LEM: a new method for nonlinear mechanics

ILEANA TOMA

ISBN -10: 1514393174
ISBN-13: 978-1514393178

DEDICATION

To my family

ACKNOWLEDGMENTS

I am grateful to the regretted professor P.P.Teodorescu, my professor and collaborator for so many years, who was the first to appreciate and promote LEM. I am also grateful to all those who tried to understand and to apply this method, especially to dr. Veturia Chiroiu, dr. Oana Firică and dr. Ştefania Donescu.

CONTENTS

INTRODUCTION

The linear equivalence method – briefly, LEM – is my original method. I created it to the purpose of determining and studying the solutions of nonlinear dynamical systems depending on parameters in a classical linear frame. A first survey of the method and its applications appeared in 1995, but I firstly published the idea of the method in 1978.

In 2008, I published in Romanian a book at Editura Tehnică (Publishing House Tehnica) of Bucharest, concerning LEM and its applications in various domains; among these applications, there were also some realized by other scientific searchers.

The present book is a revised and completed edition of the above mentioned one; it includes, besides the previous results, many other theoretical considerations and applications – made by me or other authors since 2008 – obtained by applying LEM to nonlinear models from a multitude of fields.

There are comparatively few linear models of real phenomena. The real world is rather nonlinear. Unlike the linear case, the nonlinear one involves multiple difficulties, sometimes inextricable, related to the methods of getting solutions, as well as to their qualitative interpretation. Basically, the study of these equations begins with their linearization. But linearization is often far from being satisfactory, because the nonlinear terms may become important; for a realistic perception of their influence, a great variety of approximating methods dedicated to the calculation of solution were imagined, most of them iterative and/or numeric, the degree of approximation being not always specified.

LEM overcomes this difficulty, establishing a one-to-one correspondence between the solutions of an initial polynomial problem and the analytic solutions of a certain linear partial differential equation. This is why I called it the ***L**inear **E**quivalence **M**ethod*. The method was extended to more general cases, e.g., to nonlinear ordinary differential operators with variable coefficients, that can be approximated by polynomial operators.

A characteristic LEM advantage is the *local inverse of a nonlinear operator*, which gives rise to specific LEM algorithms. It is important to notice that these algorithms are convergent, the error being estimated at every step. LEM is advantageous and effective in many problems: Cauchy, polylocal, two-point problems with multiple solutions, etc.

The *normal LEM representation* – another specific feature – is useful in the qualitative study of solutions.

The linear equivalence method is theoretically founded in the first three chapters, where there are presented various types of LEM representations and their simplified forms, as well as specific to LEM convergent algorithms; in the fourth chapter, there are realized connections of LEM with other domains, such as Fliess series, Lie derivatives, theory of graphs, theory of numbers, etc. Thus, LEM proves itself a very efficient tool for the study of a great variety of problems: the nonlinear deformation of straight bars or the motion of relativistic charged particles. In qualitative studies, there were established LEM asymptotic solutions, very useful to treat stability problems, bifurcations, catastrophes, etc. Some of these results are the buckling of a straight bar, the ecological prey-predator model leading to the Lotka-Volterra system, the existence of the solitary travelling wave solution in the model of Belousov-Zhabotinskij oscillatory chemical reaction, etc.

Among the most pertinent applications of LEM to mechanical models, one can mention those presented in chapters 5 and 6, realized in co-operation with the regretted prof. dr. rer. nat. Petre P.Teodorescu, my professor and collaborator for so many years. He was a professor of Mechanics at the faculty of Mathematics of the University of Bucharest, the head of the department of Mechanics of the Academy of Technical Sciences of Romania and a well-known author of numerous important papers and treatises concerning the mechanics of solids. Other authors successfully applied LEM to their nonlinear models, such as nonlinear double pendulum, heavy elastica or neo-Hookean models. These applications are presented in the last four chapters.

This book is addressed to all the specialists – in mathematics and/or in other domains – who deal with models of the form of nonlinear differential equations and systems. The understanding of the method requires elementary notions of differential equations, analysis and algebra.

In fact, its purpose is to provide the users with an efficient tool that maps one to one the nonlinear to the linear frame. This enables the free application of the methods of the linear, much easier handled and better founded that those of the nonlinear.

I hope that some of the readers of this book will wish to try to apply, or even to extend LEM, in connection with their own researches; I think that my expectancy has a good chance to fulfil, as the nonlinear models are in full swing of development.

ILEANA TOMA

Chapter 1

WHAT IS LEM?

As I specified in the introduction, I created LEM – the linear equivalence method – to the purpose of determining and studying, both qualitatively and numerically, the solutions of nonlinear dynamical systems depending on parameters in a classical linear frame. A first survey of the method and its applications up to 1995 was done in [143], but LEM's main idea was firstly published in 1978 [131].

LEM was initially applied to polynomial operators, but since 1983, it was extended to larger than polynomial classes of nonlinear operators [136]. LEM has the advantage of linearity, though it does not represent what is currently understood by linearization, while in the first articles it was thus called [134].

This first chapter is devoted to the theoretical foundation of LEM and to the corresponding proofs.

The central idea of LEM consists of an exponential mapping depending on parameters. After defining this mapping, its role is emphasized in the relation between the linear and the nonlinear frame; this allows building up the local LEM inverse of a nonlinear operator in matrix form. As a consequence, I could obtain specific LEM representations of the solutions of nonlinear ODSs in a standard linear frame.

Notations

Throughout this book, we shall use the following notations and abbreviations:

1) ODE – ordinary differential equation, ODS – ordinary differential system, PDE – partial differential equation, LEM – linear equivalence method;

2) For $f : I \subseteq \Re \to \Re$, the "sup" norm in $C^0(I)$ will be denoted by $\|f\|_0 = \sup_{x\in I}|f(x)|$, and the norm in $C^k(I)$, by $\|f\|_k = \sum_{m=0}^{k}\left\|\frac{d^m f}{dx^m}\right\|_0$;

3) $\mathcal{N}^* = \mathcal{N} \cup \{0\}$, where $\mathcal{N}$ is the set of naturals;

4) $D_\xi^\nu \equiv \frac{\partial^{|\nu|}}{\partial\xi_1^{\nu_1}\partial\xi_2^{\nu_2}\cdots\partial\xi_n^{\nu_n}}$ represents the partial differential operator, $\nu \equiv (\nu_j)_{j=\overline{1,n}}, \nu_j \in \mathcal{N}^*$, $\xi \equiv (\xi_j)_{j=\overline{1,n}} \in \Re^n$.

We shall introduce the LEM idea, setting up firstly an adequate functional frame. As the method was firstly applied to polynomial differential operators, we shall define more accurately this concept, pointing out its implications on LEM.

1.1. NONLINEAR ORDINARY DIFFERENTIAL OPERATORS OF POLYNOMIAL TYPE

The polynomial operators on Banach spaces were introduced by L.B.Rall [70]. Together with his disciples, Rall proved the existence of the solutions of abstract polynomial equations, also finding conditions of approximating operators with certain properties of continuity by polynomial operators; these last results mainly use the idea of the well-known theorem of Weierstrass, of approximating continuous functions by polynomials. Obviously, a polynomial differential operator defined on function spaces can be naturally introduced as a polynomial of the unknown functions and their derivatives.

This definition was extended by Rall in an adequate frame.

Let X, Y be real Banach spaces, $X^k = \underbrace{X \times X \ldots \times X}_{k \text{ ori}}$ and let diag X^k be the diagonal of X^k, isomorphic with X.

Definition 1.1. [35]. The operator $\mathsf{L} : X \to Y$ is called k-linear if L is linear with respect to each component $x_j \in X$ of $(x_1, x_2, \ldots x_k) \in X^k$, i.e.

$$\begin{aligned}\mathsf{L}(x_1, x_2, \ldots, \alpha x_j + \beta x_j', \ldots, x_k) = \alpha \mathsf{L}(x_1, x_2, \ldots, x_j, \ldots, x_k)\\ + \beta \mathsf{L}(x_1, x_2, \ldots, x_j', \ldots, x_k)\end{aligned} \tag{1.1.1}$$

for any $\alpha, \beta \in \Re$ and every $x_j, x_j' \in X$, $j = \overline{1,k}$.

The notion of k-linearity immediately leads to the general definition of a polynomial operator.

Definition 1.2 [70]. $\mathsf{P}: X \to Y$ is called *a polynomial operator of degree p* if P is of the form $\mathsf{P}x \equiv \sum_{k=0}^{p} \mathsf{L}_k x^k$, $x^k \in \operatorname{diag} X^k$, where L_0 is the "constant" operator and L_k, $k = \overline{1, p}$ are k-linear.

If X, Y are function spaces, definition 1.2 is reduced to the natural definition of polynomial differential operators, as polynomials in the unknown functions and their derivatives.

Therefore, let $I \equiv [a, b] \subset \mathfrak{R}$ and $P_j : I \times \mathfrak{R}^n \to \mathfrak{R}$, $j = \overline{1, n}$, be polynomials of degree p_j, whose coefficients $a_{j\eta}$ are of class $C^0(I)$

$$P_j(x, \zeta) \equiv \sum_{|\eta| \leq p_j} a_{j\eta}(x) \zeta^\eta, \quad \zeta \in \mathfrak{R}^n. \tag{1.1.2}$$

In formula (1.1.2), $\eta = (\eta_1, \eta_2, \ldots, \eta_n)$ are multiindices of length $|\eta|$, i.e.

$$|\eta| = \sum_{k=1}^{n} \eta_k, \quad \eta_k \in \mathfrak{N}^*; \tag{1.1.3}$$

by θ we mean the index of null components, therefore $\theta = (0, 0, \ldots, 0)$, and we put, by convention

$$\zeta^\eta = \zeta_1^{\eta_1} \zeta_2^{\eta_2} \cdots \zeta_n^{\eta_n}. \tag{1.1.4}$$

Let us denote by $p = \max\{p_j, j = \overline{1, n}\}$, $y \equiv \lfloor y_j \rfloor_{j=\overline{1,n}}$.

Let now $\mathbf{y} \equiv [y_j]_{j=\overline{1,n}} \in \left(C^1(I)\right)^n$ and consider the ODS

$$\mathcal{P}\mathbf{y} \equiv \frac{d\mathbf{y}}{dx} - \mathbf{P}(x, \mathbf{y}) = \mathbf{0}, \quad \mathbf{P} = [P_j(x, \mathbf{y})]_{j=\overline{1,n}}, \tag{1.1.5}$$

where P_j are defined in (1.1.2). According to definition 1.2, this canonical system is of polynomial type.

Let us now associate to the ODS (1.1.5) the Cauchy conditions

$$\mathbf{y}(x_0) = \mathbf{y}_0, \quad x_0 \in I, \mathbf{y}_0 \equiv [y_{j0}]_{j=\overline{1,n}} \in \mathfrak{R}^n. \tag{1.1.6}$$

The polynomial system (1.1.5) can be analyzed in the classical frame of the theory of differential operators, establishing for it theorems deduced in more general than required by its form hypotheses. Thus, for instance, due to its algebraic properties, it satisfies Lipschitz type conditions, ensuring the local existence and uniqueness of the solution of the nonlinear initial problem (1.1.5), (1.1.6), through the agency of the classical Cauchy-Picard theorem. Let us also mention that there are various numerical methods, generating numerous algorithms, which can be adapted to the polynomial initial problem (1.1.5), (1.1.6), viewed as a general nonlinear problem.

Immaterial the theoretical frame they are tackled in, their quality of polynomials is not exploited at the level of its value. In other words, the functional methods lay stress upon the properties of the domain of definition and/or the range rather than upon the specific correspondence realized by the nonlinear operators (and, particularly, by the polynomial ones) between these sets. This is why the numerical methods strongly depend on the initial data (e.g., Runge-Kutta, Adams-Bashforth-Moulton, etc.). Welcome exceptions are the semi-analytic methods, much studied and applied lastly; but they also are dependent on the initial data.

To overcome these difficulties, in the study of nonlinear problems – polynomial or not – one tries, in many cases, to establish a parallel with the linear operators; these operators, having a simpler structure and richer properties, are easier tackled.

Basically, one makes the approach with the linear operators by severing a part of the considered nonlinear operator; this is a widely used technique. This "surgical" operation is usually called *linearization*; actually, in many studies this term is identified with the approximation of the operator with its linear part.

There are also other possibilities of establishing connections between the nonlinear and linear operators: by non-singular transformations that associate to a nonlinear ODS the solution of a certain linear system. However, in general, this method depends on the specified nonlinear operator. Anyway, even realized on particular cases, one cannot completely take advantage of these algebraic techniques.

1.2. φ - IMMERSIBLE AND φ - EQUIVALENT OPERATORS

In this section, we shall consider transformations of operators, specifying the relations between them and their transformed forms on the one hand and between their corresponding kernels, on the other hand.

We shall also emphasize some properties that will be useful in what follows.

This preparatory section presents only a work frame and is neither compulsory, nor decisive, for the understanding of LEM.

The general practice shows that in many cases one can find some useful transformations that simplify the model and/or make it more accessible. To create a general frame for such transformations, we shall introduce the notion of φ-immersibility and φ-equivalence.

Consider the operators $\mathsf{P}_i : \operatorname{dom}\mathsf{P}_i \subseteq X_i \to Y_i, i = 1,2$, where X_i, Y_i, are Banach spaces and let us denote

$$\ker \mathsf{P}_j = \left\{ x \in \operatorname{dom}\mathsf{P}_j \middle| \mathsf{P}_j x = 0_{Y_j} \right\}. \tag{1.2.1}$$

Definition 1.3. [137]. We say that P_1 is φ-*immersible* in P_2 if it exists $\varphi : X_1 \to X_2$, bicontinuous from X_1 to $\varphi(X_1)$, such that $\varphi(\operatorname{dom}\mathsf{P}_1) \subset \operatorname{dom}\mathsf{P}_2$ and $\varphi(\ker\mathsf{P}_1) \subseteq \ker\mathsf{P}_2$.

Definition 1.4.[137]. P_2 is a φ-*equivalent* for P_1 if P_1 is φ-immersible in P_2 and $\varphi(\ker\mathsf{P}_1) \supseteq \ker\mathsf{P}_2$.

One can easily transpose these definitions for equations on function spaces.

Denote by Ξ_j the set of solutions of $\mathsf{E}_j x = 0, j = 1,2$. By definition 1.3, an operator E_1 of attached equation $\mathsf{E}_1 x = 0$ is φ-immersible in the operator E_2 of attached equation $\mathsf{E}_2 x = 0$, if one can find a one-to-one application $\varphi : \operatorname{dom}\mathsf{E}_1 \to \operatorname{dom}\mathsf{E}_2$ such that $\varphi(\mathsf{E}_1) \subset \varphi(\mathsf{E}_2)$. By definition 1.4, E_1 allows a φ-equivalent if, moreover, any solution of $\mathsf{E}_2 x = 0$ is the image by φ of a solution of $\mathsf{E}_1 x = 0$.

Therefore, by the property of φ-equivalence one can solve $\mathsf{E}_2 x = 0$ instead of $\mathsf{E}_1 x = 0$; naturally, we shall choose φ such that the transformed equation be easier tackled.

Specifically, let us give an example.

Example 1.1. The classical model of the elastic straight bar is represented by the Bernoulli-Euler equation

$$y'' = f(x)\left(1 + y'^2\right)^{\frac{3}{2}},\ f \in C^0([0,l]),\ y \in C^2([0,l]), \tag{1.2.2}$$

where l is the bar length and y – the bar axis displacement. As this problem will be considered further, for now we suppose that f is determined by the bar physical properties, without other specification.

The equation (1.2.2) is a second order nonlinear ODE. The order can be reduced at once if we denote $z = \frac{dy}{dx}$, or, otherwise speaking, applying the change of function $y' = z$. Thus, (1.2.2) becomes

$$\mathsf{E}_1(z) \equiv z' - f(x)\left(1 + z^2\right)^{\frac{3}{2}} = 0, \quad z \in C^1([0,l]); \tag{1.2.3}$$

then, by the transformation

$$\varphi(z) = (u,z) \equiv \left(\sqrt{1+z^2}, z\right), \tag{1.2.4}$$

it is mapped into

$$\mathsf{E}_2((u,z)) \equiv \begin{pmatrix} u' - f(x)zu^2 \\ z' - f(x)u^3 \end{pmatrix} = \begin{pmatrix} 0 \\ 0 \end{pmatrix}. \tag{1.2.5}$$

According to definition 1.3, E_1 is φ-immersible in E_2, but E_2 is not a φ-equivalent of E_1. Indeed, one can write any solution of (1.2.5) in the form $\left(\sqrt{c+z^2}, z\right)$, with c an arbitrary constant – not necessarily 1 – and z satisfies

$$\frac{dz}{dx} = f(x)\left(c + z^2\right)^{\frac{3}{2}}. \tag{1.2.6}$$

Consequently, the inclusion $\varphi(\ker \mathsf{E}_1) \subset \ker \mathsf{E}_2$ is strict.

The introduction of the initial conditions changes this situation.

Indeed, the nonlinear bending of a cantilever bar, for instance, is determined by solving the Bernoulli-Euler equation in null Cauchy conditions

$$y(0) = 0, \quad y'(0) = 0. \tag{1.2.7}$$

Hence $z(0) = 0$ for $\mathsf{E}_1(z) = 0$; $u(0) = 1$, $z(0) = 0$ are the initial conditions corresponding to the ODS $\mathsf{E}_2((u,z)) = (0,0)^T$.

Let E_1^0 be the restriction of E_1 to the set

$$\mathcal{E}_1 = \left\{ z \in C^1([0,l]), z(0) = 0 \right\}, \tag{1.2.8}$$

and E_2^0 the restriction of E_2 to

$$\mathcal{E}_2 = \left\{ (u,z)^T \in \left(C^1([0,l])\right)^2, (u(0), z(0))^T = (1,0) \right\}. \tag{1.2.9}$$

It is seen that, E_1^0 is φ-immersible in E_2^0. It is easily seen that any element belonging to $\varphi(\ker E_1) \subseteq \ker E_2$ satisfies (1.2.5) and, as (1.2.9) implies $c = 1$ for the equation $E_1(z) = 0$, it follows that $\varphi\left(\ker E_1^0\right) \supseteq \ker E_2^0$. According to Definition 1.4, this means that E_2^0 is the φ-equivalent of E_1^0, with φ given by (1.2.4).

An obvious sufficient condition of φ-equivalence is expressed by the following theorem

Theorem 1.1 [137, 143]. *If* P_1 *is φ-immersible in* P_2 *and* $\mathrm{card}(\ker P_1) = \mathrm{card}(\ker P_2) < \infty$, *then* P_2 *is a φ-equivalent of* P_1.

In other words, if P_1 is φ-immersible in P_2 and if we succeed to prove a (non-constructive) theorem of existence of the solution for the equation $P_1 x_1 = 0$ on the one hand and the uniqueness of the solution of the equation $P_2 x_2 = 0$ on the other hand, then $x_1 = \varphi^{-1}(x_2)$ can be effectively set up.

The properties of φ-immersibility and φ-equivalence are transitive. To illustrate this fact, we shall firstly extend the Definitions 1.3 and 1.4 to (finite) sequences of operators.

Definition 1.5 [137]. A finite sequence of operators $\{P_j\}_{j=\overline{1,k}}$, to which a finite sequence of applications $\{\varphi_j\}_{j=\overline{1,k-1}}$ correspond such that P_j be φ_j-immersible in P_{j+1}, $j = \overline{1, k-1}$ is called a *resonant chain.*

Definition 1.6 [137, 143]. A resonant chain of φ_j-equivalents is called *resolvent.*

Using these definitions, one can prove an immediate consequence of theorem 1.1.

Corollary 1.1. [137, 143]. *Let* $\{P_j\}_{j=\overline{1,k}}$ *be a resonant chain of operators. If* $\ker P_k$ *has at most one element and* $\ker P_1$ *at least one element, then* $\{P_j\}_{j=\overline{1,k}}$ *is a resolvent chain and*

$$\operatorname{card}\left(\ker \mathsf{P}_j\right) = 1, \quad j = \overline{1,k}. \tag{1.2.10}$$

Indeed, denoting by x_j the unique element of $\ker \mathsf{P}_j$, it results that x_j is represented as

$$x_j = \left(\underset{m=j-1}{\overset{s-1}{\circ}} \varphi_m \right)^{-1} x_s, \; s = \overline{j,k}, \tag{1.2.11}$$

for $j = \overline{1,k}$.

1.3. THE LEM MAPPING AND THE FUNDAMENTAL THEOREMS

Let us get back to the polynomial system (1.1.5). If $a_{j\eta} \in \mathrm{C}^0(I)$, then the operator P can be considered as defined from $\left(\mathrm{C}^1(I)\right)^n$ to $\left(\mathrm{C}^0(I)\right)^n$.

Let us associate to each polynomial $P_j(x,\xi)$ given by formula (1.1.2) the differential polynomial

$$\mathsf{P}_j\left(x, \mathrm{D}_\xi\right) \equiv \sum_{|\eta|=0}^{p_j} a_{j\eta}(x) \mathrm{D}_\xi^\nu, \; j = \overline{1,n}. \tag{1.3.1}$$

Thus, the vector $\mathbf{P}(x,\xi) \equiv \left(P_j(x,\xi)\right)_{j=\overline{1,n}}$ is put into correspondence with the vector having as components the differential polynomials (1.3.2)

$$\mathbf{P}\left(x, \mathrm{D}_\xi\right) \equiv \left(\mathsf{P}_j\left(x, \mathrm{D}_\xi\right)\right)_{j=\overline{1,n}}. \tag{1.3.2}$$

Consider now the exponential mapping of real parameters ξ

$$v(x,\xi) \equiv \mathrm{e}^{\langle \xi, \mathbf{y} \rangle}, \quad \xi \equiv \left(\xi_1, \xi_2, \ldots, \xi_n\right) \in \Re^n, \langle \xi, \mathbf{y} \rangle = \sum_{j=1}^{n} \xi_j y_j. \tag{1.3.3}$$

We can prove the following result

Theorem 1.2 [132, 133, 139, 143]. *The exponential mapping* (1.3.3), *in which* $\mathbf{y}$ *is the solution of the initial polynomial problem* (1.1.5), (1.1.6), *satisfies*

i) the linear PDE

$$\mathfrak{L}v \equiv \frac{\partial v}{\partial x} - \langle \boldsymbol{\xi}, \mathbf{P}(x, \mathrm{D}_{\xi}) \rangle v = 0\,, \tag{1.3.4}$$

where $\mathbf{P}(x, \mathrm{D}_{\xi})$ *is given by* (1.3.2) *and* $\langle \boldsymbol{\xi}, \mathbf{P}(x, \mathrm{D}_{\xi}) \rangle$ *is the formal scalar product*

$$\langle \boldsymbol{\xi}, \mathbf{P}(x, \mathrm{D}_{\xi}) \rangle \equiv \sum_{j=1}^{n} \xi_j \mathsf{P}_j (x, \mathrm{D}_{\xi}); \tag{1.3.5}$$

ii) the condition

$$v(x_0, \boldsymbol{\xi}) \equiv \mathrm{e}^{\langle \boldsymbol{\xi}, \mathbf{y}_0 \rangle}, \quad \boldsymbol{\xi} \in \mathfrak{R}^n\,. \tag{1.3.6}$$

Proof. Point *ii)* is obviously true. To prove *i)*, let us multiply each equation from (1.1.5) with the exponential v, given by (1.3.3); observe that the right member of each equation can be written in the form

$$vP_j(x, \mathbf{y}) = v \sum_{|\eta|=0}^{p_j} a_{j\eta}(x) \mathbf{y}^{\eta} = \sum_{|\eta|=0}^{p_j} a_{j\eta}(x) \mathrm{D}_{\xi}^{\eta} v = \mathsf{P}_j(x, \mathrm{D}_{\xi}) v\,. \tag{1.3.7}$$

But

$$\frac{\partial v}{\partial x} = \left(\sum_{j=1}^{n} \xi_j \frac{\mathrm{d}y_j}{\mathrm{d}x} \right) v \tag{1.3.8}$$

hence, taking (1.3.7) into account, it results

$$\frac{\partial v}{\partial x} = \sum_{j=1}^{n} \xi_j \mathsf{P}_j(x, \mathbf{y})\, v = \sum_{j=1}^{n} \xi_j \mathsf{P}_j(x, \mathrm{D}_{\xi})\, v\,. \tag{1.3.9}$$

We finally get

$$\frac{\partial v}{\partial x} = \langle \boldsymbol{\xi}, \mathbf{P}(x, \mathrm{D}_{\xi}) \rangle\, v\,. \tag{1.3.10}$$

Theorem 1.1. is thus completely proved.□

The exponential mapping (1.3.3) introduces a method for studying and effectively determining the solutions of polynomial ODSs in a linear frame, *without approximation*.

For this reason, I called this method *the linear equivalence method* – or, briefly, *LEM.* For the same reason, the exponential transform (1.3.3) was called *the LEM mapping*.

The LEM mapping (1.3.3) was introduced in [131, 132].

Theorem 1.2 shows that one can pass from the nonlinear to the linear frame, easier tackled. To have the benefits of the advantages offered by the linear frame, one must ensure the return to the nonlinear, still keeping the exponential form of the LEM mapping. The condition *ii)* cannot miss from the inverse of Theorem 1.2, as it is seen from the following example

Example 1.2. Consider the polynomial ODE of second order

$$y'' = y^2. \tag{1.3.11}$$

The LEM mapping is, in this case

$$v(t,\sigma,\xi) = \mathrm{e}^{\sigma y + \xi y'}. \tag{1.3.12}$$

We get, by differentiation

$$\frac{\partial v}{\partial t} = (\sigma y' + \xi y'')\mathrm{e}^{\sigma x + \xi y}. \tag{1.3.13}$$

Taking into account the equation (1.3.11), we obtain

$$\frac{\partial v}{\partial t} = \left(\sigma y' + \xi y^2\right)\mathrm{e}^{\sigma x + \xi y} = \sigma\frac{\partial}{\partial \xi}\left(\mathrm{e}^{\sigma x + \xi y}\right) + \xi\frac{\partial^2}{\partial \sigma^2}\left(\mathrm{e}^{\sigma x + \xi y}\right) \tag{1.3.14}$$

The PDE associated by LEM is

$$\frac{\partial v}{\partial x} = \sigma\frac{\partial v}{\partial \xi} + \xi\frac{\partial^2 v}{\partial \sigma^2}. \tag{1.3.15}$$

This PDE is satisfied by $v = \sigma$, among other possible solutions; though, while analytic, this function has not the exponential form (1.3.3).

Theorem 1.3 [132, 133, 139, 143]. *If* $a_{j\eta} \in \mathrm{C}^0(I)$, $\|a_{j\eta}\|_0 \le C$, *then the analytic with respect to* $\boldsymbol{\xi}$ *solution of the linear PDE* (1.3.4), *uniform with respect to* $x \in I$ *and that also* (1.3.6), *has the exponential form* (1.3.3), *where* $\mathbf{y}$ *is the solution of the polynomial initial problem* (1.1.5), (1.1.6).

Proof. Consider the function

$$H(x,\boldsymbol{\xi},\boldsymbol{\xi}') \equiv v(x,\boldsymbol{\xi}+\boldsymbol{\xi}') - v(x,\boldsymbol{\xi})v(x,\boldsymbol{\xi}'). \tag{1.3.16}$$

It is well known that any solution of the functional equation

$$H(x,\xi,\xi')=0\,, \tag{1.3.17}$$

continuous in ξ,ξ', has the exponential form

$$v(x,\xi)\equiv e^{\langle \xi,\varphi\rangle}\,, \tag{1.3.18}$$

with $\varphi : I \to \Re^n, \varphi(x)=\left[\varphi_j(x)\right]_{j=\overline{1,n}}$. On the other hand, if v satisfies (1.3.4), then H satisfies

$$\frac{\partial H}{\partial x}=\langle \xi,\mathbf{P}(x,\mathrm{D}_\xi)\rangle H+\langle \xi',\mathbf{P}(x,\mathrm{D}_{\xi'})\rangle H\,. \tag{1.3.19}$$

The condition (1.3.6) implies $H(x_0,\xi,\xi')=0$ and, in general, $\mathrm{D}_\xi^{\eta+\eta'}H(x_0,\xi,\xi')=0$, for any multiindices $\eta,\eta'\in(\mathcal{N}^*)^n$.

Let us denote by T the linear integro-differential operator

$$\mathsf{T}H(x,\xi,\xi')\equiv\int_{x_0}^{x}\left[\langle \xi,\mathbf{P}(x,\mathrm{D}_\xi)\rangle H(x,\xi,\xi')+\langle \xi',\mathbf{P}(x,\mathrm{D}_{\xi'})\rangle H(x,\xi,\xi')\right]\mathrm{d}x. \tag{1.3.20}$$

Clearly, any solution of (1.3.19) vanishing at x_0 satisfies the linear integral equation

$$H=\mathsf{T}H\,. \tag{1.3.21}$$

From the hypothesis, it results that a neighbourhood $\omega(x_0)\subseteq I$ and a positive constant M exist such that the inequality $MC\left|x-x_0\right|\sum_{j=1}^{n}\left(\left|\xi_j\right|+\left|\xi'_j\right|\right)<1$ hold in a vicinity of the origin of $\Re^n$.

Therefore, the equation (1.3.21) allows only the trivial solution.□

From now on, the PDE (1.3.4) shall be called the *linear* Exp-*equivalent* of the polynomial system (1.1.5); this denomination is fully justified by theorems 1.2 and 1.3.

These two theorems are basic for the linear equivalence method and, for this reason, one can call them *fundamental theorems*.

Most of the references of this book contain results concerning LEM, its extensions and its applications in various domains.

LEM has an important advantage in that that it faithfully reproduces the algebraic behaviour of the considered polynomial operator. Unlike other methods, e.g., standard linearization, LEM *does not approximate the operator.*

In a certain sense, the LEM mapping represents a replica of the Fourier transform in the case of nonlinear operators.

The advantages of LEM are numerous. In the first place, one must find only the analytic solution of the linear equivalent PDE (1.3.4), which considerably restrains the domain on which one serches the solution, enlarging in exchange the real practical possibilities of determining it.

Then, the restrictions regarding the shape and the order of magnitude of the nonlinearities are removed (e.g., it is not required that the nonlinearity be "small", as it happens in the theory of perturbations, or to behave as a power less than 2).

Eventually, as the correspondence is realized between the exact solutions of two kinds of problems – the polynomial one and its linear LEM equivalent – one can apply both numerical and analytical methods specific to linear differential equations, which are far better than those used in the general nonlinear case.

Unlike the algorithms deduced by standard methods (e.g., Runge-Kutta or Adams-Bashforth-Moulton, etc.), the numerical algorithms deduced by LEM have two essential qualities:

- they separate the contribution of the data from that of the coefficients of the considered ODS;
- they do not repeat for any new set of initial data.

The treatment by LEM of the two-point problems is particularly advantageous. Thus, in the case of two-point problems, a specific LEM method was set up that is completely different from the usual techniques (e.g. shooting, collocation, etc.)

Thus, the linear Exp-equivalent has from the very beginning the following advantages:

- *linearity*: according to the above fundamental theorems, one can replace a nonlinear polynomial problem by a linear one, fact which is, practically much simpler;
- it ensures *a one-to-one correspondence* with the nonlinear operator;
- it does not impose restrictions on the nonlinearity.

We shall give several examples of linear equivalents associated to some polynomial operators, currently met in applications.

Example 1.3. **The LEM equivalent of van der Pol's equation** [64]. The van der Pol oscillator with nonlinear damping evolves in time according to the second order polynomial ODE

$$\ddot{x} + \mu \dot{x}\left(x^2 - 1\right) + x = 0\,,\ \dot{x} = \frac{\mathrm{d}x}{\mathrm{d}t} \tag{1.3.22}$$

Initially deduced as a model in electrotechnics, this equation is widely used in physics, biology and seismology, being one of the first cases showing the deterministic chaos.

We firstly associate to this ODE the first order polynomial ODS

$$\begin{aligned} &\dot{x} = y, \\ &\dot{y} = -x + \mu y - \mu x^2 y, \end{aligned} \tag{1.3.23}$$

where we explicitly put into evidence the linear part.

The LEM mapping is, in this case

$$v(t, \sigma, \xi) = \mathrm{e}^{\sigma x + \xi y}\,. \tag{1.3.24}$$

By differentiation, we find

$$\frac{\partial v}{\partial t} = (\sigma \dot{x} + \xi \dot{y})\mathrm{e}^{\sigma x + \xi y}\,. \tag{1.3.25}$$

Taking into account the ODS (1.3.23), we immediately get

$$\begin{aligned} \frac{\partial v}{\partial t} &= \left[\sigma y + \xi\left(-x + \mu y - \mu x^2 y\right)\right]\mathrm{e}^{\sigma x + \xi y} = \sigma \frac{\partial}{\partial \xi}\left(\mathrm{e}^{\sigma x + \xi y}\right) \\ &+ \xi\left[-\frac{\partial}{\partial \sigma}\left(\mathrm{e}^{\sigma x + \xi y}\right) + \mu \frac{\partial}{\partial \xi}\left(\mathrm{e}^{\sigma x + \xi y}\right) - \mu \frac{\partial^3}{\partial \sigma^2 \partial \xi}\left(\mathrm{e}^{\sigma x + \xi y}\right)\right]. \end{aligned} \tag{1.3.26}$$

Hence the linear equivalent of van der Pol's equation is

$$\frac{\partial v}{\partial t} = \sigma \frac{\partial v}{\partial \xi} + \xi\left(-\frac{\partial v}{\partial \sigma} + \mu \frac{\partial v}{\partial \xi} - \mu \frac{\partial^3 v}{\partial \sigma^2 \partial \xi}\right). \tag{1.3.27}$$

Example 1.4. **The LEM equivalent of Lorenz's system** [49]. The Lorenz equations, well known in the qualitative study of meteorological phenomena, read

$$\begin{aligned} \dot{x} &= s(y - x), \\ \dot{y} &= rx - y - xz, \\ \dot{z} &= xy - \beta z. \end{aligned} \tag{1.3.28}$$

Among the parameters r, s, β of this system, we remark Prandl's number s and Raighley's number r, whose variation induces changes in the qualitative behaviour of the solution.

Here, the LEM mapping is

$$v(t, \sigma, \xi) = \mathrm{e}^{\sigma x + \xi y + \varsigma z}. \tag{1.3.29}$$

Differentiation with respect to the independent variable t yields

$$\frac{\partial v}{\partial t} = (\sigma \dot{x} + \xi \dot{y} + \varsigma \dot{z})\mathrm{e}^{\sigma x + \xi y + \varsigma \dot{z}}. \tag{1.3.30}$$

Considering the ODS (1.3.28), we obtain

$$\begin{aligned} \frac{\partial v}{\partial t} &= [\sigma s(y - x) + \xi(rx - y - xz) + \varsigma(xy - \beta z)]\mathrm{e}^{\sigma x + \xi y + \varsigma z} \\ &= \sigma s\left[\frac{\partial}{\partial \xi}\left(\mathrm{e}^{\sigma x + \xi y + \varsigma z}\right) - \frac{\partial}{\partial \sigma}\left(\mathrm{e}^{\sigma x + \xi y + \varsigma z}\right)\right] \\ &+ \xi\left[r\frac{\partial}{\partial \sigma}\left(\mathrm{e}^{\sigma x + \xi y + \varsigma z}\right) - \frac{\partial}{\partial \xi}\left(\mathrm{e}^{\sigma x + \xi y + \varsigma z}\right) - \frac{\partial^2}{\partial \sigma \partial \varsigma}\left(\mathrm{e}^{\sigma x + \xi y + \varsigma z}\right)\right] \\ &+ \varsigma\left[\frac{\partial^2}{\partial \sigma \partial \xi}\left(\mathrm{e}^{\sigma x + \xi y + \varsigma z}\right) - \beta\frac{\partial}{\partial \varsigma}\left(\mathrm{e}^{\sigma x + \xi y + \varsigma z}\right)\right]. \end{aligned} \tag{1.3.31}$$

Consequently, the LEM equivalent of Lorenz's system reads

$$\frac{\partial v}{\partial t} = \sigma s\left(\frac{\partial v}{\partial \xi} - \frac{\partial v}{\partial \sigma}\right) + \xi\left(r\frac{\partial v}{\partial \sigma} - \frac{\partial v}{\partial \xi} - \frac{\partial^2 v}{\partial \sigma \partial \varsigma}\right) + \varsigma\left(\frac{\partial^2 v}{\partial \sigma \partial \xi} - \beta\frac{\partial v}{\partial \varsigma}\right). \tag{1.3.32}$$

One must now set up an adequate functional frame for the LEM equivalent (1.3.4). Let us consider with Trèves [163] the Exp-type spaces

$$\mathcal{Q}_n^k(I) = \left\{v(x, \xi), v : I \times \mathfrak{R}^n \to \mathfrak{R}; \|v_\gamma\|_k \le M^{|\gamma|}K, |\gamma| \in \mathfrak{N}^*\right. \tag{1.3.33}$$

for functions v with exponential growth, allowing expansions in the form

$$v(x,\xi)=1+\sum_{|\gamma|\geq 1} v_\gamma(x)\frac{\xi^\gamma}{\gamma!}, \quad x\in I, \xi\in\Re^n, \tag{1.3.34}$$

uniformly valid on *I*.

Theorem 1.4 [134, 143]. *The set of the linear exp-equivalents of a polynomial operator is affine and maps $\mathcal{Q}_n^1(I)$ in $\mathcal{Q}_n^0(I)$.*

Proof. The first part is immediately proved. Indeed, let $\mathcal{L}_1, \mathcal{L}_2$ be two linear equivalents, associated to the polynomial differential operators $\mathcal{P}_1$ and $\mathcal{P}_2$ accordingly. Let c_1, c_2 be two real constant such that $c_1+c_2=1$ and take $v\in\mathcal{Q}_n^1(I)$. Then

$$\begin{aligned}(c_1\mathcal{L}_1+c_2\mathcal{L}_2)v=c_1&\left[\frac{\partial v}{\partial x}-\langle\xi,\mathbf{P}_1(x,\mathrm{D}_\xi)\rangle v\right]\\ &+c_2\left[\frac{\partial v}{\partial x}-\langle\xi,\mathbf{P}_2(x,\mathrm{D}_\xi)\rangle v\right],\end{aligned} \tag{1.3.35}$$

and, finally,

$$(c_1\mathcal{L}_1+c_2\mathcal{L}_2)v=(c_1+c_2)\frac{\partial v}{\partial x}-\langle\xi,c_1\mathbf{P}_1(x,\mathrm{D}_\xi)+c_2\mathbf{P}_2(x,\mathrm{D}_\xi)\rangle v. \quad .3.36)$$

This means that $(c_1\mathcal{L}_1+c_2\mathcal{L}_2)v=0$ is the LEM equivalent of the polynomial system

$$\mathcal{P}\mathbf{y}\equiv\frac{\mathrm{d}\mathbf{y}}{\mathrm{d}x}-[c_1\mathbf{P}_1(x,\mathbf{y})+c_2\mathbf{P}_2(x,\mathbf{y})]=\mathbf{0}. \tag{1.3.37}$$

To prove the last part of the theorem, it suffices to show that $\mathcal{Q}_n^0(I)$ is invariant to any differential polynomial with respect to ξ variables.

Therefore, let $\mathsf{P}(x,\mathrm{D}_\xi)$ be such an operator, corresponding to a polynomial of degree *p* in ξ, with coefficients depending on *x* bounded on *I*, hence

$$|a_\mu(x)|\leq C, |\mu|\leq p. \tag{1.3.38}$$

For $v\in\mathcal{Q}_n^0(I)$ we have

$$\mathsf{P}(x, \mathrm{D}_\xi)v = \sum_{|\mu| \le p} a_\mu(x) \mathrm{D}_\xi^\mu v = \sum_{|\mu| \le p} a_\mu(x) \sum_{|\gamma-\mu| \ge 0} v_\gamma(x) \frac{\xi^{\gamma-\mu}}{(\gamma-\mu)!}$$
$$= \sum_{|\mu| \le p} a_\mu(x) \sum_{|\gamma| \ge 0} v_{\gamma+\mu}(x) \frac{\xi^\gamma}{\gamma!}. \tag{1.3.39}$$

But

$$|\mathsf{P}(x, \mathrm{D}_\xi)v| \le \sum_{|\mu| \le p} |a_\mu(x)| \sum_{|\gamma| \ge 0} |v_{\gamma+\mu}(x)| \frac{|\xi|^{|\gamma|}}{\gamma!}$$
$$\le C \sum_{|\mu| \le p} \sum_{|\gamma| \ge 0} K M^{|\gamma|+|\mu|} \frac{|\xi|^{|\gamma|}}{\gamma!} \tag{1.3.40}$$
$$\le CK \left(\sum_{|\mu| \le p} M^{|\mu|} \right) \mathrm{e}^{M|\xi|}.$$

Thus, $\mathsf{P}(x, \mathrm{D}_\xi)v$ allows an expansion in the form

$$\mathsf{P}(x, \mathrm{D}_\xi)v = \sum_{|\gamma| \ge 0} b_\gamma(x) \frac{\xi^\gamma}{\gamma!}, \tag{1.3.41}$$

where

$$b_\gamma(x) = \sum_{|\mu| \le p} a_\mu(x) v_{\gamma+\mu}(x). \tag{1.3.42}$$

Taking into account the inequalities (1.3.40), we observe that $b_\gamma(x)$ satisfies the inequality

$$|b_\gamma(x)| \le \sum_{|\mu| \le p} |a_\mu(x)| \, |v_{\gamma+\mu}(x)| \le CK \sum_{|\mu| \le p} M^{|\gamma|+|\mu|}, \tag{1.3.43}$$

i.e.

$$|b_\gamma(x)| \le Q M^{|\gamma|}, \qquad Q = CK \sum_{|\mu| \le p} M^{|\mu|}. \tag{1.3.44}$$

Consequently, if $v \in \mathcal{Q}_n^0(I)$, then $\mathsf{P}(x, \mathrm{D}_\xi)v \in \mathcal{Q}_n^0(I)$ too.■

As a conclusion, the LEM mapping (1.3.3) leads to the following diagram

$$\begin{array}{l} \mathcal{P} : \left(\mathrm{C}^1(I)\right)^n \to \left(\mathrm{C}^0(I)\right)^n \\ \mathrm{e}^{\langle \xi, \bullet \rangle} \downarrow \\ \mathcal{L} : \mathcal{A}_n^1(I) \to \mathcal{A}_n^0(I) \end{array} \tag{1.3.45}$$

In this frame, the fundamental theorems 1.2 and 1.3 can be expressed in other terms, taking into consideration the definitions and the facts from Sections 1.1 and 1.2. The following theorem hold true

Theorem 1.5 [134, 143]. *Any polynomial differential operator with continuous coefficients is globally exp-immersible in a linear operator. Any restriction of a polynomial differential operator with continuous coefficients to a set of functions satisfying given Cauchy conditions locally allows an affine exp-equivalent.*

1.4. THE LEM EQUIVALENT SYSTEM ASSOCIATED TO A POLYNOMIAL DIFFERENTIAL OPERATOR

Even if the analyticity of the solution of the LEM equivalent PDE in ξ does not suffice to return to the solutions of the polynomial system, it remains however an important condition. We shall show that analyticity not only determines a functional working frame, but it also plays an important part in the effective solving of the LEM equivalent. Let us take for $v = v(x, \xi)$ the previously mentioned development

$$v(x, \xi) = 1 + \sum_{|\gamma| \geq 1} v_\gamma(x) \frac{\xi^\gamma}{\gamma!}, \quad x \in I, \xi \in \Re^n. \tag{1.4.1}$$

Formally replacing this expression in the first LEM equivalent (1.3.4), we obtain the linear infinite ODS

$$\frac{\mathrm{d}v_\gamma}{\mathrm{d}x} = \sum_{j=1}^{n} \gamma_j \sum_{|\mu| \leq p_j} a_{j\mu}(x) v_{\gamma+\mu-\mathbf{e}_j}, \quad |\gamma| \in \mathfrak{N}, \tag{1.4.2}$$

where

$$\mathbf{e}_j = \left(\delta_j^k\right)_{k=\overline{1,j}} \equiv \left(0,0,\ldots,0,\underset{j}{1},0,\ldots,0\right), \quad j = \overline{1,n}. \tag{1.4.3}$$

The conditions (1.3.6) become

$$v_\gamma(x_0) = \mathbf{y}_0^\gamma, \quad |\gamma| \in \mathfrak{N}. \tag{1.4.4}$$

The ODS (1.4.2) is infinite, but it is linear, and its special form is very useful in applications.

Let us give some examples of how the LEM equivalent ODS is associated to a polynomial operator.

Example 1.5. Consider the **Riccati** equation

$$\mathcal{P}y \equiv y' - \left(a + by + cy^2\right) = 0, \tag{1.4.5}$$

where $a, b, c \in C^0(I)$, $I \subseteq \mathfrak{R}$.

In this case, $n = 1$, hence the LEM mapping will depend only on one parameter

$$v(x,\xi) = e^{\xi y}. \tag{1.4.6}$$

Consequently, the first LEM equivalent of (1.4.5) is

$$\frac{\partial v}{\partial x} = \xi\left(av + b\frac{\partial v}{\partial \xi} + c\frac{\partial^2 v}{\partial \xi^2}\right). \tag{1.4.7}$$

The expansion of v will be

$$v(x,\xi) = 1 + \sum_{i \geq 1} v_i(x)\frac{\xi^i}{i!}; \tag{1.4.8}$$

thus, we obtain for the coefficients v_i the LEM equivalent ODS

$$\frac{\mathrm{d}v_i}{\mathrm{d}x} = i\left(av_{i-1} + bv_i + cv_{i+1}\right), \quad i \in \mathfrak{N}. \tag{1.4.9}$$

Example 1.6. Consider the polynomial ODE of second order

$$\mathcal{P}y \equiv y'' - \left(ay + by' + Ay^2 + Byy' + Cy'^2\right) = 0, \tag{1.4.10}$$

with $a, b, A, B, C \in C^0(I)$, $I \subseteq \mathfrak{R}$.

Let us associate to this equation the first order ODS, also polynomial

$$\begin{aligned} y' &= z, \\ z' &= ay + by' + Ay^2 + Byy' + Cy'^2. \end{aligned} \tag{1.4.11}$$

With the above notations, we observe that in this case $n = 2$, hence the LEM mapping depends on two parameters

$$v(x, \xi_1, \xi_2) = e^{(\xi_1 y + \xi_2 z)}. \tag{1.4.12}$$

The first linear equivalent of (1.4.11) reads

$$\frac{\partial v}{\partial x} = \xi_1 \frac{\partial v}{\partial \xi_2} + \xi_2 \left(a \frac{\partial v}{\partial \xi_1} + b \frac{\partial v}{\partial \xi_2} + A \frac{\partial^2 v}{\partial \xi_1^2} + B \frac{\partial^2 v}{\partial \xi_1 \partial \xi_2} + C \frac{\partial^2 v}{\partial \xi_2^2} \right). \tag{1.4.13}$$

Using for v a similar to (1.4.1) expansion

$$v(x, \xi_1, \xi_2) = 1 + \sum_{i+j \geq 1} v_{ij}(x) \frac{\xi_1^i}{i!} \frac{\xi_2^j}{j!}, \tag{1.4.14}$$

we get for the coefficients v_{ij} the linear equivalent system

$$\frac{\mathrm{d}v_{ij}}{\mathrm{d}x} = i v_{i-1,j+1} + j\left(a v_{i+1,j-1} + b v_{i,j} + A v_{i+2,j-1} + B v_{i+1,j} + C v_{i,j+1}\right) \quad i, j \in \mathfrak{N}. \tag{1.4.15}$$

Example 1.7. In the case of **van der Pol's equation** (1.3.22), we already set up its first linear equivalent (1.3.27). Applying to $v(t, \sigma, \xi)$ the corresponding expansion of the type (1.4.14), we also get the linear equivalent ODS

$$\dot{v}_{ij} = i v_{i-1,j+1} + j\left(-v_{i+1,j-1} + \mu v_{i,j} - \mu v_{i+2,j}\right), \qquad i, j \in \mathfrak{N}. \tag{1.4.16}$$

Example 1.8. We showed that the first LEM equivalent of **Lorenz's system** (1.3.28) is (1.3.32). Considering for $v(x, \sigma, \xi, \varsigma)$ the expansion

$$v(t, \sigma, \xi, \varsigma) = 1 + \sum_{i+j+k \geq 1} v_{ijk}(t) \frac{\sigma^i}{i!} \frac{\xi^j}{j!} \frac{\varsigma^k}{k!}, \tag{1.4.17}$$

we obtain the LEM equivalent system for Lorenz's equations

$$\begin{aligned} \dot{v}_{ijk} &= is\left(v_{i-1,j+1,k} - v_{ijk}\right) + j\left(r v_{i+1,j-1,k} - v_{ijk} - v_{i+1,j-1,k+1}\right) \\ &\quad + k\left(v_{i+1,j+1,k-1} - \beta v_{ijk}\right) \qquad i, j \in \mathfrak{N}. \end{aligned} \tag{1.4.18}$$

From the previous examples, as well as from the construction pattern of the linear equivalents, it is easily seen that for n-th order ODEs we always have

$$\frac{\mathrm{d}v_{\mathbf{e}_j}}{\mathrm{d}x} = v_{\mathbf{e}_{j+1}}, \quad j = \overline{1,n}. \tag{1.4.19}$$

To write the linear equivalent system in matrix form, we firstly consider the infinite vector

$$\mathbf{V}(x) = [\mathbf{V}_k(x)]_{k\in\mathfrak{N}}, \qquad \mathbf{V}_k(x) \equiv [v_\gamma(x)]_{|\gamma|=k}, \tag{1.4.20}$$

whose components are also the coefficients of the expansion (1.4.1) of v.

A natural functional frame for $\mathbf{V}$ and, implicitly, for the LEM equivalent system, is given by the spaces

$$\mathcal{B}_n^k(I) = \left\{ \mathbf{V}(x) \equiv [v_\gamma(x)]_{|\gamma|\in\mathfrak{N}}; v_\gamma \in C^k(I), \|v_\gamma\|_k \le M^{|\gamma|}K, |\gamma| \in \mathfrak{N}^* \right\} \tag{1.4.21}$$

Let us note that $\mathcal{A}_n^k(I), \mathcal{B}_n^k(I)$ are isomorphic, with the application τ defined as

$$\tau : \mathcal{A}_n^k(I) \to \mathcal{B}_n^k(I), \quad \tau(v(x,\xi)) = \mathbf{V}(x) \equiv [v_\gamma(x)]_{|\gamma|\in\mathfrak{N}}. \tag{1.4.22}$$

1.4.1. THE LEM MATRIX

With the preceding specifications, the system (1.4.2) can be written in matrix form, using $\mathbf{V}$ as a vector whose components are the unknown functions:

$$\mathcal{S}\mathbf{V} \equiv \frac{\mathrm{d}\mathbf{V}}{\mathrm{d}x} - \mathbf{A}(x)\mathbf{V} - \mathbf{B}(x) = \mathbf{0}. \tag{1.4.23}$$

The conditions (1.4.4) become initial conditions for $\mathbf{V}$

$$\mathbf{V}(x_0) = \left[v_0^\gamma \right]_{|\gamma|\in\mathfrak{N}}. \tag{1.4.24}$$

Thus, we put into evidence the matrix associated to the LEM linear equivalent system (1.4.23), which depends only on the coefficients of the polynomial ODS. The matrix $\mathbf{A}$, generated by LEM, has a cell-diagonal structure, being row- and column-finite. On the main diagonal there are square cells $\mathbf{A}_{kk}(x)$, generated by the coefficients of the linear part of the operator defined by (1.1.5); actually, $\mathbf{A}_{11}(x)$ is its linear part. The first parallel to the main diagonal contains rectangular cells, generated by the coefficients of the second-degree terms. The coefficients of higher degree terms produce the next diagonal cell-bands, the p-th being the last one (remember that p is the degree of the polynomial ODS). There is only one sub-diagonal band, generated by the vector **B**, representing the free term of the system (1.4.23)

$$\mathbf{B}(x)=\left[a_{1\theta}(x)\quad a_{2\theta}(x)\quad a_{3\theta}(x)\ldots a_{n\theta}(x)\quad 0\quad 0\ldots\right]^{\mathrm{T}},\qquad \theta=(0\quad 0\quad \ldots 0). \tag{1.4.25}$$

More precisely, $\mathbf{A}(x)$ can be written emphasizing the cells $\mathbf{A}_{ij}(x)$:

$$\mathbf{A}(x)=\begin{bmatrix} \mathbf{A}_{11}(x) & \mathbf{A}_{12}(x) & \mathbf{A}_{13}(x) & \ldots & \mathbf{A}_{1p}(x) & \mathbf{0} & \ldots \\ \mathbf{A}_{21}(x) & \mathbf{A}_{22}(x) & \mathbf{A}_{23}(x) & \ldots & \mathbf{A}_{2p}(x) & \mathbf{A}_{2,p+1}(x) & \ldots \\ \mathbf{0} & \mathbf{A}_{32}(x) & \mathbf{A}_{33}(x) & \ldots & \mathbf{A}_{3p}(x) & \mathbf{A}_{3,p+1}(x) & \ldots \\ \mathbf{0} & \mathbf{0} & \mathbf{A}_{43}(x) & \ldots & \mathbf{A}_{4p}(x) & \mathbf{A}_{4,p+1}(x) & \ldots \\ \ldots & \ldots & \ldots & \ldots & \ldots & \ldots & \ldots \\ \mathbf{0} & \mathbf{0} & \mathbf{0} & \ldots & \mathbf{0} & \mathbf{A}_{p+2,p+1}(x) & \ldots \\ \ldots & \ldots & \ldots & \ldots & \ldots & \ldots & \ldots \end{bmatrix} \tag{1.4.26}$$

In what follows, we shall use the term *LEM matrix*, with reference to $\mathbf{A}(x)$.

The polynomial operator is therefore completely represented by the LEM matrix.

If $a_{j\theta}=0,\ j=\overline{1,n}$, then **B** is the null vector and the subdiagonal LEM cells $\mathbf{A}_{k,k\text{-}1}$ have only null entries. The LEM (1.4.23) becomes

$$\mathfrak{S}\mathbf{V}\equiv\frac{\mathrm{d}\mathbf{V}}{\mathrm{d}x}-\mathbf{A}(x)\mathbf{V}=\mathbf{0}, \tag{1.4.27}$$

where the LEM matrix reads

$$\mathbf{A}(x)=\begin{bmatrix} \mathbf{A}_{11}(x) & \mathbf{A}_{12}(x) & \mathbf{A}_{13}(x) & \ldots & \mathbf{A}_{1p}(x) & \mathbf{0} & \ldots \\ \mathbf{0} & \mathbf{A}_{22}(x) & \mathbf{A}_{23}(x) & \ldots & \mathbf{A}_{2p}(x) & \mathbf{A}_{2,p+1}(x) & \ldots \\ \mathbf{0} & \mathbf{0} & \mathbf{A}_{33}(x) & \ldots & \mathbf{A}_{3p}(x) & \mathbf{A}_{3,p+1}(x) & \ldots \\ \mathbf{0} & \mathbf{0} & \mathbf{0} & \ldots & \mathbf{A}_{4p}(x) & \mathbf{A}_{4,p+1}(x) & \ldots \\ \ldots & \ldots & \ldots & \ldots & \ldots & \ldots & \ldots \\ \mathbf{0} & \mathbf{0} & \mathbf{0} & \ldots & \mathbf{0} & \mathbf{A}_{p+1,p+1}(x) & \ldots \\ \ldots & \ldots & \ldots & \ldots & \ldots & \ldots & \ldots \end{bmatrix}. \tag{1.4.28}$$

Remark. The general form of the LEM equivalent ODS (1.4.23), as well as the form of its associated LEM matrix, prove that the only ODSs for which the associated LEM systems can be solved by separate blocks are the linear ones.

Let us take again some of the previous examples. in order to set up their corresponding LEM matrices.

Example 1.9. The LEM matrix for **Lorenz's system** is

$$\mathbf{A}(x)=\begin{bmatrix} \mathbf{A}_{11} & \mathbf{A}_{12} & \mathbf{0} & \mathbf{0} & \mathbf{0} & \mathbf{0} & \cdots \\ \mathbf{0} & \mathbf{A}_{22} & \mathbf{A}_{23} & \mathbf{0} & \mathbf{0} & \mathbf{0} & \cdots \\ \mathbf{0} & \mathbf{0} & \mathbf{A}_{33} & \mathbf{A}_{34} & \mathbf{0} & \mathbf{0} & \cdots \\ \mathbf{0} & \mathbf{0} & \mathbf{0} & \mathbf{A}_{44} & \mathbf{A}_{45} & \mathbf{0} & \cdots \\ \cdots & \cdots & \cdots & \cdots & \cdots & \cdots & \cdots \\ \mathbf{0} & \mathbf{0} & \mathbf{0} & \cdots & \mathbf{0} & \mathbf{A}_{p,p} & \cdots \\ \cdots & \cdots & \cdots & \cdots & \cdots & \cdots & \cdots \end{bmatrix}, \tag{1.4.29}$$

where the matrix of the linear part is

$$\mathbf{A}_{11}=\begin{bmatrix} -s & s & 0 \\ r & -1 & 0 \\ 0 & 0 & -\beta \end{bmatrix}, \tag{1.4.30}$$

and the first cell generated by the second-degree terms reads

$$\mathbf{A}_{12}=\begin{bmatrix} 0 & 0 & 0 & 0 & 0 & 0 \\ 0 & 0 & 0 & -1 & 0 & 0 \\ 0 & 1 & 0 & 0 & 0 & 0 \end{bmatrix}. \tag{1.4.31}$$

Example 1.10. The LEM matrix for **van der Pol's equation** can be "contracted", because the equation is odd in y; it will be

$$\mathbf{A}=\begin{bmatrix} \mathbf{A}_{11} & \mathbf{A}_{13} & \mathbf{0} & \mathbf{0} & \mathbf{0} & \mathbf{0} & \cdots \\ \mathbf{0} & \mathbf{A}_{33} & \mathbf{A}_{35} & \mathbf{0} & \mathbf{0} & \mathbf{0} & \cdots \\ \mathbf{0} & \mathbf{0} & \mathbf{A}_{55} & \mathbf{A}_{57} & \mathbf{0} & \mathbf{0} & \cdots \\ \mathbf{0} & \mathbf{0} & \mathbf{0} & \mathbf{A}_{77} & \mathbf{A}_{79} & \mathbf{0} & \cdots \\ \cdots & \cdots & \cdots & \cdots & \cdots & \cdots & \cdots \\ \mathbf{0} & \mathbf{0} & \mathbf{0} & \cdots & \mathbf{0} & \mathbf{A}_{2p-1,2p-1} & \cdots \\ \cdots & \cdots & \cdots & \cdots & \cdots & \cdots & \cdots \end{bmatrix}. \tag{1.4.32}$$

where the linear part of the operator is

$$\mathbf{A}_{11} = \begin{bmatrix} 0 & 1 \\ -1 & \mu \end{bmatrix}. \tag{1.4.33}$$

The other square cells from the main diagonal read

$$\mathbf{A}_{2j-1,2j-1} = \begin{bmatrix} 0 & 2j-1 & 0 & \dots & 0 & 0 \\ -1 & \mu & 2j-2 & \dots & 0 & 0 \\ 0 & -2 & 2\mu & \dots & 0 & 0 \\ \dots & \dots & \dots & \dots & 0 & 0 \\ 0 & 0 & 0 & \dots & (2j-2)\mu & 1 \\ 0 & 0 & 0 & \dots & -(2j-1) & (2j-1)\mu \end{bmatrix}. \tag{1.4.34}$$

The super-diagonal cells have the form

$$\mathbf{A}_{2j-1,2j+1} = -\mu \begin{bmatrix} 0 & 0 & 0 & \dots & 0 & 0 & 0 \\ 0 & 1 & 0 & \dots & 0 & 0 & 0 \\ 0 & 0 & 2 & \dots & 0 & 0 & 0 \\ \dots & \dots & \dots & \dots & 0 & 0 & 0 \\ 0 & 0 & 0 & \dots & 2j-1 & 0 & 0 \end{bmatrix}. \tag{1.4.35}$$

1.4.2. THE LOCAL LEM INVERSE OF A POLYNOMIAL OPERATOR

The above facts emphasize the idea that, once LEM applied, one must also find specific methods of tackling both the LEM linear equivalents: the PDE and the ODS. These methods must permit us to turn back to the solutions of the polynomial system we started from. This is put into evidence by the following theorem

Theorem 1.6 [135, 143]. *Let* $a_{j\nu} \in C^{\infty}(I)$, $a_{j\theta} = 0, j = \overline{1,n}$ *and suppose that* $\|a_{j\nu}\|_k \leq C, k \in \mathfrak{N}^*$. *Then the solution of the linear Cauchy problem* (1.4.27), (1.4.24) *(formally) allows the representation*

$$\mathbf{V}(x) = \mathbf{\Pi}(x - x_0)\mathbf{V}(x_0), \tag{1.4.36}$$

where the infinite matrix $\mathbf{\Pi}$ *is given by*

$$\Pi(x-x_0)=\sum_{k\geq 0}\mathbf{A}^{(k)}(x_0)\frac{(x-x_0)^k}{k!}. \qquad (1.4.37)$$

The matrices $\mathbf{A}^{(k)}$ *are determined by recurrence*

$$\mathbf{A}^{(k)}=\frac{\mathrm{d}\,\mathbf{A}^{(k-1)}}{\mathrm{d}x}+\mathbf{A}^{(k-1)}(x)\mathbf{A}(x),\quad \mathbf{A}^{(0)}=\mathbf{E}\equiv\left[\delta^i_j\right]_{i,j\in\mathfrak{N}}. \qquad (1.4.38)$$

The first n components of $\mathbf{V}$ *are consistent on the interval*

$$I_1=\{x\in I;|x-x_0|<h\},\quad h=\frac{1}{CQn(p+1)},\quad Q=\sum_{j=1}^{n}\sum_{|\mu|\leq p}|\mathbf{y}_0|^{|\gamma|}, \qquad (1.4.39)$$

and represent the Taylor series of the solution of the initial polynomial problem (1.1.5), (1.1.6) *around* x_0.

Proof. From (1.4.2) it immediately follows that

$$\left|\frac{\mathrm{d}v_\gamma}{\mathrm{d}x}\right|\leq CQn|\gamma||\mathbf{y}_0|^{|\gamma|}, \qquad (1.4.40)$$

and then, step by step, we have

$$\left|\frac{\mathrm{d}^m v_\gamma}{\mathrm{d}x^m}\right|\leq|\gamma|\left(|\gamma|+p\right)\left(|\gamma|+2p\right)\cdots\left(|\gamma|+(m-1)p\right)(CQn)^m|\gamma||\mathbf{y}_0|^{|\gamma|}. \qquad (1.4.41)$$

This leads to the inequality

$$\begin{aligned}|v_\gamma(x)|&\leq\sum_{m\geq 0}|\gamma|\left(|\gamma|+p\right)\left(|\gamma|+2p\right)\cdots\left(|\gamma|+(m-1)p\right)\frac{(CQn)^m}{m!}|x-x_0|^m|\mathbf{y}_0|^{|\gamma|}\\&<\frac{1}{1-(CQn)\left(|\gamma|+p\right)|x-x_0|}|\mathbf{y}_0|^{|\gamma|},\qquad x\in I_{|\gamma|}.\end{aligned} \qquad (1.4.42)$$

For the first n components of $\mathbf{V}$, $|\gamma|=1$, such that the inequality (1.4.42) ensures the consistency on I_1 of each $v_{\mathbf{e}_j}$. By theorems 1.3 and 1.5, we conclude that $\left[v_{\mathbf{e}_j}\right]_{j=\overline{1,n}}$ is the local solution of the initial polynomial problem (1.1.5), (1.1.6). ◘

Corollary 1.2 [134, 135, 143]. *Under the assumptions of theorem* 1.6, *the components* $v_\gamma(x)$ *of the vector* $\mathbf{V}(x)$ *given by* (1.4.36) *are each of them consistent on the intervals* $I_{|\gamma|}$ *defined by the inequalities*

$$I_{|\gamma|} = \{x \in I, |x - x_0| < CQn(|\gamma| + p)\}. \tag{1.4.43}$$

Proof. We firstly set up

$$v_\gamma(x) = \sum_{m=0}^{\infty} \frac{\mathrm{d}^m v_\gamma}{\mathrm{d}x^m}(x_0)\frac{(x - x_0)^m}{m!}; \tag{1.4.44}$$

from (1.4.41) it results that around x_0 the series $v_\gamma(x)$ satisfy

$$|v_\gamma(x)| \le \sum_{m \ge 0} (|\gamma| + p)^m (CQn|x - x_0|)^m |\mathbf{y}_0|^{|\gamma|}. \tag{1.4.45}$$

This implies the consistency of $v_\gamma(x)$ on $I_{|\gamma|}$.■

Following the same proof pattern, we can prove an extension of theorem 1.6 for non-homogeneous systems.

Theorem 1.7 [135, 143]. *Let* $a_{j\mu} \in \mathrm{C}^\infty(I)$ *and* $\|a_{j\mu}\|_k \le C, k \in \mathfrak{N}^*$. *Then the solution of the polynomial Cauchy problem* (1.4.23), (1.4.24) *(formally) allows the representation*

$$\mathbf{V}(x) = \mathbf{\Pi}(x - x_0)\mathbf{V}(x_0) + \mathbf{\Theta}(x - x_0), \tag{1.4.46}$$

where the infinite matrix $\mathbf{\Pi}$ *is given by* (1.4.37), (1.4.38) *and the vector* $\mathbf{\Theta}$ *– by the formula*

$$\mathbf{\Theta}(x - x_0) = \sum_{k \ge 0} \mathbf{B}^{(k-1)}(x_0)\frac{(x - x_0)^k}{k!}. \tag{1.4.47}$$

The vectors $\mathbf{B}^{(k)}$ *are determined by recurrence*

$$\mathbf{B}^{(k)} = \frac{\mathrm{d}\mathbf{B}^{(k-1)}}{\mathrm{d}x} + \mathbf{B}^{(k-1)}(x)\mathbf{A}(x), \quad \mathbf{B}^{(0)} = \mathbf{B}. \tag{1.4.48}$$

The first n components of $\mathbf{V}$ *are consistent on the interval* (1.4.39) *and coincide with the Taylor series of the polynomial initial problem* (1.1.5), (1.1.6) *around* x_0.

Remark. In the particular case of homogeneous ODSs with constant coefficients, the $\mathbf{\Pi}$ matrix takes an exponential form.

Corollary 1.3 [135, 143]. *In the case of polynomial homogeneous ODSs* ($a_{j\theta} = 0, j = \overline{1,n}$) *with constant coefficients, the solution of the* *polynomial initial problem* (1.1.5), (1.1.6) *locally coincides with the first n components of the vector*

$$\mathbf{V}(x) = e^{\mathbf{A}(x-x_0)}\mathbf{V}_0, \quad e^{\mathbf{A}(x-x_0)} = \mathbf{E} + \sum_{m=1}^{\infty} \mathbf{A}^m \frac{(x-x_0)^m}{m!}. \qquad (1.4.49)$$

This representation coincide in form with the well known result of the theory of linear ODSs with constant coefficients; the powers of the LEM matrix – constant in this case – are computed by block partitioning. As in the linear case, the knowledge of the eigenvalues of the square cells from the main diagonal is very useful.

A direct algebraic calculation of the corresponding determinants leads to

Theorem 1.8 [132, 139, 143]. *Let* $\lambda_j, j = \overline{1,n}$ *be the eigenvalues of the* $n \times n$ *matrix* $\mathbf{A}_{11}$*,the linear part of the polynomial operator given by* (1.1.5). *Then the eigenvalues of any diagonal cell* $\mathbf{A}_{mm}$ *is expressed in terms of* λ_j *in the form of the linear combinations*

$$\langle \mu, \lambda \rangle \equiv \sum_{j=1}^{n} \mu_j \lambda_j, \quad |\mu| = m, \qquad (1.4.50)$$

where μ_j *are components of some multi-indices of length m.*

In fact, the last two theorems identify the polynomial initial problem (1.1.5), (1.1.6) with an initial problem associated to a linear, while infinite, ODS. A suggestive expression of this affirmation can be obtained by using the definitions and the results of Section 1.2.

Indeed, let us index by 0 the restrictions of $\mathcal{P}, \mathcal{L}, \mathcal{S}$ to admissible sets of functions/vector functions satisfying the conditions (1.1.6), (1.3.6) and (1.4.24) respectively. Then the theorems 1.2, 1.3 and 1.7 imply

Theorem 1.9 [137, 143]. *The chain* $\{\mathcal{P}, \mathcal{L}, \mathcal{S}\}$ *is resonant, and the chain of their restrictions* $\{\mathcal{P}_0, \mathcal{L}_0, \mathcal{S}_0\}$ *is resolvent.*

This theorem completes the preceding diagram (1.3.45) as follows

$$
\begin{aligned}
&\mathcal{P}:\left(\mathrm{C}^1(I)\right)^n \to \left(\mathrm{C}^0(I)\right)^n \\
&\mathrm{e}^{\langle \xi, \bullet \rangle} \downarrow \\
&\mathcal{L}:\mathcal{A}_n^1(I) \to \mathcal{A}_n^0(I) \\
&\tau \ \ \downarrow \\
&\mathcal{S}:\mathcal{B}_n^1(I) \to \mathcal{B}_n^0(I)
\end{aligned}
\tag{1.4.51}
$$

The diagram (1.4.51) is not closed; yet, by the previous theorem, one can get back to the local solution of the polynomial Cauchy problem.

The same mentioned above theorems also involve that the matrix $\mathbf{\Pi}$ defined by (1.4.37) is essential for the effective determination of the LEM solutions of the polynomial system (1.1.5).

The first n rows of the matrix $\mathbf{\Pi}$ represent the *local LEM inverse of the polynomial operator*. The LEM inverse does not depend on the initial data, it only depends on the coefficients of the polynomial system. This is one of the most effective specific advantages of LEM. From formulae (1.4.36) and (1.4.46) we see that, within the LEM solution, the contribution of the coefficients of the polynomial ODS is clearly separated from that of the initial data. Therefore, it is important to get $\mathbf{\Pi}$ numerically or analytically, or, at least, to find proper techniques of getting approximately its first n rows. Once we have obtained these first n rows, we can set up the transport matrix of the solution of a polynomial ODS, that can be applied to any set of initial data.

In a certain way, $\mathbf{\Pi}$ corresponds to the *fundamental matrix* associated to a linear ODS.

Let us also mention that, from the numerical point of view, the local LEM inverse leads to specific LEM algorithms, that, unlike those corresponding to usual standard methods (e.g. Euler, Runge-Kutta, Adams-Bashforth-Moulton, etc.), do not repeat for any new set of initial data; thus, the LEM algorithms save a lot of computation, both in time and memory.

Another advantage is the possibility of realizing the matrix products required by the recurrence formulae (1.4.38) or (1.4.48) by block partitioning, as a consequence of the row-finiteness of the LEM matrix; moreover, if we truncate the series (1.4.37) up to a degree m, all the necessary computation for getting the first n components of $\mathbf{V}$ are finite.

We will retake all these facts in detail in the next chapter.

Chapter 2

LEM APPROXIMATIONS AND EXTENSIONS OF THE METHOD

In the preceding chapter, we defined LEM and set up the LEM solution for a polynomial Cauchy problem, computed by block partitioning. In this chapter, we shall present the local LEM inverse for a polynomial operator, which gave rise to the transport matrix of the solution and to specific LEM numerical algorithms. Along with the extensions of LEM to more general nonlinear operators, presented in section 2.3, these results proved to be very useful in applications.

2.1. A CONVERGENT LEM ALGORITHM

Starting from the linear equivalent system, one can set up a convergent algorithm for the solutions of a polynomial ODS.

Consider, therefore, the polynomial system

$$\mathcal{G}\mathbf{y} \equiv \frac{d\mathbf{y}}{dx} - \mathbf{P}(x,\mathbf{y}) = \mathbf{0}, \quad \mathbf{P} = \left[P_j(x,\mathbf{y})\right]_{j=\overline{1,n}}, \qquad P_j(x,\mathbf{y}) \equiv \sum_{|\mu| \le p_j} a_{j\mu}(x)\mathbf{y}^{\mu}, \tag{2.1.1}$$

that must be solved in the Cauchy conditions

$$\mathbf{y}(x_0) = \mathbf{y}_0, \quad x_0 \in I, \mathbf{y}_0 \equiv \left[y_{j0}\right]_{j=\overline{1,n}} \in \Re^n. \tag{2.1.2}$$

Suppose that $a_{j\mu}(x)$ – the coefficients of the polynomial system (2.1.1) – are polynomials of degree q_j in x

$$a_{j\mu}(x) = \sum_{s=0}^{q_j} a_{j\mu}^s (x - x_0)^s, \tag{2.1.3}$$

and let p, q be defined by

$$p=\max\left\{p_j, j=\overline{1,n}\right\} \quad q=\max\left\{q_j, j=\overline{1,n}\right\}. \tag{2.1.4}$$

Also suppose that the coefficients $v_\gamma(x)$ of the series (1.4.1) locally allow the following expansion

$$v_\gamma(x)=\sum_{i=0}^{\infty} v_\gamma^i (x-x_0)^i ; \tag{2.1.5}$$

later on, we shall specify the neighbourhood of x_0 on which (2.1.5) converges. Applying LEM, we obtain the corresponding linear equivalent system; taking it into account, we obtain for the coefficients v_γ^i by identification

$$(i+1)v_\gamma^{i+1}=\sum_{j=1}^{n}\gamma_j \sum_{|\mu|\le p_j} \sum_{\substack{m=0,q_i \\ m\le i}} a_{j\mu}^m v_{\gamma+\mu-\mathbf{e}_j}^{i-m}, \quad |\gamma|\in \mathfrak{N}, \quad i\in\mathfrak{N}^*. \tag{2.1.6}$$

The relations (2.1.6) represent, in fact, a linear explicit algorithm for the coefficients v_γ^i, which can be determined for any i, starting from $i=0$.

For $i=0$, the values v_γ^0 are determined from the initial conditions (1.4.4)

$$v_\theta^0=\mathbf{y}_0^\gamma \quad |\gamma|\in\mathfrak{N}. \tag{2.1.7}$$

As in the series (1.4.1) of v the free term is 1, we shall also have

$$v_\theta^0=1, \qquad v_\theta^m=0, \qquad m\in\mathfrak{N}; \tag{2.1.8}$$

let us recall that θ is the multi-index of n null components.

Following the same pattern as in the proof of theorem 1.6, we shall establish the convergence of the above algorithm

Theorem 2.1 [134, 135]. *The algorithm given by formulae* (2.1.6)-(2.1.8) *converges for*

$$|x-x_0|<\frac{1}{cn\widetilde{Q}(p+1)}, \tag{2.1.9}$$

where

$$\widetilde{Q}=\sum_{|\mu|\le p}\left|\mathbf{y}_0^{\mu-\mathbf{e}_j}\right|, \quad c=\max\left\{\left|a_{j\mu}^s\right|, |\mu|\le p_j, s\le q_j, j=\overline{1,n}\right\}, \tag{2.1.10}$$

and the series

$$\sum_{m=1}^{\infty} v_{\mathbf{e}_j}^m (x-x_0)^m, \quad j=\overline{1,n}, \tag{2.1.11}$$

coincide with the Taylor series of the components $y_j(x)$, $j=\overline{1,n}$, *of the solution of the polynomial Cauchy problem* (2.1.1), (2.1.2).

Proof. We shall follow the same way as in theorem 1.6. For the first iteration, one has

$$\left|v_\gamma^1\right| \le c\sum_{j=1}^{n} \gamma_j \sum_{|\mu|\le p_j} \sum_{s=0}^{q_j} \left|\mathbf{y}_0^{\mu-\mathbf{e}_j}\right| \left|\mathbf{y}_0^\gamma\right| \le c|\gamma|Qn\left|\mathbf{y}_0^\gamma\right|; \tag{2.1.12}$$

at the m-th step, we get, by induction,

$$\left|v_\gamma^m\right| \le \frac{|\gamma|(|\gamma|+p)(|\gamma|+2p)\cdots(|\gamma|+(m-1)p)}{m!}(cQn)^m\left|\mathbf{y}_0^\gamma\right|. \tag{2.1.13}$$

Finally, we deduce the following majorizing for the coefficients of the series (2.1.5)

$$\begin{aligned} |v_\gamma(x)| &\le \sum_{m\ge 0} \left|v_\gamma^m\right| |x-x_0|^m \\ &\le \sum_{m\ge 0} \frac{|\gamma|(|\gamma|+p)(|\gamma|+2p)\cdots(|\gamma|+(m-1)p)}{m!}(cQn|x-x_0|)^m\left|\mathbf{y}_0^\gamma\right| \\ &\le \sum_{m\ge 0} (|\gamma|+p)^m(cQn|x-x_0|)^m\left|\mathbf{y}_0^\gamma\right|. \end{aligned} \tag{2.1.14}$$

This means that the above algorithm converges for $|\gamma|=1$ and for x belonging to the interval defined by (2.1.9). By virtue of theorems 1.3 and 1.6, it converges to the solution $\mathbf{y}$ of the polynomial initial problem (1.1.5), (1.1.6).□

From theorem 2.1, it also results that the components y_j, $j=\overline{1,n}$, of this solution locally allow the representation

$$y_j(x) = \sum_{m=1}^{\infty} v_{\mathbf{e}_j}^m (x-x_0)^m, \quad j=\overline{1,n}. \tag{2.1.15}$$

In conclusion, theorem 2.1 validates the correspondence between the polynomial initial problem and its associated linear one. At the same time, it offers an effective representation of the solution in the case of polynomial coefficients.

2.2. TRUNCATED LEM SYSTEMS AND THE LEM TRANSPORT MATRIX OF THE SOLUTIONS

Let us note that, due to theorems 1.2 and 1.3, it is not necessary to study the linear equivalent in more general assumptions on the behaviour of v; it suffices to consider v analytic with respect to ξ and then to truncate the corresponding series. Yet, this truncation resulting in the truncation of the linear equivalent system, cannot be arbitrarily done up to any term, because this could lead to an unsatisfactory result.

We shall show how this truncation is realized in order to obtain good approximates of the solutions of polynomial ODSs.

2.2.1. TRUNCATED LEM SYSTEMS

We start from the truncation of the LEM exponential

$$v^{(m)}(x,\xi)=1+\sum_{|\gamma|=1}^{m} v_{\gamma}^{(m)}(x)\frac{\xi^{\gamma}}{\gamma!}, \quad \xi\in\Re^{n}. \tag{2.2.1}$$

From the second LEM equivalent (1.4.2) we get a finite linear ODS

$$\frac{\mathrm{d}v_{\gamma}^{(m)}}{\mathrm{d}x}=\sum_{j=1}^{n}\gamma_j\sum_{|\mu|\le p_j} a_{j\mu}(x)v_{\gamma+\mu-\mathbf{e}_j}^{(m)}{}_{j\,\gamma+\mu-\mathbf{e}_j}, \quad |\gamma|\le m, \tag{2.2.2}$$

whose solution is the closer to the exact solution of the polynomial initial problem as m increases.

Let us express this by using the form (1.4.23) of the LEM system. The above truncation can be considered as a result of the application of a projector $\boldsymbol{\pi}^{(m)}:\mathcal{B}_n^0(I)\to\mathcal{B}_n^{0,m}(I)$ associating to any infinite vector $\mathbf{V}$ a vector $\mathbf{V}^{(m)}$ from the finite dimensional $\mathcal{B}_n^{0,m}(I)$, obtained by cancelling all components $v_{\gamma}(x)$ for which $|\gamma|>m$. Obviously, if $\mathbf{V}\in\mathcal{B}_n^1(I)$, then $\boldsymbol{\pi}^{(m)}\mathbf{V}\in\mathcal{B}_n^1(I)$ too. Denote now by $\ker\mathfrak{S}$, $\ker\boldsymbol{\pi}^{(m)}\mathfrak{S}$ the kernels of the operators $\mathfrak{S}$, $\boldsymbol{\pi}^{(m)}\mathfrak{S}$ respectively. In [143] I proved that

$$\ker\mathfrak{S}=\left(\bigcap_{m}\ker\boldsymbol{\pi}^{(m)}\mathfrak{S}\right)\cap\mathcal{B}_n^1(I). \tag{2.2.3}$$

Remark. In the case of constant coefficients, the above double inclusion and theorem 1.8 emphasize a new way of approximating the solutions of polynomial problems: by exponential or trigonometric functions, occasionally multiplied by polynomials, in connection with the linear part of the operator. This method is completely distinct from Taylor series expansion, currently in use.

The above remark also leads to the idea of the direct approximation, by exponentials, of the solution of a nonlinear system, an approximation in which one previously knows the exact exponents, unlike those considered in some numerical methods [38], in which one approximates them afterwards.

Let us specify that the LEM approximations do not coincide with the solutions thus configured.

The projector also cancels the entries of the cells of the LEM matrix, starting with the $m+1$-th block upwards. The truncated system (2.2.2), written in matrix form, reads

$$\boldsymbol{\pi}^{(m)}\mathfrak{S}\mathbf{V} \equiv \frac{\mathrm{d}\mathbf{V}^{(m)}}{\mathrm{d}x} - \mathbf{A}^{(m)}\mathbf{V}^{(m)} - \boldsymbol{\pi}^{(m)}\mathbf{B} = \mathbf{0}. \tag{2.2.4}$$

In the case of homogeneous systems, $a_{j\theta} = 0,\ j = \overline{1,n}$. The approximating LEM matrix $\mathbf{A}^{(m)}(x)$ becomes

$$\mathbf{A}^{(m)}(x) = \begin{bmatrix} \mathbf{A}_{11}(x) & \mathbf{A}_{12}(x) & \dots & \mathbf{A}_{1p}(x) & \mathbf{0} & \dots & \mathbf{0} \\ \mathbf{0} & \mathbf{A}_{22}(x) & \dots & \mathbf{A}_{2p}(x) & \mathbf{A}_{2,p+1}(x) & \dots & \mathbf{0} \\ \dots & \dots & \dots & \dots & \dots & \dots & \dots \\ \mathbf{0} & \mathbf{0} & \mathbf{0} & \mathbf{0} & \mathbf{0} & \dots & \mathbf{A}_{mm}(x) \end{bmatrix}, \tag{2.2.5}$$

hence the finite approximating system (2.2.2) can be written by blocks

$$\begin{aligned} &\frac{\mathrm{d}\mathbf{V}_1^{(m)}}{\mathrm{d}x} = \mathbf{A}_{11}(x)\mathbf{V}_1^{(m)} + \mathbf{A}_{12}(x)\mathbf{V}_2^{(m)} + \dots \mathbf{A}_{1m}(x)\mathbf{V}_m^{(m)}, \\ &\frac{\mathrm{d}\mathbf{V}_2^{(m)}}{\mathrm{d}x} = \mathbf{A}_{22}(x)\mathbf{V}_2^{(m)} + \mathbf{A}_{23}(x)\mathbf{V}_3^{(m)} + \dots \mathbf{A}_{2m}(x)\mathbf{V}_m^{(m)}, \\ &\dots\dots\dots\dots\dots\dots\dots\dots\dots\dots\dots\dots\dots\dots \\ &\frac{\mathrm{d}\mathbf{V}_{m-1}^{(m)}}{\mathrm{d}x} = \mathbf{A}_{m-1,m-1}(x)\mathbf{V}_{m-1}^{(m)} + \mathbf{A}_{m-1,m}(x)\mathbf{V}_m^{(m)}, \\ &\frac{\mathrm{d}\mathbf{V}_m^{(m)}}{\mathrm{d}x} = \mathbf{A}_{mm}(x)\mathbf{V}_m^{(m)}, \end{aligned} \tag{2.2.6}$$

where we used again the finite segments $\mathbf{V}_m$, defined in (1.4.20).

The finite linear ODS (2.2.6) must be solved in the Cauchy conditions

$$\mathbf{V}_k^{(m)}(x_0)=\left[\mathbf{y}_0^{\gamma}\right]_{|\gamma|=k},\ k=\overline{1,m}. \tag{2.2.7}$$

The form of this system clearly suggests the mode of determining the solution. One starts from the last block that represents a linear homogeneous ODS, to which one associates the corresponding initial conditions deduced from (2.2.7)

$$\frac{\mathrm{d}\mathbf{V}_m^{(m)}}{\mathrm{d}x}=\mathbf{A}_{mm}\mathbf{V}_m^{(m)},\qquad \mathbf{V}_m^{(m)}(x_0)=\left[\mathbf{y}_0^{\mu}\right]_{|\mu|=m}. \tag{2.2.8}$$

Using classical methods, we find the (unique) solution of this initial problem.

Indeed, considered the n-dimensional ODS associated to the linear part of the operator with the corresponding Cauchy condition

$$\begin{aligned}&\frac{\mathrm{d}\mathbf{V}_1}{\mathrm{d}x}=\mathbf{A}_{11}(x)\mathbf{V}_1,\\&\mathbf{V}_1(x_0)=\mathbf{y}_0.\end{aligned} \tag{2.2.9}$$

As the coefficients of this system are at least continuous, the solution of (2.2.9) exists at least locally and it is unique. We can prove the following general result, shaping the solution of the linear initial problem (2.2.8) for any m:

Theorem 2.2 [153]. *If* $\mathbf{Y}$ *is the unique solution of* (2.2.9)*, then* $\left(\mathbf{Y}^{\mu}\right)_{|\mu|=m}$ *is the unique solution of the finite linear Cauchy problem* (2.2.8) *for any m.*

Proof. We apply the LEM mapping to the linear system (2.2.9). The associated LEM matrix will contain only the diagonal cells $\mathbf{A}_{mm}(x)$ and the system (2.2.6) will have only de-coupled blocks that one can easily solve separately. According to theorems 1.2 and 1.3, the analytic solution of the linear LEM equivalent (1.3.4), also satisfying the condition

$$v(x_0,\xi)=\mathrm{e}^{\langle\xi,\mathbf{y}_0\rangle}, \tag{2.2.10}$$

has the exponential form

$$v(x,\xi)=\mathrm{e}^{\langle\xi,\mathbf{Y}\rangle}, \tag{2.2.11}$$

where $\mathbf{Y}$ is the solution of the Cauchy problem (2.2.9); taking into account the expansion (1.4.1), it is seen that $\mathbf{V}_k(x)=\left(\mathbf{Y}^{\mu}(x)\right)_{|\mu|=m}$. ◘

Once $\mathbf{V}_m^{(m)}$ determined, we get on with the $(m-1)$-th block, which, by the replacement of $\mathbf{V}_m^{(m)}$, now known, becomes a non-homogeneous ODS, still linear. After $m-1$ steps, we obtain the finite vector $\mathbf{V}_1^{(m)}$, whose components approximate the solution $\mathbf{y}$ of the nonlinear initial problem (1.1.5), (1.1.6).

Remark. This approximation does no more coincide with the truncation of the Taylor series of the solution; it was obtained on a completely different way.

Let us also specify that, in order to take into account the influence of all the coefficients $a_{j\mu}$, it is necessary to choose a truncating order $m \geq p$.

2.2.2. THE LEM TRANSPORT MATRIX OF THE SOLUTION

From theorems 1.6 and 1.7 it follows that the matrix $\mathbf{\Pi}$, defined in (1.4.37), (1.4.38), plays an important part in getting solutions of a nonlinear ODS by LEM.

This matrix gives the general for of the solution of a polynomial initial problem with arbitrary data; we previously specified that, for this reason, the first n rows of $\mathbf{\Pi}$ represent in fact the *local inverse* of the considered polynomial operator.

This particularity of LEM is extremely useful; it leads to specific schemes of getting the solution independently of the initial data. In fact, the LEM representations themselves separate the contribution of the coefficients of the nonlinear ODS from that of the data.

Clearly, it is important to get this local inverse – more precisely, the first n rows of the matrix $\mathbf{\Pi}$ – on intervals of convergence of the type (1.4.39). By doing this, we obtain a transport matrix of the solution, which can be applied to any admissible set of initial data.

This advantage of LEM, commonly met in the linear, but never in the nonlinear frame, leads to numerical algorithms completely different from the standard ones. The current numerical methods for nonlinear Cauchy problems completely depend on the initial data; this is the reason why the corresponding algorithms must be repeated for any new set of initial data, a shortcoming overcome by LEM.

Starting from the above locally valid representation formulae, we can set up a transport matrix for the solutions of a polynomial ODS.

Indeed, theorem 1.6 ensures the absolute (and uniform) convergence of the solution of the polynomial initial problem (1.1.5), (1.1.6) on the interval I_1 defined by (1.4.39). Suppose that we wish to obtain the inverse of the polynomial operator on an interval $I' = [x_0, x_0 + a)$, $I' \subset I \subset \Re$, for which we could establish a majorant M for solutions starting from a known range of initial values around $\mathbf{y}_0$. To this purpose, we build up an equidistant network on I', say, $\{x_0, x_1, \ldots, x_n\} \subset I'$, $x_{k-1} < x_k$, $k = \overline{1,q}$, of step h strictly less than h_0, where

$$h_0 = \frac{1}{C(p+1)\sum_{|\mu|=0}^{p} M^{|\mu|}}. \tag{2.2.12}$$

The interval I' is thus divided in q subintervals $I'_k = [x_{k-1}, x_k)$, $k = \overline{1,q}$, each of them of length less than h.

With the notations from theorems 1.6 and 1.7, the first n rows of the matrix $\Pi(x - x_{k-1})$, as well as the first n components of the vector $\Theta(x - x_{k-1})$, converge on each I'_k, $k = \overline{1,q}$ and do not depend on the initial data. By virtue of theorem 1.7, we can approximate any solution of (1.1.5), considering only a finite number m of blocks of $\mathbf{\Pi}$, determined by a finite number of operations. Denote by $\mathbf{\Pi}^{(m)}$ the corresponding truncated matrix and by $\mathbf{\Theta}^{(m)}$ the truncated vector, corresponding to $\mathbf{\Theta}$ from (1.4.47). We can apply the projector $\boldsymbol{\pi}^{(1)}$, which cancels all the rows of $\mathbf{\Pi}^{(m)}$ and $\mathbf{\Theta}^{(m)}$, except for the first n ones. We then determine the approximations $\boldsymbol{\pi}^{(1)}\left(\mathbf{\Pi}^{(m)}(x_k - x_{k-1})\right)$, $\boldsymbol{\pi}^{(1)}\left(\mathbf{\Theta}^{(m)}(x_k - x_{k-1})\right)$ for $k = \overline{1,q}$; obviously, this calculation will require only a finite number of steps. We thus built up a new matrix, of rows $\boldsymbol{\pi}^{(1)}\left(\mathbf{\Pi}^{(m)}(x_k - x_{k-1})\right)$, $k = \overline{1,q}$, having m column-blocks

$$\mathbf{T}^{(m)} = \begin{bmatrix} \boldsymbol{\pi}^{(1)}\left(\mathbf{\Pi}^{(m)}(x_1 - x_0)\right) \\ \boldsymbol{\pi}^{(1)}\left(\mathbf{\Pi}^{(m)}(x_2 - x_1)\right) \\ \boldsymbol{\pi}^{(1)}\left(\mathbf{\Pi}^{(m)}(x_3 - x_2)\right) \\ \vdots \\ \boldsymbol{\pi}^{(1)}\left(\mathbf{\Pi}^{(m)}(x_q - x_{q-1})\right) \end{bmatrix}. \tag{2.2.13}$$

We call the matrix $\mathbf{T}^{(m)}$ *the transport matrix of the solutions of the polynomial ODS*. The number of rows of the transport matrix is qn.

Denote by $\mathbf{\Psi}^{(m)}$ the vector of qn components, obtained by the successive components of $\boldsymbol{\pi}^{(1)}\left(\mathbf{\Theta}^{(m)}(x_k - x_{k-1})\right)$ for $k=\overline{1,q}$

$$\mathbf{\Psi}^{(m)} = \begin{bmatrix} \boldsymbol{\pi}^{(1)}\left(\mathbf{\Theta}^{(m)}(x_1 - x_0)\right) \\ \boldsymbol{\pi}^{(1)}\left(\mathbf{\Theta}^{(m)}(x_2 - x_1)\right) \\ \boldsymbol{\pi}^{(1)}\left(\mathbf{\Theta}^{(m)}(x_3 - x_2)\right) \\ \vdots \\ \boldsymbol{\pi}^{(1)}\left(\mathbf{\Theta}^{(m)}(x_q - x_{q-1})\right) \end{bmatrix}. \tag{2.2.14}$$

To use the transport matrix for a set of initial data $\mathbf{y}_0$, we firstly set up $\boldsymbol{\pi}^{(m)}(\mathbf{V}(x_0))$. To get the solution at the point $x_1 = x_0 + h$ we shall compute the following finite scalar products and sums

$$\begin{aligned} \boldsymbol{\pi}^{(1)}\left(\mathbf{V}^1\right) \equiv \boldsymbol{\pi}^{(1)}\mathbf{V}(x_1) = \boldsymbol{\pi}^{(1)}\left(\mathbf{\Pi}^{(m)}(x_1 - x_0)\ \boldsymbol{\pi}^{(m)}\mathbf{V}(x_0)\right) + \\ + \boldsymbol{\pi}^{(1)}\left(\mathbf{\Theta}^{(m)}(x_1 - x_0)\right) \end{aligned} \tag{2.2.15}$$

The starting point for the next step is $\boldsymbol{\pi}^{(1)}\left(\mathbf{V}^1\right)$, by means of which we set up an approximation for $\boldsymbol{\pi}^{(m)}(\mathbf{V}(x_1))$, repeating this cycle up to the step q.

Remark. The transport matrix will not remain in a purely numerical frame, as we also have the benefit of analytical expressions for the approximations of $\mathbf{\Pi}(x - x_{k-1})$, as well as for $\mathbf{\Theta}(x - x_{k-1})$; thus, LEM leads to a semi-analytic method, independent of the initial data.

This idea was applied to set up a transport matrix for relativistic electron beams (REB)[162]; before this, such a transport matrix could only be set up in the non-relativistic case.

We used the LEM transport matrix for REBs for the fast calculation of a great number of trajectories. This allowed us to determine in real time the shape of the beam for a given potential configuration, as well as to compute other physical quantities of interest, such as the cross-over or the beam density distribution.

Another pertinent application of the LEM transport matrix is the study of the Bernoulli-Euler bar in various hypostases and for various bar shapes [94-97,100-108, 112, 116, 122, 125, 126].

All these applications will be treated in detail in chapters 5 and 7.

2.3. LEM EXTENSIONS

The linear equivalence method can be extended to other classes of operators, more general than the polynomial ones. Using polynomials in the unknown functions, we can approximate a nonlinear operator by a polynomial one.

The problem in this case is to estimate how close is the solution of the approximating polynomial ODS to the exact one.

Consider the nonlinear ODS

$$\mathfrak{F}\mathbf{y} \equiv \frac{\mathrm{d}\mathbf{y}}{\mathrm{d}x} - \mathbf{f}(x, \mathbf{y}) = \mathbf{0}, \quad \mathbf{f}(x, \mathbf{y}) = \left[f_j(x, \mathbf{y}) \right]_{j=\overline{1,n}}, \tag{2.3.1}$$

currently met in the nonlinear dynamics; here, we take $f_j \in \mathcal{A}_n^0(I)$, in other words, $f_j(x, \xi)$ allows an expansion of the form

$$f_j(x, \xi) = \sum_{|\mu| \geq 0} f_{j\mu}(x) \xi^{\mu}, \quad x \in I, \xi \in \mathfrak{R}^n, \tag{2.3.2}$$

valid on I absolutely and uniformly.

As in the preceding cases, let us associate to the system (2.3.1) the Cauchy conditions

$$\mathbf{y}(x_0) = \mathbf{y}_0, \quad x_0 \in I, \; \mathbf{y}_0 \equiv \left[y_{j0} \right]_{j=\overline{1,n}} \in \mathfrak{R}^n. \tag{2.3.3}$$

Applying the LEM mapping, we obtain a formal PDE

$$\mathfrak{L}_{\mathbf{f}} v \equiv \frac{\partial v}{\partial x} - \left\langle \xi, \mathbf{f}(x, \mathrm{D}_{\xi}) \right\rangle v = 0, \tag{2.3.4}$$

in which $\left\langle \xi, \mathbf{f}(x, \mathrm{D}_{\xi}) \right\rangle$ is, as in the case of polynomial operators, the formal scalar product

$$\left\langle \xi, \mathbf{f}(x, \mathrm{D}_{\xi}) \right\rangle \equiv \sum_{j=1}^{n} \xi_j f_j(x, \mathrm{D}_{\xi}), \tag{2.3.5}$$

and

$$f_j(x, \mathrm{D}_{\xi}) = \sum_{|\mu|=0}^{\infty} f_{j\mu}(x) \mathrm{D}_{\xi}^{\mu}. \tag{2.3.6}$$

One can say that (2.3.5) is linear, in a generalized sense. It represents the first LEM equivalent of the nonlinear system (2.3.1). Yet, in order to benefit by its linearity, we must first prove its consistency.

Theorem 2.3 [143]. *If* $f_j \in \mathcal{C}_n^0(I), j=\overline{1,n}$, *then* $f_j(x, \mathrm{D}_\xi), j=\overline{1,n}$ *are consistent on* $\mathcal{C}_n^0(\mathrm{I})$.

Proof. As $f_j \in \mathcal{C}_n^0(I), j=\overline{1,n}$, it follows that one can find $C_j, j=\overline{1,n}$ such that

$$\sup_{x\in I}\left| f_{j\mu}(x) \right| \le C_j, \quad |\mu| \in \mathfrak{N}^*, \quad j=\overline{1,n}. \tag{2.3.7}$$

Let $C=\max\left\{C_j, j=\overline{1,n}\right\}$. Then, if $v \in \mathcal{C}_n^0(I)$, it results that v allows a development of the form (1.4.1) and that a positive M exists such that

$$\sup_{x\in I}\left| v_\gamma(x)\right| \le M^{|\gamma|}, \quad |\gamma| \in \mathfrak{N}. \tag{2.3.8}$$

Consequently,

$$\begin{aligned}\left|f_j(x,\mathrm{D}_\xi)v\right| &\equiv \left|\sum_{|\mu|=0}^{\infty} f_{j\mu}(x)\mathrm{D}_\xi^\mu v\right| \equiv \left|\sum_{|\mu|=0}^{\infty} f_{j\mu}(x)\sum_{|\gamma|\ge 0} v_{\gamma+\mu}(x)\frac{\xi^\mu}{\mu!}\right| \\ &\le \sum_{|\mu|=0}^{\infty}\sum_{|\gamma|\ge 0}\frac{C^{|\mu|}}{|\mu|!}(Mn)^{|\gamma|+|\mu|}\frac{|\xi|^{|\gamma|}}{|\gamma|!} \le \mathrm{e}^{(C+|\xi|)Mn}.\end{aligned} \tag{2.3.9}$$

This last equality proves the consistency of the linear equivalent PDE (2.3.4) on $\mathcal{C}_n^0(I)$. ◘

Therefore, the linear equivalence of the ODS (2.3.1) is a well posed problem and the corresponding LEM PDE is consistent on the Exp-type spaces previously defined.

To make use of this result, we must conveniently approximate $f_j(x,\mathbf{y})$, e.g. by using the ideas of L.B.Rall and his disciples [70]. The most natural way of approximation is to consider the partial sums of the series of f_j, as $f_j \in \mathcal{C}_n^0(I)$.. Let $m=m(\varepsilon)$ be such that

$$\left|R_j(\varepsilon)\right| \equiv \left|\sum_{|\mu|=m(\varepsilon)+1}^{\infty} f_{j\mu}(x)\xi^\mu\right| \le \sum_{|\mu|=m(\varepsilon)+1}^{\infty}\left|f_{j\mu}(x)\right||\xi|^{|\mu|} < \varepsilon \tag{2.3.10}$$

and consider the polynomial ODS

$$\mathcal{P}_\varepsilon \mathbf{y} \equiv \frac{\mathrm{d}\mathbf{y}}{\mathrm{d}x} - \mathbf{P}_\varepsilon(x,\mathbf{y}) = \mathbf{0}, \quad \mathbf{P}_\varepsilon(x,\mathbf{y}) = \left[P_{j,\varepsilon}(x,\mathbf{y})\right]_{j=\overline{1,n}}, \tag{2.3.11}$$

where

$$P_{j,\varepsilon}(x,\xi)=\sum_{|\mu|=0}^{m(\varepsilon)} f_{j\mu}(x)\xi^{\mu}. \tag{2.3.12}$$

Applying the LEM mapping to (2.3.11), we find its linear equivalent PDE

$$\mathcal{L}_{\varepsilon} v \equiv \frac{\partial v}{\partial x} - \left\langle \xi, \mathbf{P}_{\varepsilon}(x, \mathrm{D}_{\xi}) \right\rangle v = 0. \tag{2.3.13}$$

We then apply theorem 1.6 if $f_{j\theta}(x)=0,\ j=\overline{1,n}$, or theorem 1.7, for a non-homogeneous system.

Let us also approximate by polynomials in x the coefficients $f_{j\mu}(x)$. More precisely, for a given ρ one can find $q=q(\rho)$ such that

$$\left| f_{j\mu}(x) - p_{j\mu}^{q}(x) \right| \le \rho, \quad x \in I, |\mu| = \overline{1, m(\varepsilon)}; \tag{2.3.14}$$

thus $f_{j\mu}(x)$ are uniformly approximated on I by the polynomials $p_{j\mu}^{q}(x)$ of degrees at most q. Consequently, the polynomial system (2.3.1) can be expressed by

$$\mathcal{P}_{\varepsilon,\rho}\mathbf{y} \equiv \frac{\mathrm{d}\mathbf{y}}{\mathrm{d}x} - \mathbf{P}_{\varepsilon,\rho}(x,\mathbf{y}) = \mathbf{0}, \tag{2.3.15}$$

where

$$\mathbf{P}_{\varepsilon,\rho}(x,\mathbf{y}) = \left[P_{j,\varepsilon}^{\rho}(x,\mathbf{y})\right]_{j=\overline{1,n}}, \quad P_{j,\varepsilon}^{\rho}(x,\xi)=\sum_{|\mu|=0}^{m(\varepsilon)} p_{j\mu}^{q}(x)\xi^{\mu}. \tag{2.3.16}$$

Note that the polynomial ODS (2.3.15) satisfies the hypotheses of theorem 2.1. Its linear equivalent is

$$\mathcal{L}_{\varepsilon,\rho} v \equiv \frac{\partial v}{\partial x} - \left\langle \xi, \mathbf{P}_{\varepsilon,\rho}(x, \mathrm{D}_{\xi}) \right\rangle v = 0. \tag{2.3.17}$$

Obviously, in order to use the LEM approximations $\mathcal{L}_{\varepsilon}$, $\mathcal{L}_{\varepsilon,\rho}$, it is necessary to prove their convergence to $\mathcal{L}_{\mathbf{f}}$.

Theorem 2.4 [143]. *Let* $f_j \in \mathcal{C}_n^0(\mathrm{I}),\ j=\overline{1,n}$. *Then* $\mathcal{L}_{\varepsilon}$, $\mathcal{L}_{\varepsilon,\rho}$ *converge to* $\mathcal{L}_{\mathbf{f}}$ *for* $\varepsilon \to 0$ ($\varepsilon, \rho \to 0$ *respectively*), *as a consequence of the inequalities*

$$\left|\mathfrak{L}_{\mathbf{f}}v-\mathfrak{L}_{\varepsilon}v\right|\leq\varepsilon\varphi(M,\xi), \tag{2.3.18}$$

and

$$\left|\mathfrak{L}_{\mathbf{f}}v-\mathfrak{L}_{\varepsilon,\rho}v\right|\leq\left(\varepsilon+\rho\sum_{s=1}^{m(\varepsilon)}M^{s}\right)\varphi(M,\xi), \tag{2.3.19}$$

that hold true for $v\in\mathcal{Q}_n^1(I)$. *In the above* inequalities,

$$\varphi(M,\xi)\equiv\left(\sum_{j=1}^{n}\left|\xi_j\right|\right)\mathrm{e}^{Mn\sum_{j=1}^{n}\left|\xi_j\right|}, \tag{2.3.20}$$

and M is the majorant of v from the definition of $\mathcal{Q}_n^0(\mathrm{I})$.

Proof. To prove the first inequality, we observe that, as $v\in\mathcal{Q}_n^1(\mathrm{I})$, it allows an expansion of the form

$$v(x,\xi)=1+\sum_{|\gamma|\geq 0}v_\gamma(x)\frac{\xi^\gamma}{\gamma!}, \tag{2.3.21}$$

whose coefficients satisfy

$$\sup_{x\in I}\left|v_\gamma(x)\right|\leq M^{|\gamma|}. \tag{2.3.22}$$

It follows, step by step,

$$\begin{aligned}
\left|\mathfrak{L}_{\mathbf{f}}v-\mathfrak{L}_{\varepsilon}v\right| &\leq\sum_{j=1}^{n}\left|\xi_j\right|\sum_{|\mu|>p(\varepsilon)}\left|f_{j\mu}(x)\mathrm{D}^{\mu}v\right|\\
&\leq\sum_{j=1}^{n}\left|\xi_j\right|\sum_{|\mu|>p(\varepsilon)}\left|f_{j\mu}(x)\right|\sum_{|\gamma|\geq 0}\left|v_{\gamma+\mu-\mathbf{e}_j}(x)\right|\frac{|\xi|^{|\gamma|}}{|\gamma|!}n^{|\gamma|}\\
&\leq\sum_{j=1}^{n}\left|\xi_j\right|\sum_{|\mu|>p(\varepsilon)}\left|f_{j\mu}(x)\right|\sum_{|\gamma|\geq 0}M^{|\gamma|+|\mu|}\frac{|\xi|^{|\gamma|}}{|\gamma|!}n^{|\gamma|}\\
&\leq\sum_{j=1}^{n}\left|\xi_j\right|\mathrm{e}^{Mn\sum_{j=1}^{n}\left|\xi_j\right|}\sum_{|\mu|>p(\varepsilon)}\left|f_{j\mu}(x)\right|M^{|\mu|}\leq\varepsilon\,\varphi(M,\xi).
\end{aligned} \tag{2.3.23}$$

The inequality (2.3.18) is thus proved.

We apply a similar procedure for the second inequality. Taking (2.3.14) into account, we have

$$\left|\mathcal{L}_{\mathbf{f}} v - \mathcal{L}_{\varepsilon,\rho} v\right| \leq \sum_{j=1}^{n} \left|\xi_j\right| \sum_{|\mu|>p(\varepsilon)} \left|f_{j\mu}(x)\right| M^{|\gamma|+|\mu|} n^{|\gamma|} \frac{|\xi|^{|\gamma|}}{|\gamma|!}$$

$$+ \sum_{j=1}^{n} \left|\xi_j\right| \sum_{|\mu|\leq p(\varepsilon)} \left|f_{j\mu}(x) - p_{j\mu}^{q}(x)\right| \sum_{|\gamma|\geq 0} M^{|\gamma|+|\mu|} \frac{|\xi|^{|\gamma|}}{|\gamma|!} n^{|\gamma|} \tag{2.3.24}$$

$$\leq \varepsilon\varphi(M,\xi) + \rho\left(\sum_{s\leq p(\varepsilon)} M^{s}\right)\varphi(M,\xi) \leq \left(\varepsilon + \rho\sum_{s\leq p(\varepsilon)} M^{s}\right)\varphi(M,\xi).$$

Thus, the theorem 2.4 is completely proved. ◘

An immediate consequence of this theorem, on the one hand, and of theorems 1.2 and 1.3, on the other hand, is

Theorem 2.5 [137, 139]. *Any differential operator of class $\mathcal{Q}_n^0(I)$ is globally* exp-*immersible in a consistent on $\mathcal{Q}_n^1(I)$ linear operator. Any restriction of a differential operator of class $\mathcal{Q}_n^0(I)$ to a set of admissible functions, satisfying Cauchy conditions, locally allows a linear* Exp-*equivalent, consistent on $\mathcal{Q}_n^1(I)$.*

Using the above transformation τ that associates to v the coefficients $v_\gamma(x)$ from the expansion

$$v(x,\xi) = 1 + \sum_{|\gamma|\geq 1} v_\gamma(x)\frac{\xi^\gamma}{\gamma!}, \quad x\in I, \xi\in\Re^n, \tag{2.3.25}$$

we get the second LEM equivalent, that is, the linear infinite ODS

$$\frac{\mathrm{d}v_\gamma}{\mathrm{d}t} = \sum_{j=1}^{n} \gamma_j \sum_{|\mu|=1}^{\infty} f_{j\mu}(t) v_{\gamma+\mu-\mathbf{e}_j}, \quad \mathbf{e}_j = \left(\delta_i^j\right)_{i=\overline{1,n}}. \tag{2.3.26}$$

It can be written in matrix form

$$\mathfrak{S}\mathbf{V} \equiv \frac{\mathrm{d}\mathbf{V}}{\mathrm{d}x} - \mathbf{A}(x)\mathbf{V} - \mathbf{B}(x) = \mathbf{0}\ . \tag{2.3.27}$$

The associated LEM matrix is generated similar to the case of polynomial operators and it reads

$$\mathbf{A}(x)=\begin{bmatrix} \mathbf{A}_{11}(x) & \mathbf{A}_{12}(x) & \mathbf{A}_{13}(x) & \ldots & \mathbf{A}_{1,m-1}(x) & \mathbf{A}_{1,m}(x)\ldots \\ \mathbf{A}_{21}(x) & \mathbf{A}_{22}(x) & \mathbf{A}_{23}(x) & \ldots & \mathbf{A}_{2,m-1}(x) & \mathbf{A}_{2,m}(x)\ldots \\ \mathbf{0} & \mathbf{A}_{32}(x) & \mathbf{A}_{33}(x) & \ldots & \mathbf{A}_{3,m-1}(x) & \mathbf{A}_{3,m}(x)\ldots \\ \ldots & \ldots & \ldots & \ldots & \ldots & \ldots \quad \ldots \\ \mathbf{0} & \mathbf{0} & \mathbf{0} & \ldots & \mathbf{A}_{m,m-1}(x) & \mathbf{A}_{m,m}(x)\ldots \\ \ldots & \ldots & \ldots & \ldots & \ldots & \ldots \quad \ldots \end{bmatrix}. \quad (2.3.28)$$

We observe that in this case the matrix is no more row-finite, but is still column-finite, hence the calculation by block partitioning is still working. The cells $\mathbf{A}_{ss}(x)$ of the main diagonal are square, being generated by the coefficients of the linear part of the operator – namely, by those $f_{j\mu}(x)$ for which $|\mu|=1$; in fact, $\mathbf{A}_{11}(x)$ is the Jacobian matrix of the system. All the other cells are rectangular; the cells $\mathbf{A}_{k,k+s}(x)$ are set up by using those coefficients $f_{j\mu}(x)$ for which $|\mu|=s+1$, as in the case of polynomial operators. For instance, on the first parallel to the main diagonal we find cells generated by the coefficients of second degree terms in **y**, etc.

The relations between the nonlinear operators and their linear equivalents are explicitly given by the following diagram

$$\begin{array}{ll} \mathcal{F}:\left(\mathrm{C}^1(I)\right)^n \to \left(\mathrm{C}^0(I)\right)^n & \\ \mathrm{e}^{\langle \xi,\bullet\rangle} \;\downarrow & \\ \mathcal{L}:\mathcal{A}_n^1(I) \;\to \mathcal{A}_n^0(I) & , \\ \tau \quad\downarrow & \\ \mathcal{S}:\mathcal{B}_n^1(I) \;\to \mathcal{B}_n^0(I) & \end{array} \quad (2.3.29)$$

which represents the natural generalization of (1.4.51) – the similar diagram, set up for polynomial operators.

Let us note that, as in the case of polynomial ODSs, the diagram is not closed; yet, it can be used to "see" the solutions through a linear frame. With the same ideas for the proofs, one can extend the LEM representations from the polynomial case to the ODS (2.3.1). In this respect, we have

Theorem 2.6 [139, 143]. If $f_j \in \mathcal{A}_n^\infty(I)$, *then the solution of the nonlinear initial problem* (2.3.1), (2.3.3) *formally allows the representation*

$$\mathbf{V}(x)=\mathbf{\Phi}(x-x_0)\mathbf{V}(x_0)+\mathbf{\Theta}(x-x_0), \tag{2.3.30}$$

where the infinite matrix $\mathbf{\Phi}$ *is given by*

$$\mathbf{\Phi}(x-x_0)\equiv\sum_{k\geq 0}\mathbf{A}^{(k)}(x_0)\frac{(x-x_0)^k}{k!}, \tag{2.3.31}$$

and the infinite vector $\mathbf{\Theta}$ *by*

$$\mathbf{\Theta}(x-x_0)=\sum_{k\geq 0}\mathbf{B}^{(k-1)}(x_0)\frac{(x-x_0)^k}{k!}. \tag{2.3.32}$$

The matrices $\mathbf{A}^{(k)}$ *are determined by recurrence*

$$\mathbf{A}^{(k)}(x)=\frac{\mathrm{d}\mathbf{A}^{(k-1)}}{\mathrm{d}x}(x)+\mathbf{A}^{(k-1)}(x)\mathbf{A}(x), \quad \mathbf{A}^{(0)}(x)=\mathbf{E}, \tag{2.3.33}$$

where $\mathbf{E}$ *is the unit infinite matrix and the vectors* $\mathbf{B}^{(k)}$ *also by recurrence, using the formulae*

$$\mathbf{B}^{(k)}=\frac{\mathrm{d}\mathbf{B}^{(k-1)}}{\mathrm{d}x}+\mathbf{B}^{(k-1)}(x)\mathbf{A}(x), \quad \mathbf{B}^{(0)}=\mathbf{B}. \tag{2.3.34}$$

Particularly, the first n components of $\mathbf{V}$ are consistent at least on an interval I centered at x_0 and they coincide with the Taylor series of the solution of the nonlinear initial problem (2.3.1), (2.3.3). The length of I depends on the bounds of the coefficients $f_{j\mu}(x)$ and on $\mathbf{y}_0$.

In this case too, the first n rows of $\mathbf{\Phi}$ represent in fact *the local inverse matrix* of the nonlinear operator $\mathfrak{F}$. This also means that the LEM solution (2.3.30) straightforwardly separates the contribution of the coefficients form that of the data, an advantage already obtained in the previous section for polynomial differential operators.

Let us mention that, if the series (2.3.30) is truncated at some order k, all the involved operations are in finite number, as the LEM matrix is column-finite.

Example 2.1. Consider the ODE

$$y''=f(x,y), \quad x\in I\subseteq\Re, \tag{2.3.35}$$

often met in applications. Suppose that the equation is homogeneous and odd with respect to y and that $f\in\mathcal{C}_n^0(I)$, i.e.

$$f(x,y)=\sum_{k=1}^{\infty} f_{2k-1}(x)\, y^{2k-1}, \tag{2.3.36}$$

uniformly for $x \in I$. The LEM mapping depends here on two parameters; the first linear equivalent reads

$$\mathfrak{L}v \equiv \frac{\partial v}{\partial x} - \left(\xi_1 \frac{\partial v}{\partial \xi_2} + \xi_2 \sum_{k=1}^{\infty} f_{2k-1}(x) \frac{\partial^{2k-1} v}{\partial \xi_1^{2k-1}} \right) = 0, \tag{2.3.37}$$

and, according to theorem 2.2, it is consistent on $\mathcal{A}_n^0(I)$.

The associated LEM matrix is

$$\mathbf{A}(x) = \begin{bmatrix} \mathbf{A}_{11}(x) & \mathbf{0} & \mathbf{A}_{13}(x) & \dots & \mathbf{0} & \mathbf{A}_{1,2m-1}(x) & \dots \\ \mathbf{0} & \mathbf{A}_{22}(x) & \mathbf{0} & \dots & \mathbf{A}_{2,2m-2}(x) & \mathbf{0} & \dots \\ \mathbf{0} & \mathbf{0} & \mathbf{A}_{33}(x) & \dots & \mathbf{0} & \mathbf{A}_{3,2m-1}(x) & \dots \\ \dots & \dots & \dots & \dots & \dots & \dots & \dots \\ \mathbf{0} & \mathbf{0} & \mathbf{0} & \dots & \mathbf{0} & \mathbf{A}_{2m-1,2m-1}(x) & \dots \\ \dots & \dots & \dots & \dots & \dots & \dots & \dots \end{bmatrix}. \tag{2.3.38}$$

We observe that all the operations indicated in theorem 2.6 to the purpose of getting solutions of (2.3.35) show that the local inverse will not contain cells of even indices. Therefore, from the very beginning, we can work with the "compressed" LEM matrix

$$\mathbf{A}(x) = \begin{bmatrix} \mathbf{A}_{11}(x) & \mathbf{A}_{13}(x) & \mathbf{A}_{15}(x) & \dots & \mathbf{A}_{1,2m-3}(x) & \mathbf{A}_{1,2m-1}(x) & \dots \\ \mathbf{0} & \mathbf{A}_{33}(x) & \mathbf{A}_{35}(x) & \dots & \mathbf{A}_{3,2m-3}(x) & \mathbf{A}_{3,2m-1}(x) & \dots \\ \mathbf{0} & \mathbf{0} & \mathbf{A}_{55}(x) & \dots & \mathbf{A}_{5,2m-3}(x) & \mathbf{A}_{5,2m-1}(x) & \dots \\ \dots & \dots & \dots & \dots & \dots & \dots & \dots \\ \mathbf{0} & \mathbf{0} & \mathbf{0} & \dots & \mathbf{0} & \mathbf{A}_{2m-1,2m-1}(x) & \dots \\ \dots & \dots & \dots & \dots & \dots & \dots & \dots \end{bmatrix}. \tag{2.3.39}$$

Consider now ODSs with constant coefficients

$$\begin{gathered} \mathcal{F}(\mathbf{y}) \equiv \frac{\mathrm{d}\mathbf{y}}{\mathrm{d}x} - \mathbf{f}(\mathbf{y}) = \mathbf{0}, \\ \mathbf{f}(\mathbf{y}) = \left[f_j(\mathbf{y}) \right]_{j=\overline{1,n}}, \quad f_j(\mathbf{y}) = \sum_{|\mu|=1}^{\infty} f_{j\mu}\, \mathbf{y}^{\mu}, \end{gathered} \tag{2.3.40}$$

where $f_{j\mu} \in \mathfrak{R},\ j = \overline{1,n},\ |\mu| \in \mathfrak{N}$. Theorem 2.5, applied in this case, leads to the following result

Corollary 2.1. [136, 139, 143] *The solution of the nonlinear initial problem* (2.3.40), (2.3.3) *locally coincides with the first n components of the infinite vector*

$$\mathbf{V}(x) = \mathrm{e}^{\mathbf{A}(x-x_0)}\,\mathbf{V}_0, \tag{2.3.41}$$

the exponential matrix being introduced similarly to the finite case, i.e.

$$\mathrm{e}^{\mathbf{A}(x-x_0)} = \mathbf{E} + \frac{(x-x_0)}{1!}\mathbf{A} + \frac{(x-x_0)^2}{2!}\mathbf{A}^2 + \dots + \frac{(x-x_0)^k}{k!}\mathbf{A}^k + \dots \quad , \tag{2.3.42}$$

because all the involved step by step matrix computation is finite.

One can consider Corollary 2.1 as a generalization of corollary 1.3, which presented a similar result, obtained for polynomial differential operators.

From the above considerations, it follows that the linear equivalence method can be extended to other classes of operators, larger than those considered until now. One of these possibilities could be, e.g., the extension of LEM to operators continuously depending on the unknown functions; such operators should be conveniently approximated by polynomial operators in the frame of Weierstrass-like theorems, though, in this case, the approximating operator must be carefully handled and controlled.

Chapter 3

LEM TECHNIQUES AND REPRESENTATIONS

From the previous chapters, it is seen that there are at least two possibilities to apply LEM in order to get solutions of nonlinear ODSs: by using the first linear equivalent (the PDE) or the linear equivalent system; we shall examine each of them.

3.1. LEM REPRESENTATIONS BASED ON THE FIRST LINEAR EQUIVALENT

Consider the polynomial n-dimensional ODS

$$\mathcal{P}\mathbf{y} \equiv \frac{\mathrm{d}\mathbf{y}}{\mathrm{d}x} - \mathbf{P}(\mathbf{y}) = \mathbf{0}, \quad \mathbf{P} = \left[P_j(\mathbf{y})\right]_{j=\overline{1,n}}, \tag{3.1.1}$$

with constant coefficients, that is

$$P_j(\zeta) \equiv \sum_{|\mu| \le p_j} a_{j\mu}\zeta^{\eta}, \quad \zeta \in \mathfrak{R}^n, \quad a_{j\mu} \in \mathfrak{R}, |\mu| \le p_j, j = \overline{1,n}. \tag{3.1.2}$$

Let us associate to (3.1.1) the Cauchy conditions

$$\mathbf{y}(x_0) = \mathbf{y}_0 . \tag{3.1.3}$$

According to theorems 1.2 and 1.3, the linear equivalent of the polynomial system is

$$\mathcal{L}v \equiv \frac{\partial v}{\partial x} - \langle \xi, \mathbf{P}(\mathrm{D}_\xi) \rangle v = 0, \ \langle \xi, \mathbf{P}(\mathrm{D}_\xi) \rangle \equiv \sum_{j=1}^{n} \xi_j \mathrm{P}_j(\mathrm{D}_\xi); \tag{3.1.4}$$

v must also satisfy the exponential condition

$$v(x_0, \xi) \equiv \mathrm{e}^{\langle \xi, \mathbf{y}_0 \rangle}, \quad \xi \in \mathfrak{R}^n . \tag{3.1.5}$$

Let us consider solutions v of the form

$$v(x, \xi) = \varphi(x)\psi(\xi), \tag{3.1.6}$$

in which we separated the newly introduced variables from the independent variable of the nonlinear system. Introducing (3.1.6) in (3.1.4), we find

$$\frac{d\varphi}{dx}\psi - \left(\sum_{j=1}^{n} \xi_j P_j(D_\xi)\psi\right)\varphi = 0 ; \tag{3.1.7}$$

due to the linearity, the variables separate from each other, hence φ satisfies

$$\frac{d\varphi}{dx} = \lambda\varphi , \tag{3.1.8}$$

and ψ is solution of the linear PDE with variable coefficients

$$\sum_{j=1}^{n} \xi_j P_j(D_\xi)\psi = \lambda\psi . \tag{3.1.9}$$

But, as v must be analytic with respect to ξ, it follows that the solution $\psi(\xi,\lambda)$ of (3.1.9) has to be searched for through the analytic with respect to ξ functions.

If (3.1.9) allows a unique analytic solution for every λ belonging to a set $\Lambda \subseteq C$, then, applying the superposition principle, we can write the solution of (3.1.4) in the form

$$v(x,\xi) = \int e^{\lambda x}\psi(\xi,\lambda)d\lambda , \tag{3.1.10}$$

in which the integration domain is established taking Λ and the condition (3.1.5) into account.

Getting analytic solution for the PDE (3.1.9) seems to be a difficult task. For this reason, formula (3.1.10) does not allow, at first sight, too many practical applications. Yet, (3.1.10) has the advantage of separating the "dynamical element" of the polynomial operator – that is, the differentiation with respect to x – from the "stationary" one, which gives the algebraic behaviour of the operator.

Example 3.1. Consider the polynomial initial problem

$$y' = y^2 , y(0) = y_0 . \tag{3.1.11}$$

Applying the LEM mapping $v(x,\sigma) = e^{\sigma y}$, we obtain its linear equivalent

$$\frac{\partial v}{\partial x} - \sigma \frac{\partial^2 v}{\partial \sigma^2} = 0 \tag{3.1.12}$$

and the condition

$$v(0, \sigma) = \mathrm{e}^{\sigma y_0} . \tag{3.1.13}$$

If $v(x, \sigma) = \varphi(x)\psi(\sigma)$, then φ satisfies (3.1.8), and ψ is solution of the linear PDE

$$\sigma \frac{\partial^2 \psi}{\partial \sigma^2} = \lambda \psi . \tag{3.1.14}$$

As ψ must be analytic in σ, we shall search for it in the form

$$\psi(\sigma) = \sum_{j=1}^{\infty} \psi_j \frac{\sigma^j}{j!} , \tag{3.1.15}$$

wherefrom we find the recurrence relation for the coefficients

$$j \psi_{j+1} = \lambda \psi_j , \quad j \in \mathfrak{N} , \tag{3.1.16}$$

involving

$$\psi_j = \frac{\lambda^{j-1}}{(j-1)!} \psi_1 , \quad j \in \mathfrak{N} . \tag{3.1.17}$$

It results that v allows the general representation formula

$$v(x, \sigma) = 1 + \int_{-\infty}^{0} \mathrm{e}^{\lambda x} \sum_{j=1}^{\infty} \frac{\lambda^{j-1}}{(j-1)!} \frac{\sigma^j}{j!} \psi_1 \, \mathrm{d}\lambda , \tag{3.1.18}$$

for $x \geq 0$. Imposing now the initial condition $y(0) = y_0$, we get

$$\int_{-\infty}^{0} \frac{\lambda^{j-1}}{(j-1)!} \psi_1 \, \mathrm{d}\lambda = y_0^j , \quad j \in \mathfrak{N} . \tag{3.1.19}$$

The two last relations are satisfied for

$$\psi_1 = \mathrm{e}^{-\frac{\lambda}{y_0}} , \tag{3.1.20}$$

which implies

$$v(x,\sigma)=1+\int_{-\infty}^{0} e^{\lambda\left(x-\frac{1}{y_0}\right)} \sum_{j=1}^{\infty} \frac{\lambda^{j-1}}{(j-1)!} \frac{\sigma^j}{j!} d\lambda = e^{\frac{y_0}{1-xy_0}\sigma} . \tag{3.1.21}$$

The solution of the Cauchy problem is then

$$y(x)=\frac{y_0}{1-xy_0} . \tag{3.1.22}$$

Among the applications based on the first LEM equivalent, we remark the coupled pendulum, studied by Ştefania Donescu in [21, 22, 24, 57].

3.2. LEM REPRESENTATIONS BASED ON THE LINEAR EQUIVALENT SYSTEM

For polynomial ODSs with constant coefficients, one can find LEM solutions based on the second linear equivalent. One can prove

Theorem 3.1 [143]. *The solution of the polynomial initial problem* (3.1.1), (3.1.3) *locally allows the following representation*

$$y_j(x)= y_{0j} +$$

$$+\sum_{|\mu^0|\le p} a_{j\mu^0} \sum_{s=1}^{\infty} \sum_{j_1,j_2,\dots j_{s-1}=1}^{n} A^s_{j_1 j_2 \dots j_{s-1}} (y_0)^{\sum_{m=1}^{s-1}\left(\mu^m - e_{j_m}\right)+\mu^0} , \tag{3.2.1}$$

where μ^k *is a n-dimensional multi-index,* $\mu^k_{j_r}$ *– its* j_r *component,* $\delta^k_{j_r}$ *– Kronecker's delta and we used the following notations*

$$A^s_{j_1 j_2 \dots j_{s-1}} = \sum_{\substack{|\mu^k|\le p \\ k=\overline{1,s-1}}} \prod_{r=1}^{s-1} \left\{\left[\mu^0_{j_r} + \sum_{m=1}^{r-1}\left(\mu^m_{j_r} - \delta^m_{j_r}\right)\right] a_{j_r \mu^r}\right\} . \tag{3.2.2}$$

Proof. Let us consider the form (1.4.2) of the linear equivalent system corresponding to the polynomial operator defined by (3.1.1); we have

$$\frac{\mathrm{d}v_\gamma}{\mathrm{d}x}=\sum_{j=1}^{n}\gamma_j\sum_{|\mu|\le p_j}a_{j\mu}v_{\gamma+\mu-\mathbf{e}_j},\quad |\gamma|\in\mathfrak{N},\tag{3.2.3}$$

as well as

$$v_\gamma(x_0)=\mathbf{y}_0^\gamma,\quad \gamma\in\mathfrak{N}.\tag{3.2.4}$$

According to theorem 1.9, the solution of the solution of the initial polynomial problem (3.1.1), (3.1.3) coincides with $v_{\mathbf{e}_j}$, $j=\overline{1,n}$. We first have

$$\frac{\mathrm{d}v_{\mathbf{e}_{j_0}}}{\mathrm{d}x}=\sum_{|\mu^0|\le p}a_{j\mu^0}v_{\mu^0},\qquad j_0=\overline{1,n},\tag{3.2.5}$$

which allows us to determine $\frac{\mathrm{d}^s v_{\mathbf{e}_j}}{\mathrm{d}x^s}$; this calculation is much easier done, due to the linearity of the system. Step by step, it follows

$$\frac{\mathrm{d}^s v_{\mathbf{e}_j}}{\mathrm{d}x^s}=\sum_{|\mu^0|\le p}a_{j\mu^0}\sum_{j_1,j_2,\dots j_{s-1}=1}^{n}\sum_{\substack{|\mu^k|\le p\\ k=\overline{1,s-1}}}A^{\mu^1\mu^2\dots\mu^{s-1}}_{j_1j_2\dots j_{s-1}}v_{\mu^0+\sum_{l=1}^{s-1}(\mu^l-\mathbf{e}_{j_l})},\tag{3.2.6}$$

where the A coefficients from the right member are given by

$$A^{\mu^1\mu^2\dots\mu^{s-1}}_{j_1j_2\dots j_{s-1}}=\prod_{r=1}^{s-1}a_{j_r\mu^r}\left[\mu^0_{j_r}+\sum_{m=1}^{r-1}\left(\mu^m_{j_r}-\delta^m_{j_r}\right)\right].\tag{3.2.7}$$

This immediately yields (3.2.1). ◘

In the particular case $n=1$ we obtain

Corollary 3.1 [143]. *The solution of the initial polynomial problem*

$$y'=\sum_{l=0}^{p}a_l y^l,\quad y(x_0)=y_0,\tag{3.2.8}$$

a_l being real constants, locally allows the representation

$$y(x)=y_0+\sum_{s=1}^{\infty}\sum_{\substack{0\le l_j\le p\\ 1\le j\le s}} A_{l_1 l_2 \dots l_s} y_0^{l_1+l_2+\dots+l_{s-1}-s+1}\frac{(x-x_0)^s}{s!}, \tag{3.2.9}$$

where

$$A_{l_1 l_2 \dots l_s}=a_{l_1}a_{l_2}\dots a_{l_s} l_1(l_1+l_2-1)..(l_1+l_2+\dots+l_{s-1}-s+2). \tag{3.2.10}$$

There is a good chance to get this representation otherwise, but, by all means, LEM offered an organized and systematic way of obtaining higher derivatives of the solution.

Even if the LEM solution (3.2.9) has no more the property of separating the contribution of the data from that of the coefficients, it still has the advantage of an easy computer implementation and, in certain particular cases, it even leads to closed form solutions, as it can be seen from the following examples.

Example 3.2. Consider the linear ODE

$$y'=y, \tag{3.2.11}$$

to be solved in the Cauchy condition

$$y(x_0)=y_0. \tag{3.2.12}$$

With classical means, we immediately get

$$y(x)=y_0 e^{x-x_0} \tag{3.2.13}$$

Let us apply the preceding corollary. In this case, $p=1, a_0=0$, $a_1=1$ and formula (3.2.9) becomes

$$\begin{aligned} y(x)&=y_0+\sum_{s=1}^{\infty}\left[1^s\cdot 1\cdot(2-1)\cdot(3-2)\cdots(s-1-s+2)y_0^{s+1-s}\right.\\ &=y_0\left[1+\sum_{s=1}^{\infty}\frac{(x-x_0)^s}{s!}\right]=y_0 e^{x-x_0}. \end{aligned} \tag{3.2.14}$$

Example 3.3. Consider the problem of solving the nonlinear ODE

$$y'=y^2, \tag{3.2.15}$$

with the initial condition

$$y(0) = y_0 \,. \tag{3.2.16}$$

Applying corollary 3.1, we have $p = 2, a_0 = 0$, $a_1 = 0$, $a_2 = 1$ and $l_j = 2, j = \overline{1,s}$, for any $s \in \mathfrak{N}$. From formula (3.2.9) it is obtained

$$\begin{aligned} y(x) &= y_0 + \sum_{s=0}^{\infty} \left[1^{s+1} \cdot 2 \cdot (2 \cdot 2 - 1) \cdot (3 \cdot 2 - 2) \cdots (2s - s + 1) \frac{x^{s+1}}{(s+1)!} y_0^{2(s+1)-s} \right] \\ &= y_0 + y_0 \sum_{s=0}^{\infty} (s+1)! \frac{x^{s+1}}{(s+1)!} y_0^{s+1}; \end{aligned} \tag{3.2.17}$$

this obviously yields

$$y(x) = \frac{y_0}{1 - x y_0} \,. \tag{3.2.18}$$

3.3. NORMAL LEM REPRESENTATIONS

We recall that by LEM one can obtain the local inverse of a nonlinear operator in matrix form; we showed that this leads to specific representations of the solutions of nonlinear ODEs and ODSs, separating the data and the coefficients.

The previous paragraph offers a partial release from the local character of the representations. More precisely, the obtained representation is no more a Taylor series and, as it can be easily seen, it is valid on a larger interval; there are cases in which the LEM representation holds true on the whole interval on which the problem is considered.

A thorough analysis of the LEM matrix leads, in the case of constant coefficients, to certain qualitative representations, extremely useful in the study of the long term behaviour of the solutions of nonlinear dynamical systems.

Consider again the homogeneous nonlinear ODS with constant coefficients

$$\mathfrak{F}\mathbf{y} \equiv \frac{\mathrm{d}\mathbf{y}}{\mathrm{d}x} - \mathbf{f}(\mathbf{y}) = \mathbf{0}, \quad \mathbf{f}(\mathbf{y}) = \left[f_j(\mathbf{y}) \right]_{j=\overline{1,n}}, f_j \in \mathcal{A}_n^0(I), \tag{3.3.1}$$

where

$$f_j(\mathbf{y})=\sum_{|\mu|=1}^{\infty} f_{j\mu}\mathbf{y}^{\mu}, \quad |f_{j\mu}|\le K^{|\mu|}M, \ |\mu|\in\mathfrak{N}. \tag{3.3.2}$$

Let us associate to this ODS the following Cauchy conditions

$$\mathbf{y}(x_0)=\mathbf{y}_0, \quad x_0\in I, \mathbf{y}_0\equiv\left[y_{j0}\right]_{j=\overline{1,n}}\in\Re^n. \tag{3.3.3}$$

The first LEM equivalent is

$$\mathfrak{L}_{\mathbf{f}}v\equiv\frac{\partial v}{\partial x}-\left\langle\xi,\mathbf{f}\left(\mathrm{D}_{\xi}\right)\right\rangle v=0, \tag{3.3.4}$$

and the corresponding LEM matrix reads

$$\mathbf{A}=\begin{bmatrix} \mathbf{A}_{11} & \mathbf{A}_{12} & \mathbf{A}_{13} & \dots & \mathbf{A}_{1,m} & \dots \\ \mathbf{0} & \mathbf{A}_{22} & \mathbf{A}_{23} & \dots & \mathbf{A}_{2,m} & \dots \\ \mathbf{0} & \mathbf{0} & \mathbf{A}_{33} & \dots & \mathbf{A}_{3,m} & \dots \\ \dots & \dots & \dots & \dots & \dots & \dots \\ \mathbf{0} & \mathbf{0} & \mathbf{0} & \dots & \mathbf{A}_{m,m} & \dots \\ \dots & \dots & \dots & \dots & \dots & \dots \end{bmatrix}; \tag{3.3.5}$$

let us remind that the cells $\mathbf{A}_{m,m+k-1}$ are generated by those coefficients $f_{j\mu}$ such that $|\mu|=k$. By virtue of theorems 2.3 and 2.4, as well as of corollary 2.1, the solution of the nonlinear Cauchy problem (3.3.1), (3.3.3) is given by the first n components of the vector

$$\mathbf{V}(x)=\mathrm{e}^{\mathbf{A}(x-x_0)}\mathbf{V}_0, \tag{3.3.6}$$

where

$$\mathbf{V}_0=\mathbf{V}(x_0)=\left[\mathbf{y}_0^{\gamma}\right]_{|\gamma|\in\mathfrak{N}}. \tag{3.3.7}$$

Therefore, in order to determine the solution of the above mentioned problem, we only need the first n rows of $\mathbf{A}^m$. We thus get

Theorem 3.2 [139]. *The LEM solution of the nonlinear problem* (3.3.1), (3.3.3) *is*

$$\mathbf{y}(x)\equiv\mathbf{V}_1(x)=\mathbf{e}^{\mathbf{A}_{11}(x-x_0)}\mathbf{V}_1(x_0)+\sum_{m,k}\frac{(x-x_0)^m}{m!}\mathbf{S}_{km}\mathbf{V}_k(x_0), \tag{3.3.8}$$

the finite vectors $\mathbf{V}_k$ *being given by*

$$\mathbf{V}_k(x)=[v_\gamma(x)]_{|\gamma|=k}. \tag{3.3.9}$$

The finite matrices $\mathbf{S}_{km}$ *are determined as sums of matrix products*

$$\mathbf{S}_{km}=\sum_{|\gamma|=m-k+1}\mathbf{A}_{11}^{\gamma_1}\mathbf{A}_{12}\mathbf{A}_{22}^{\gamma_2}\mathbf{A}_{23}\ldots\mathbf{A}_{k-1,k}\mathbf{A}_{kk}^{\gamma_k}. \tag{3.3.10}$$

Proof. We already mentioned that the form (3.3.5) of the LEM matrix allows the calculation of its successive powers by block partitioning. Thus, the first n rows of $\mathbf{A}^2$ may be written by blocks

$$\begin{array}{l}\mathbf{A}_{11}^2\\ \mathbf{A}_{11}\mathbf{A}_{12}+\mathbf{A}_{12}\mathbf{A}_{22}\\ \mathbf{A}_{11}\mathbf{A}_{13}+\mathbf{A}_{12}\mathbf{A}_{23}+\mathbf{A}_{13}\mathbf{A}_{33}\\ \ldots\ldots\ldots\ldots\ldots\ldots\ldots\ldots\ldots\ldots\ldots\ldots\ldots\ldots\ldots\\ \sum_{j=1}^{k}\mathbf{A}_{1j}\mathbf{A}_{jk}\\ \ldots\ldots\ldots\ldots\ldots\ldots\ldots\ldots\ldots\ldots\ldots\ldots\ldots\ldots\ldots\end{array} \tag{3.3.11}$$

The k-th block must be multiplied by $\mathbf{V}_k(x_0)$.

This first calculation is sufficient to conclude that the k-th block from $\mathbf{A}^m$ is written in the form (3.3.10). ◘

The above LEM representation has at least two advantages:

- it separates the data from the coefficients;
- each term is obtained by a finite matrix calculation.

Theorem 3.2 is an intermediate step on the way of obtaining the normal LEM representation.

Theorem 3.3 [139]. *The solution of the nonlinear ODS* (3.3.1), *also satisfying the initial conditions* (3.3.3), *allows* ***the normal LEM representation***

$$y_j(x)=\sum_{|\mu|=1}^{\infty}u_\mu^j(x)\mathbf{y}_0^\mu,\quad j=\overline{1,n}, \tag{3.3.12}$$

where $u_\mu^j(x),\ j=\overline{1,n}$ *are components of the finite vectors* $\mathbf{U}_m^j(x)\equiv[u_\mu^j(x)]_{|\mu|=m}$ *satisfying the linear finite ODSs with constant coefficients*

$$\frac{d\mathbf{U}_1^j}{dx} = \mathbf{A}_{11}^{\mathrm{T}}\mathbf{U}_1^j,$$

$$\frac{d\mathbf{U}_2^j}{dx} = \mathbf{A}_{22}^{\mathrm{T}}\mathbf{U}_2^j + \mathbf{A}_{12}^{\mathrm{T}}\mathbf{U}_1^j, \tag{3.3.13}$$

....

$$\frac{d\mathbf{U}_m^j}{dx} = \mathbf{A}_{mm}^{\mathrm{T}}\mathbf{U}_m^j + \mathbf{A}_{m-1,m}^{\mathrm{T}}\mathbf{U}_{m-1}^j + \ldots + \mathbf{A}_{2m}^{\mathrm{T}}\mathbf{U}_2^j + \mathbf{A}_{1m}^{\mathrm{T}}\mathbf{U}_1^j,$$

as well as the Cauchy conditions

$$\mathbf{U}_1^j(x_0) = \mathbf{e}_j, \quad \mathbf{U}_s^j(x_0) = \mathbf{0},\ s = \overline{2,m}, \quad \mathbf{e}_j = \left[\delta_j^m\right]_{m=\overline{1,n}}. \tag{3.3.14}$$

In the above equations (3.3.13), T *stands for transposed matrix.*

Proof. In the general classical theory of linear ODSs, the notion of normal system of solutions means a fundamental system of solutions whose associated matrix coincides with the unit matrix at the initial point (in this case). Suppose that we determined such a system for a given ODS. Then its general solution can be written as a linear combination of the functions of the normal system, having as coefficients arbitrary Cauchy data, given at x_0.

This property is also valid for the normal LEM solution of a nonlinear ODS, only, in this case the sum will be infinite.

The calculation of the coefficients $\mathbf{y}_0^{\mu}, |\mu| = k$, can be done by classical methods applied to finite ODSs with constant coefficients.

Comparing (3.3.12) with (3.3.8), we obtain for $k = 1$ the normal solution of the linear part of the ODS (3.3.1). From the previous theorem, we see that each $\mathbf{y}_0^{\mu}$, $|\mu| = k$, which are, in fact, components of $\mathbf{V}_k(x_0)$, must be multiplied by

$$\sum_m \frac{(x - x_0)^m}{m!}\mathbf{S}_{km}, \tag{3.3.15}$$

where $\mathbf{S}_{km}$ are given in (3.3.10). From the last expression we observe that every step involves only a finite number of LEM cells, more precisely, those cells belonging to the "triangle" of vertices $\mathbf{A}_{11}, \mathbf{A}_{1k}, \mathbf{A}_{kk}$. Searching for the coefficient of $\mathbf{y}_0^{\mu}$, $|\mu| = k$ at the step *m*, we note that it depends on the previous step by the transposed of the truncated LEM matrix

$$\begin{bmatrix} \mathbf{A}_{11} & \mathbf{A}_{12} & \dots & \mathbf{A}_{1k} \\ \mathbf{0} & \mathbf{A}_{22} & \dots & \mathbf{A}_{2k} \\ \dots & \dots & \dots & \dots \\ \mathbf{0} & \mathbf{0} & \dots & \mathbf{A}_{kk} \end{bmatrix}, \tag{3.3.16}$$

and this dependence does not change with m. Combining this remark with the general normal solution for finite linear ODSs, we see that all the coefficients of $\mathbf{y}_0^{\mu}$ must satisfy the finite linear ODSs with constant coefficients (3.3.13), as well as the initial conditions (3.3.14). ◘

We called (3.3.12) *normal LEM representation* by analogy with the classical normal solution met in the linear case [42]. As a matter of fact, if the considered ODS is linear, its standard normal solution becomes a particular case of the normal LEM representation.

Remark. By theorem 3.2, the solutions of a nonlinear ODS may be regarded as elements of some Clifford algebra with respect to the Cauchy data.

To solve the linear ODS (3.3.13), which represents, in a way, the core of the normal LEM representation, it should be useful to know the eigenvalues of the diagonal cells $\mathbf{A}_{kk}$. Let us recall that this was done in theorem 1.8.

In what follows we shall see how these representations were applied to the stability study of the Bernoulli-Euler bar, of the elastic nonlinear frame, as well as of the nonlinear rigid pendulum.

3.4. THE LAPLACE TRANSFORM OF THE LEM SOLUTION

A convenient method to solve the finite initial problem (3.3.13), (3.3.14) is to apply the Laplace transformation.

Let us denote by

$$\mathsf{U}_k(p) = \mathcal{L}(\mathbf{U}_k(x)) \tag{3.4.1}$$

the Laplace transform of $\mathbf{U}_k(t)$. Then

$$\mathsf{U}_1(p) = \left(p\mathbf{E}_1 - \mathbf{A}_{11}^{\mathrm{T}}\right)^{-1}\mathbf{e}_j, \qquad \mathbf{e}_j = \left\lfloor \delta_i^j \right\rfloor_{i=\overline{1,n}}, \tag{3.4.2}$$

where $\mathbf{E}_1 \in \mathfrak{M}(n \times n)$ is the unit matrix.

As the inverse matrix from the right member of (3.4.2) is typical for this application, we shall use the notation

$$\mathbf{\Phi}_k(p) = \left(p\mathbf{E}_k - \mathbf{A}_{kk}^{\mathrm{T}}\right)^{-1}, \tag{3.4.3}$$

$\mathbf{E}_k$ being the unit matrix of same dimensions as $\mathbf{A}_{kk}^{\mathrm{T}}$. Thus,

$$\mathbf{U}_1(p) = \mathbf{\Phi}_1(p)\mathbf{e}_j. \tag{3.4.4}$$

The second step implies

$$\mathbf{U}_2(p) = \left(p\mathbf{E}_2 - \mathbf{A}_{22}^{\mathrm{T}}\right)^{-1}\mathbf{A}_{12}^{\mathrm{T}}\,\mathbf{U}_1(p), \tag{3.4.5}$$

hence

$$\mathbf{U}_2(p) = \mathbf{\Phi}_2(p)\mathbf{A}_{12}^{\mathrm{T}}\,\mathbf{\Phi}_1(p)\mathbf{e}_j. \tag{3.4.6}$$

Later on,

$$\mathbf{U}_3(p) = \left[\mathbf{\Phi}_3(p)\mathbf{A}_{13}^{\mathrm{T}}\,\mathbf{\Phi}_1(p) + \mathbf{\Phi}_3(p)\mathbf{A}_{23}^{\mathrm{T}}\,\mathbf{\Phi}_2(p)\mathbf{A}_{12}^{\mathrm{T}}\,\mathbf{\Phi}_1(p)\right]\mathbf{e}_j, \tag{3.4.7}$$

etc. In general, to determine $\mathbf{U}_k(p)$, one must add the matrices

$$\begin{array}{l} \mathbf{\Phi}_k(p)\mathbf{A}_{1k}^{\mathrm{T}}\,\mathbf{\Phi}_1(p), \\ \mathbf{\Phi}_k(p)\mathbf{A}_{2l}^{\mathrm{T}}\,\mathbf{\Phi}_2(p)\mathbf{A}_{1k}^{\mathrm{T}}\,\mathbf{\Phi}_1(p), \\ \dots\dots\dots\dots\dots\dots\dots\dots\dots\dots \\ \mathbf{\Phi}_k(p)\mathbf{A}_{k-1,k}^{\mathrm{T}}\,\mathbf{\Phi}_{k-1}(p)\mathbf{A}_{k-2,k-1}^{\mathrm{T}}\,\mathbf{\Phi}_{k-2}(p)..\mathbf{\Phi}_2(p)\mathbf{A}_{1k}^{\mathrm{T}}\,\mathbf{\Phi}_1(p) \end{array} \tag{3.4.8}$$

Knowing $\mathbf{U}_k(p)$, i.e. the Laplace transform of $\mathrm{U}_k(t)$, we immediately get the corresponding originals and, consequently, the normal LEM representation.

3.4.1. LEM BASES FOR A CLASS OF ODEs

Consider the second order nonlinear homogeneous ODE

$$y'' = f(y), \quad f(y) = \sum_{j=1}^{\infty} a_j y^j, \quad a_j \in \Re, \forall j \in \mathfrak{N}; \tag{3.4.9}$$

admitting for f the above expansion, we also supposed that $f \in \mathcal{C}_2^1(I), I \subseteq \Re$.

Let us apply LEM to the associated first order homogeneous ODS, equivalent with (3.4.9)

$$y' = z, \qquad z' = f(y), \tag{3.4.10}$$

The LEM mapping depends in this case on two parameters

$$v(\sigma, \xi, x) = \mathrm{e}^{\sigma y + \xi z}. \tag{3.4.11}$$

The first LEM equivalent is then

$$\mathfrak{L}(y, z) \equiv \frac{\partial v}{\partial x} - \sigma\, \mathrm{D}_\xi v - \xi\, f(\mathrm{D}_\sigma) v = 0, \tag{3.4.12}$$

the operators D have the significance of partial derivatives with respect to σ and ξ, accordingly. In the previous chapter, we proved that the operator

$$f(\mathrm{D}_\sigma) \equiv \sum_{j=1}^{\infty} a_j \mathrm{D}_\sigma^j, \tag{3.4.13}$$

is consistent on $\mathcal{C}_2^1(I)$.

Let us take for v a development in the form

$$v(x, \sigma, \xi) = 1 + \sum_{j+k=1}^{\infty} v_{jk}(x) \frac{\sigma^j}{j!} \frac{\xi^k}{k!}, \tag{3.4.14}$$

which, introduced in (3.4.12), leads to the second LEM equivalent, also homogeneous, written in matrix form

$$\mathfrak{S}\mathbf{V}(x) \equiv \frac{\mathrm{d}\mathbf{V}}{\mathrm{d}x} - \mathbf{A}\mathbf{V} = \mathbf{0}. \tag{3.4.15}$$

In (3.4.15), the unknown vector **V** will be divided, as previously, in finite vectors

$$\mathbf{V}(x) = [\mathbf{V}_m(x)]_{m \in \mathfrak{N}}, \quad \mathbf{V}_m(x) = \lfloor v_{jk}(x) \rfloor_{j+k=m}. \tag{3.4.16}$$

The LEM matrix **A** is constant and is of the form (3.3.5). The diagonal blocks read

$$\mathbf{A}_{11}=\begin{bmatrix}0 & 1\\ a_1 & 0\end{bmatrix},\quad \mathbf{A}_{kk}=\begin{bmatrix}0 & k & 0 & \dots & 0 & 0\\ a_1 & 0 & k-1 & \dots & 0 & 0\\ 0 & 2a_1 & 0 & \dots & 0 & 0\\ \dots & \dots & \dots & \dots & \dots & \dots\\ 0 & 0 & 0 & \dots & 0 & 1\\ 0 & 0 & 0 & \dots & ka_1 & 0\end{bmatrix}. \tag{3.4.17}$$

Observe that the rectangular blocks can be written as

$$\mathbf{A}_{12}=a_2\begin{bmatrix}0 & 0 & 0\\ 1 & 0 & 0\end{bmatrix},\quad \mathbf{A}_{k,k+s}=a_{s+1}\mathbf{B}_{k,k+s}, \tag{3.4.18}$$

where the **B** matrices do not depend on the coefficients of the ODE (3.4.9); more precisely,

$$\mathbf{B}_{k,k+s}=\begin{bmatrix}0 & 0 & 0 & \dots & 0 & 0 & \dots & 0\\ 1 & 0 & 0 & \dots & 0 & 0 & \dots & 0\\ 0 & 2 & 0 & \dots & 0 & 0 & \dots & 0\\ \dots & \dots & \dots & \dots & \dots & \dots & \dots & \dots\\ 0 & 0 & 0 & \dots & k & 0 & \dots & 0\end{bmatrix}. \tag{3.4.19}$$

This remark forms the basis of a general procedure of determining by LEM the solutions of ODEs of the form (3.4.9). The cells $\mathbf{A}_{k,k+s}$ have $(k+1)$ rows and $(k+s+1)$ columns.

The nature of the LEM solution depends on the linear part of the operator.

Indeed, if $a_1=-\omega^2<0$, then the eigenvalues of $\mathbf{A}_{11}$ – the linear part of the operator – are $\pm\mathrm{i}\omega$, i.e., purely imaginary. According to theorem 1.8, all the diagonal cells $\mathbf{A}_{kk}$ have either purely imaginary or null eigenvalues. By virtue of theorem 3.2, it follows that the normal LEM representation will contain only trigonometric functions, occasionally multiplied by polynomials.

If $a_1=\omega^2>0$, then $\mathbf{A}_{11}$ has real and opposite eigenvalues $\pm\omega$ and the corresponding normal LEM representation will contain only hyperbolic functions and polynomials.

To obtain LEM solutions, we shall consider the corresponding to (3.3.13) finite linear ODS, to which we apply the Laplace transformation.

Let us observe that in order to get the transforms $\mathbf{U}_k(p)$, one must compute the following matrices for every k

$$\begin{aligned}
&\mathbf{W}_1(p)=\mathbf{\Phi}_1(p),\\
&\mathbf{W}_{1k}(p)=\mathbf{\Phi}_k(p)\mathbf{B}_{1k}^{\mathrm{T}}\ \mathbf{\Phi}_1(p),\\
&\mathbf{W}_{1(k-1)2k}(p)=\mathbf{\Phi}_k(p)\mathbf{B}_{2k}^{\mathrm{T}}\ \mathbf{\Phi}_2(p)\mathbf{B}_{1,k-1}^{\mathrm{T}}\ \mathbf{\Phi}_1(p),\\
&\cdots\cdots\cdots\cdots\cdots\cdots\cdots\cdots\cdots\cdots\cdots\cdots\cdots\cdots\cdots\cdots\cdots\cdots\cdots\\
&\mathbf{W}_{1223\ldots(k-1)k}(p)=\mathbf{\Phi}_k(p)\mathbf{B}_{k-1,k}^{\mathrm{T}}\ \mathbf{\Phi}_{k-1}(p)\mathbf{B}_{k-2,k-1}^{\mathrm{T}}\ldots\mathbf{\Phi}_2(p)\mathbf{B}_{12}^{\mathrm{T}}\ \mathbf{\Phi}_1(p);
\end{aligned}\tag{3.4.20}$$

these matrices depend only on a_1 and on no other coefficient of (3.4.9). Hence, according to theorem 3.2, the corresponding originals form a *LEM basis* for the solutions of the ODE (3.4.9).

The first matrix (3.4.20) has to be multiplied by a_k, the second – by $a_{k-1}a_2$, etc., the last one being multiplied by a_2^{k-1}.

For instance, we have

$$\begin{aligned}
&\mathbf{U}_1(p)=\mathbf{\Phi}_1(p)\,\mathbf{e}_j=\mathbf{W}_1(p)\,\mathbf{e}_j,\\
&\mathbf{U}_2(p)=a_2\mathbf{W}_{12}(p)\,\mathbf{e}_j,\\
&\mathbf{U}_3(p)=\left[a_3\mathbf{W}_{13}(p)+a_2^2\mathbf{W}_{1223}(p)\right]\mathbf{e}_j,
\end{aligned}\tag{3.4.21}$$

where

$$\mathbf{W}_1(p)=\frac{1}{\Delta_1}\begin{bmatrix} p & a_1\\ 1 & p\end{bmatrix},\quad \mathbf{W}_{12}(p)=\frac{1}{\Delta_{12}}\begin{bmatrix} p^2-2a_1 & p\left(p^2-2a_1\right)\\ 2p & 2p^2\\ 2 & 2p\end{bmatrix},\tag{3.4.22}$$

are used to get $\mathbf{U}_1(p)$, $\mathbf{U}_2(p)$ and

$$\begin{aligned}
&\mathbf{W}_{13}(p)=\frac{1}{\Delta_1\Delta_3}\begin{bmatrix} p\left(p^2-7a_1\right) & p^2\left(p^2-7a_1\right)\\ 10p\left(p^2-3a_1\right) & 10p^2\left(p^2-3a_1\right)\\ 20p^2 & 20p^3\\ 20p & 20p^2\end{bmatrix},\\
&\mathbf{W}_{1223}(p)=\frac{1}{\Delta_1\Delta_2\Delta_3}\begin{bmatrix} 2\left(p^2+a_1\right)\left(p^2-7a_1\right) & 2p\left(p^2+a_1\right)\left(p^2-7a_1\right)\\ 3\left(p^2-3a_1\right) & 3p\left(p^2-3a_1\right)\\ 6p & 6p^2\\ 6 & 6p\end{bmatrix},
\end{aligned}\tag{3.4.23}$$

serve to obtain $\mathbf{U}_3(p)$. In the above formulae,

$$\Delta_k(p)=\det\left(p\mathbf{E}_k-\mathbf{A}_{kk}^{\mathrm{T}}\right),\quad k\in\mathfrak{N}. \tag{3.4.24}$$

Remark. The **W** matrices of the LEM basis – except for the first – have an interesting particularity; their second column is obtained by multiplying with p the first column, such that they can also be written in the form

$$\begin{aligned}
\mathbf{W}_{12}(p)&=\frac{1}{\Delta_{12}}\left[\mathbf{D}_{13}(p)\quad p\mathbf{D}_{13}(p)\right],\\
\mathbf{W}_{13}(p)&=\frac{1}{\Delta_1\Delta_3}\left[\mathbf{D}_{13}(p)\quad p\mathbf{D}_{13}(p)\right],\\
\mathbf{W}_{1223}(p)&=\frac{1}{\Delta_1\Delta_2\Delta_3}\left[\mathbf{D}_{1223}(p)\quad p\mathbf{D}_{1223}(p)\right],
\end{aligned} \tag{3.4.25}$$

where the **D** columns read

$$\mathbf{D}_{12}(p)=\begin{bmatrix} p^2-2a_1\\ 2p\\ 2\end{bmatrix},\qquad \mathbf{D}_{13}(p)=\begin{bmatrix} p\left(p^2-7a_1\right)\\ 10p\left(p^2-3a_1\right)\\ 20p^2\\ 20p\end{bmatrix},$$
$$\mathbf{D}_{1223}(p)=\begin{bmatrix} 2\left(p^2+a_1\right)\left(p^2-7a_1\right)\\ 3\left(p^2-3a_1\right)\\ 6p\\ 6\end{bmatrix}. \tag{3.4.26}$$

3.4.2. NONLINEAR ODD ODEs

In this case, (3.4.9) is of the form

$$y''=f(y),\quad f(y)=\sum_{k=1}^{\infty}a_{2k-1}y^{2k-1},\quad a_{2k-1}\in\Re,\ \forall k\in\mathfrak{N}. \tag{3.4.27}$$

Applying LEM, we obtain the first equivalent as in (3.4.12), only, in this case

$$f(\mathrm{D}_\sigma)\equiv\sum_{j=1}^{\infty}a_{2j-1}\mathrm{D}_\sigma^{2j-1}. \tag{3.4.28}$$

Considering now for v a development of the type (3.4.14), we get the linear equivalent system, that can be "compressed", taking in the corresponding LEM matrix only cells with odd indices. More precisely, the second LEM equivalent can be written as

$$\mathfrak{S}\mathbf{V}(x) \equiv \frac{\mathrm{d}\mathbf{V}}{\mathrm{d}t} - \mathbf{A}\mathbf{V} = \mathbf{0},$$
$$\mathbf{V}(x) = \left[\mathbf{V}_{2m-1}(x)\right]_{m\in\mathfrak{N}}, \quad \mathbf{V}_{2m-1}(x) = \left\lfloor v_{jk}(x)\right\rfloor_{j+k=2m-1}. \tag{3.4.29}$$

The diagonal blocks read

$$\mathbf{A}_{11} = \begin{bmatrix} 0 & 1 \\ a_1 & 0 \end{bmatrix},$$
$$\mathbf{A}_{2k-1,2k-1} = \begin{bmatrix} 0 & 2k-1 & 0 & \dots & 0 & 0 \\ a_1 & 0 & 2k-2 & \dots & 0 & 0 \\ 0 & 2a_1 & 0 & \dots & 0 & 0 \\ \dots & \dots & \dots & \dots & \dots & \dots \\ 0 & 0 & 0 & \dots & 0 & 1 \\ 0 & 0 & 0 & \dots & (2k-1)a_1 & 0 \end{bmatrix}, \tag{3.4.30}$$

and the rectangular ones are

$$\mathbf{A}_{13} = a_3 \begin{bmatrix} 0 & 0 & 0 \\ 1 & 0 & 0 \end{bmatrix}, \quad \mathbf{A}_{2k-1,2k+2s-1} = a_{2s+1}\mathbf{B}_{2k-1,2k+2s-1}, \tag{3.4.31}$$

where the $\mathbf{B}$ matrices do not depend on the coefficients a_{2k-1}, being given by

$$\mathbf{B}_{2k-1,2k+2s-1} = \begin{bmatrix} 0 & 0 & 0 & \dots & 0 & 0 & \dots & 0 \\ 1 & 0 & 0 & \dots & 0 & 0 & \dots & 0 \\ 0 & 2 & 0 & \dots & 0 & 0 & \dots & 0 \\ \dots & \dots & \dots & \dots & \dots & \dots & \dots & \dots \\ 0 & 0 & 0 & \dots & 2k-1 & 0 & \dots & 0 \end{bmatrix}. \tag{3.4.32}$$

The matrices $\mathbf{A}_{2k-1,2k+2s-1}$ have $2k$ rows and $(2k+2s)$ columns.

We shall obtain the normal LEM solution by using the method and the notations from the preceding section. To get $\mathbf{U}_{2k-1}(p)$, one must firstly compute the matrices giving the LEM basis

$$
\begin{aligned}
&\mathbf{W}_1(p)=\mathbf{\Phi}_1(p),\\
&\mathbf{W}_{1,2k-1}(p)=\mathbf{\Phi}_{2k-1}(p)\mathbf{B}_{12k-1}^{\mathrm{T}}\,\mathbf{\Phi}_1(p),\\
&\mathbf{W}_{1,2k-3,3,2k-1}(p)=\mathbf{\Phi}_{2k-1}(p)\mathbf{B}_{3,2k-1}^{\mathrm{T}}\,\mathbf{\Phi}_3(p)\mathbf{B}_{1,2k-3}^{\mathrm{T}}\,\mathbf{\Phi}_1(p),\\
&\dots\dots\dots\dots\dots\dots\dots\dots\dots\dots\dots\dots\\
&\mathbf{W}_{1335\ldots 2k-3,2k-1}(p)=\mathbf{\Phi}_{2k-1}(p)\mathbf{B}_{2k-3,2k-1}^{\mathrm{T}}\,\mathbf{\Phi}_{2k-3}(p)..\mathbf{\Phi}_3(p)\mathbf{B}_{13}^{\mathrm{T}}\,\mathbf{\Phi}_1(p);
\end{aligned}
\tag{3.4.33}
$$

they depend only on a_1. The first one is multiplied by a_{2k-1}, the second – by $a_{2k-1}a_3$, etc., and the last one – by a_3^{k-1}.

As the nature of the LEM solution depends on the linear part of the equation, let us make a choice, supposing that $a_1=-\omega^2$.

Our purpose is to find solutions of (3.4.27) also satisfying the Cauchy conditions

$$
y(0)=\alpha,\quad \dot{y}(0)=0\,. \tag{3.4.34}
$$

The corresponding LEM basis (3.4.33), going as far as $k=3$ (i.e., for fifth order effects), read

$$
\begin{aligned}
&\mathbf{W}_1(p)=\frac{1}{\Delta_1}\begin{bmatrix} p & -\omega^2\\ 1 & p\end{bmatrix},\\
&\mathbf{W}_{15}(p)=\frac{1}{\Delta_1\Delta_5}\left[\mathbf{D}_{15}\quad p\mathbf{D}_{15}\right],\\
&\mathbf{W}_{13}(p)=\frac{1}{\Delta_1\Delta_3}\left[\mathbf{D}_{13}\quad p\mathbf{D}_{13}\right],\\
&\mathbf{W}_{1335}(p)=\frac{1}{\Delta_1\Delta_3\Delta_5}\left[\mathbf{D}_{1335}\quad p\mathbf{D}_{1335}\right],
\end{aligned}
\tag{3.4.35}
$$

where

$$
\mathbf{D}_{13}(p)=\begin{bmatrix} p\left(p^2+7\omega^2\right)\\ 3\left(p^2+3\omega^2\right)\\ 6p\\ 6\end{bmatrix},\quad
\mathbf{D}_{15}(p)=\begin{bmatrix} p\left(p^4+30\omega^2p^2+149\omega^4\right)\\ 5\left(p^4+22\omega^2p^2+45\omega^4\right)\\ 20p\left(p^2+13\omega^2\right)\\ 60\left(p^2+5\omega^2\right)\\ 120p\\ 120\end{bmatrix},
\tag{3.4.36}
$$

and

$$\mathbf{D}_{1335}(p)=\begin{bmatrix} 3p\left(p^2+9\omega^2\right)\left(p^4+28\omega^2p^2+87\omega^4\right) \\ 27\left(p^2+\omega^2\right)\left(p^4+24\omega^2p^2+75\omega^4\right) \\ 18p\left(p^2+13\omega^2\right)\left(7p^2+15\omega^2\right) \\ 54\left(p^2+5\omega^2\right)\left(9p^2+25\omega^2\right) \\ 108p\left(7p^2+15\omega^2\right) \\ 108\left(7p^2+15\omega^2\right) \\ \end{bmatrix}. \tag{3.4.37}$$

The determinants Δ_{2k-1} are those from formula (3.4.24).

These results will be used especially in the study of the nonlinear rigid pendulum.

3.5. LEM SOLUTIONS FOR POLYNOMIAL ODSs WITH POLYNOMIAL COEFFICIENTS

Consider the polynomial ODS

$$\mathcal{P}\mathbf{y} \equiv \frac{\mathrm{d}\mathbf{y}}{\mathrm{d}x} - \mathbf{P}(x,\mathbf{y}) = \mathbf{0}, \quad \mathbf{P} = \left[P_j(x,\mathbf{y})\right]_{j=\overline{1,n}}, \tag{3.5.1}$$

with

$$P_j(x,\zeta) \equiv \sum_{|\mu|\le p_j} p_{j\mu}(x)\zeta^{\mu}, \quad \zeta \in \Re^n. \tag{3.5.2}$$

Suppose that the coefficients $p_{j\mu}$ are polynomials in x, having the maximum degree q

$$p_{j\mu}(x) = \sum_{k=0}^{q} p_{j\mu}^{k}(x-x_0)^k, \quad j=\overline{1,n},\ |\mu| = \overline{1,m}; \tag{3.5.3}$$

we wrote this polynomials putting into evidence the initial point x_0, at which we impose standard arbitrary Cauchy conditions

$$\mathbf{y}(x_0) = \mathbf{y}_0. \tag{3.5.4}$$

The LEM system can be written in this case in the form

$$\frac{\mathrm{d}\mathbf{V}}{\mathrm{d}x} = \left[\mathbf{A}_0 + (x-x_0)\mathbf{A}_1 + (x-x_0)^2\mathbf{A}_2 + \ldots + (x-x_0)^q\mathbf{A}_q\right]\mathbf{V}, \quad \mathbf{V} = \left[\mathbf{V}_j\right]_{j\in\mathfrak{N}}, \quad \mathbf{V}_j = \left[v_\gamma\right]_{\cdot|\gamma|=j}, \tag{3.5.5}$$

the matrices $\mathbf{A}_k$ being constant and of LEM type, i.e., similarly generated; each of these matrices contains only the coefficients $p_{j\mu}^k$. One can even affirm that the matrix obtained from $\mathbf{A}_k$ by replacing correspondingly the coefficients $p_{j\mu}^k$ with $p_{j\mu}^m$ is $\mathbf{A}_m$.

The initial conditions (3.5.4) become, as previously,

$$\mathbf{V}(x_0) = \mathbf{V}_0, \qquad \mathbf{V}_0 = \left[v_0^{\gamma} \right]_{|\gamma| \in \mathfrak{N}}. \tag{3.5.6}$$

The LEM system can be also formally written as an integral vector equation

$$\mathbf{V}(x) = \mathbf{V}_0 + \int_{x_0}^{x} \left[\sum_{j=0}^{q} (t - x_0)^q \mathbf{A}_q \right] \mathbf{V}(t) \mathrm{d}t; \tag{3.5.7}$$

being linear, one can apply in an easier and controlled manner the method of successive approximations, starting from $\mathbf{V}_0$

$$\begin{aligned} &\mathbf{V}^{(0)} = \mathbf{V}_0, \\ &\mathbf{V}^{(l)}(x) = \mathbf{V}_0 + \int_{x_0}^{x} \left[\sum_{j=0}^{q} (t - x_0)^q \mathbf{A}_q \right] \mathbf{V}^{(l-1)}(t) \mathrm{d}t. \end{aligned} \tag{3.5.8}$$

With these preparations, also considering theorems 1.2, 1.3, one can prove

Theorem 3.4 [153, 153, 161]. *The solution of the initial polynomial problem* (3.5.1), (3.5.4) *coincides with the first n components of the vector*

$$\mathbf{V}(x) = \mathbf{V}_0 + \sum_{l=1}^{\infty} \mathbf{A}^{(l)}(x - x_0) \mathbf{V}_0, \tag{3.5.9}$$

the matrices $\mathbf{A}^{(l)}(t)$ *being determined as follows*

$$\mathbf{A}^{(l)}(t) \equiv \sum_{k_1, k_2, \ldots, k_l = 1}^{q} \mathbf{A}_{k_1} \mathbf{A}_{k_2} \ldots \mathbf{A}_{k_l} \varphi(t, l; k_1, k_2, \ldots k_l), \tag{3.5.10}$$

where

$$\varphi(t, l; k_1, k_2, \ldots k_l) = \frac{t^{k_1 + k_2 + \ldots + k_l + l}}{(k_1 + k_2 + \ldots + k_l + l)(k_2 + \ldots + k_l + l - 1) \ldots (k_l + 1)}; \tag{3.5.11}$$

the first n components of $\mathbf{V}$ *are consistent at least on the interval* (1.4.39), *where* $C = \max_{j=\overline{1,n}} \left\{ \left| p^k_{j\mu} \right|, k = \overline{0,q}, |\mu| = \overline{1,m} \right\}$ *with* $p^k_{j\mu}$ *given by* (3.5.3).

This theorem treats the same problem as the one tackled in theorem 1.6 and, in particular, in theorem 2.1, which presented a convergent LEM algorithm for the solution. In theorem 3.4, LEM is still applied, but the way of treating the linear equivalent system is distinct. If the algorithm from theorem 2.1 has the advantage of being rapidly transposed on a computer, the procedure proposed in theorem 3.4, along with this advantage, mainly due to the constant matrices $\mathbf{A}_k$, brings in the following important contributions:

- to have approximations of order $(x - x_0)^s$, one must effectuate s steps by theorem 1.6, while by applying theorem 3.4, the same number of steps, easier realized, leads to terms containing $(x - x_0)^{s(m+1)}$;
- the iterative procedure does not affect the previously obtained results;
- unlike theorem 1.6, products, differentiation, addition of matrices with variable entries are not required and this saves a lot of time and memory;
- if one sets up a programming code based on theorem 1.6, it may be partly used also in a programming code based on theorem 3.4.

3.6. LEM SOLUTIONS BY SUCCESSIVE APPROXIMATIONS

The $\mathbf{\Phi}$ matrix (2.3.31), whose first rows coincide with the local inverse of a nonlinear operator, was obtained starting from the Taylor series solution around x_0 of the ODS (2.3.1). Yet, there are many other ways of solving this system, as it was shown, e.g., in the previous section.

We can apply the method of successive approximations to other classes of LEM systems, associated to more general than (3.5.1) nonlinear ODSs.

Consider the nonlinear homogeneous ODS of form (2.3.1)

$$\mathfrak{F}\mathbf{y} \equiv \frac{\mathrm{d}\mathbf{y}}{\mathrm{d}x} - \mathbf{f}(x, \mathbf{y}) = \mathbf{0}, \quad \mathbf{f}(x, \mathbf{y}) = \left[f_j(x, \mathbf{y}) \right]_{j=\overline{1,n}}, \tag{3.6.1}$$

hence $f_j \in \mathcal{C}_n^0(I), j=\overline{1,n}$, and $f_j(x,\mathbf{0}) = f_{j0} = 0, j=\overline{1,n}$. Applying LEM, we obtain the homogeneous linear equivalent system

$$\mathbb{S}\mathbf{V} \equiv \frac{\mathrm{d}\mathbf{V}}{\mathrm{d}x} - \mathbf{A}(x)\mathbf{V} = \mathbf{0}\ , \tag{3.6.2}$$

Associating to the system (3.6.1) the initial conditions (3.5.4), we obtain for **V** the conditions (3.5.6).

The LEM matrix has in this case the simpler than (2.3.28) form

$$\mathbf{A}(x) = \begin{bmatrix} \mathbf{A}_{11}(x) & \mathbf{A}_{12}(x) & \dots & \mathbf{A}_{1m}(x) & \dots \\ \mathbf{0} & \mathbf{A}_{22}(x) & \dots & \mathbf{A}_{2m}(x) & \dots \\ \dots & \dots & \dots & \dots & \dots \\ \mathbf{0} & \mathbf{0} & \dots & \mathbf{A}_{mm}(x) & \dots \\ \dots & \dots & \dots & \dots & \dots \end{bmatrix}. \tag{3.6.3}$$

The associated LEM system is (formally) equivalent with the linear integral equation

$$\mathbf{V}(x) = \int_{x_0}^{x} \mathbf{A}(t)\mathbf{V}(t)\mathrm{d}t + \mathbf{V}_0 . \tag{3.6.4}$$

Considering the successive approximating sequence

$$\mathbf{V}_{(n)}(x) = \int_{x_0}^{x} \mathbf{A}(t)\mathbf{V}_{(n-1)}(t)\mathrm{d}t + \mathbf{V}_0, \quad \mathbf{V}_{(0)} = \mathbf{V}_0 , \tag{3.6.5}$$

one can prove

Theorem 3.5 [153]. *The (formal) solution of the linear initial problem* (3.6.2), (3.5.6) *allows the representation*

$$\mathbf{V}(x) = \mathbf{V}_0 + \left(\sum_{m=1}^{\infty} \int_{x_0}^{x} \mathbf{A}(t_1) \int_{x_0}^{t_1} \mathbf{A}(t_2) \dots \int_{x_0}^{t_{m-1}} \mathbf{A}(t_m)\mathrm{d}t_m \dots \mathrm{d}t_2 \mathrm{d}t_1 \right) \mathbf{V}_0 , \tag{3.6.6}$$

The first n components of **V** *represents the LEM solution of the nonlinear initial problem* (3.6.1), (3.5.4).

In the particular case of polynomial ODSs, frequently met in practice,

$$\mathbf{y}' = f(x)\mathbf{P}(\mathbf{y}), \qquad \mathbf{P}(\mathbf{y}) \equiv [\mathrm{P}_1(\mathbf{y}), \mathrm{P}_2(\mathbf{y}), \dots, \mathrm{P}_n(\mathbf{y})]^{\mathrm{T}}, \quad f \in \mathrm{C}^0(I) \tag{3.6.7}$$

the following result is obtained

Corollary 3.2 [**125**, 148, 153]. *The first n components of the vector*

$$\mathbf{V}(x) = \mathrm{e}^{\mathbf{A}\int_{x_0}^{x} f(t)\mathrm{d}t}\,\mathbf{V}_0, \tag{3.6.8}$$

where **A** *is the corresponding to* **P** *LEM matrix, locally coincide with the solution of the initial polynomial problem* (3.5.4), (3.6.7).

Proof. Indeed, the LEM matrix associated to the system (3.6.7) reads

$$\overline{\mathbf{A}}(x) = f(x)\mathbf{A}, \tag{3.6.9}$$

where **A** is the constant LEM matrix, generated by the coefficients of **P**

$$\mathbf{A} = \begin{bmatrix} \mathbf{A}_{11} & \mathbf{A}_{12} & \mathbf{A}_{13} & \dots & \mathbf{A}_{1p} & \mathbf{0} & \dots \\ \mathbf{0} & \mathbf{A}_{22} & \mathbf{A}_{23} & \dots & \mathbf{A}_{2p} & \mathbf{A}_{2,p+1} & \dots \\ \mathbf{0} & \mathbf{0} & \mathbf{A}_{33} & \dots & \mathbf{A}_{3p} & \mathbf{A}_{3,p+1} & \dots \\ \mathbf{0} & \mathbf{0} & \mathbf{0} & \dots & \mathbf{A}_{4p} & \mathbf{A}_{4,p+1} & \dots \\ \dots & \dots & \dots & \dots & \dots & \dots & \dots \\ \mathbf{0} & \mathbf{0} & \mathbf{0} & \dots & \mathbf{0} & \mathbf{A}_{p+1,p+1} & \dots \\ \dots & \dots & \dots & \dots & \dots & \dots & \dots \end{bmatrix}. \tag{3.6.10}$$

Applying now formula (3.6.6), we find, step by step

$$\begin{gathered} \int_{x_0}^{x} \overline{\mathbf{A}}(t)\mathrm{d}t = \mathbf{A}\int_{x_0}^{x} f(t)\mathrm{d}t = \mathbf{A}h(x), \qquad h(x) = \int_{x_0}^{x} f(t)\mathrm{d}t, \\ \int_{x_0}^{x} \overline{\mathbf{A}}(t)\mathrm{d}t \int_{x_0}^{t} \overline{\mathbf{A}}(s)\mathrm{d}s = \left[\int_{x_0}^{x} f(t)\mathrm{d}t \int_{x_0}^{t} f(s)\mathrm{d}s\right]\mathbf{A}^2 = \frac{1}{2!}h^2(x)\mathbf{A}^2, \end{gathered} \tag{3.6.11}$$

$$\int_{x_0}^{x} \mathbf{A}(t_1)\int_{x_0}^{t_1} \mathbf{A}(t_2)\dots\int_{x_0}^{t_{m-1}} \mathbf{A}(t_m)\mathrm{d}t_m\dots\mathrm{d}t_2\mathrm{d}t_1 = \frac{1}{m!}h^m(x)\mathbf{A}^m; \tag{3.6.12}$$

this completes the proof. ◘

Using the calculation by block partitioning, one can prove the validity of the above LEM representation, at least on the interval I_1 given by formula (1.4.39). Actually, I_1 is hardly the interval of maximum length on which the LEM solutions are consistent, as it is seen from applications.

Remark. A nonlinear ODS, explicitly depending on the independent variable x, can be transformed in a nonlinear ODS with constant coefficients by introducing the equation $\frac{\mathrm{d}x}{\mathrm{d}t} = 1$; we can find LEM solutions for this new system, or we can apply corollary 1.3, extended to non-homogeneous ODSs with constant free term.

Chapter 4

LEM LINKS

In this chapter, we shall present several connections of LEM with other domains, some of them unexpected at first sight and leading to unforeseen results.

4.1. LEM AND LIE DERIVATIVES

Let $\mathbf{f}(\mathbf{y}) \equiv [f_j(\mathbf{y})]_{j=\overline{1,n}}$ be a vector of analytic with respect to $\mathbf{y}$ components. The Lie derivative with respect to $\mathbf{f}$ of the analytic function λ is [173]

$$\mathsf{L}_{\mathbf{f}}\lambda \equiv \langle \mathbf{f}, \operatorname{grad}\lambda \rangle . \qquad (4.1.1)$$

Consider the nonlinear ODS with constant coefficients

$$\mathfrak{F}\mathbf{y} \equiv \frac{\mathrm{d}\mathbf{y}}{\mathrm{d}x} - \mathbf{f}(\mathbf{y}) = \mathbf{0}, \quad \mathbf{f}(\mathbf{y}) = [f_j(\mathbf{y})]_{j=\overline{1,n}},\ f_j \in \mathcal{A}_n^0(I), \qquad (4.1.2)$$

with f_j given by (3.3.2), to be solved in the Cauchy conditions

$$\mathbf{y}(x_0) = \mathbf{y}_0, \quad x_0 \in I, \mathbf{y}_0 \equiv [y_{j0}]_{j=\overline{1,n}} \in \mathfrak{R}^n . \qquad (4.1.3)$$

We can prove

Theorem 4.1 [140, 141, 153]. *The LEM equivalent of* (4.1.2) *can be expressed in terms of Lie derivatives as*

$$\mathfrak{L}v \equiv \frac{\partial v}{\partial x} - \mathsf{L}_{\mathbf{f}} v = 0 . \qquad (4.1.4)$$

Proof. Applying the Lie derivative to the LEM mapping, it results

$$\mathsf{L}_{\mathbf{f}}\, \mathrm{e}^{\langle \xi, \mathbf{y}\rangle} = \left\langle \mathbf{f}, \operatorname{grad} \mathrm{e}^{\langle \xi, \mathbf{y}\rangle} \right\rangle = \left\langle \boldsymbol{\xi}, \mathbf{f}(\mathbf{y}) \mathrm{e}^{\langle \xi, \mathbf{y}\rangle} \right\rangle = \langle \boldsymbol{\xi}, \mathbf{f}(\mathrm{D}_{\xi}) \rangle \mathrm{e}^{\langle \xi, \mathbf{y}\rangle}; \qquad (4.1.5)$$

the theorem is thus proved. ◘

Let us see how this fact is reflected in the structure of the solution of the above nonlinear problem

Theorem 4.2 [140, 141, 153]. *The solution of the nonlinear initial problem*(4.1.2), (4.1.3) *locally allows the following equivalent representations*

$$\mathbf{y}(x) = \operatorname{grad}_{\xi}\left[\sum_{k=1}^{\infty}\langle \xi, \mathbf{f}(\mathrm{D}_{\xi})\rangle^{k} \mathrm{e}^{\langle \xi, \mathbf{y}_0\rangle} \frac{(x-x_0)^k}{k!}\right]_{\xi=0}, \tag{4.1.6}$$

$$\mathbf{y}(x) = \operatorname{grad}_{\xi}\left[\sum_{k=1}^{\infty}\mathsf{L}_{\mathbf{f}}^{k} \mathrm{e}^{\langle \xi, \mathbf{y}_0\rangle} \frac{(x-x_0)^k}{k!}\right]_{\xi=0}. \tag{4.1.7}$$

Proof. The theorem 4.2 is a consequence of the central idea of LEM. Indeed, the first LEM equivalent of the nonlinear ODS is (3.3.4), as shown in chapter 3. The Taylor series expansion of $v(x,\xi)$ around x_0 reads

$$v(x,\xi) = 1 + \sum_{k=1}^{\infty}\langle \xi, \mathbf{f}(\mathrm{D}_{\xi})\rangle^{k} \mathrm{e}^{\langle \xi, \mathbf{y}_0\rangle} \frac{(x-x_0)^k}{k!}. \tag{4.1.8}$$

By virtue of theorem 2.4, it follows that v is of the exponential form (1.3.3), with $\mathbf{y}$ satisfying the nonlinear ODS (4.1.2) and the conditions (4.1.3).

Further, according to theorem 4.1, we obviously obtain for v the expansion

$$v(x,\xi) = 1 + \sum_{k=1}^{\infty}\mathsf{L}_{\mathbf{f}}^{k} \mathrm{e}^{\langle \xi, \mathbf{y}_0\rangle} \frac{(x-x_0)^k}{k!}. \tag{4.1.9}$$

On the other hand, from (4.1.2) it straightforwardly follows

$$y_j^{(m)}(x_0) = \mathsf{L}_{\mathbf{f}}^{m}(y_{j0}), \tag{4.1.10}$$

hence

$$y_j(x) = \sum_{m=0}^{\infty}\mathsf{L}_{\mathbf{f}}^{m}(y_{j0})\frac{(x-x_0)^m}{m!}, \quad j=\overline{1,n}, \tag{4.1.11}$$

i.e., precisely the Taylor series of the solution of the nonlinear problem (4.1.2), (4.1.3).◘

Remark. The representation (4.1.7) based on Lie derivatives may be considered as a particular case of a Fliess series. Theorem 4.2 thus offer the perspective of a comparison between Fliess series and LEM representations in the case of control systems.

4.2. LEM SOLUTIONS AND FLIESS EXPANSIONS

In order to realize such a parallel, a short introduction in M. Fliess' theory is necessary [33]. This presentation is inspired by A. Isidori's book [39].

4.2.1. FLIESS REPRESENTATIONS

Fliess's representations were deduced for nonlinear control systems of the form

$$\dot{\mathbf{y}} = \mathbf{f}^0(\mathbf{y}) + \sum_{i=1}^{m} p_i(t)\, \mathbf{f}^i(\mathbf{y}), \quad \dot{\mathbf{y}} = \frac{\mathrm{d}\mathbf{y}}{\mathrm{d}t}, \tag{4.2.1}$$

where p_i, $i=\overline{1,m}$, are piecewise continuous on the real interval I and $\mathbf{f}^i(\mathbf{y})$, $i=\overline{1,m}$, are analytic with respect to $\mathbf{y} \in \Omega$, the domain $\Omega \subset \Re^n$ being open. More precisely, $\mathbf{y}(t) \in \Omega, \forall t \in I$. As the time is the independent variable, we used the point for the time derivative, a common notation in mechanics and engineering.

Let us associate to (4.2.1) the initial conditions

$$\mathbf{y}(t_0) = \mathbf{y}_0, \quad \mathbf{y}_0 = [y_{10}, y_{20}, \ldots, y_{n0}]^{\mathrm{T}} \in \Re^n. \tag{4.2.2}$$

The outputs φ_j must satisfy

$$\varphi_j = h_j(\mathbf{y}), \quad j = \overline{1,l}, \tag{4.2.3}$$

where h_j are known and analytic with respect to $\mathbf{y}$.

The relations (4.2.3) are standard for the optimal control problems; they will be taken into account only on such occasions.

As M. Fliess' method was introduced in the functional frame of formal series, we begin by presenting them.

Let $J = \{0, 1, 2, \ldots, m\}$ be a set of indices and denote by J_k the set of all finite sequences $(i_k, i_{k-1}, \ldots, i_1)$ formed by the elements of *J*. For consistency, one introduces also the set J_0, composed of a unique element of length 0, i.e. $J_0 = (\emptyset)$.

The union $J^* = \bigcup_k J_k$ becomes a free monoid with the composition law

$$(i_k, i_{k-1}, ..., i_1)(j_s, j_{s-1}, ..., j_1) = (i_k, i_{k-1}, ..., i_1, j_s, j_{s-1}, ..., j_1). \quad (4.2.4)$$

An application $c : J^* \to \Re$ is *a formal series* in $(m+1)$ non-commutative indeterminates with real coefficients. The value of c at the point $(i_k, i_{k-1}, ..., i_1)$ is denoted by $c(i_k, i_{k-1}, ..., i_1)$.

Another notion, used in order to introduce the Fliess series, is that of *iterated integral* on a set of functions. For the functions p_i from (4.2.1), the iterated integral is defined as

$$\int_{t_0}^{t} \mathrm{d}q_{i_k} \mathrm{d}q_{i_{k-1}} ... \mathrm{d}q_{i_0}, \quad (4.2.5)$$

by recurrence with respect to the multi-index length, i.e.

$$q_0(t) = t,$$
$$q_i(t) = \int_{t_0}^{t} p_i(t')\mathrm{d}t', \quad i = \overline{1, m}, \quad (4.2.6)$$

and

$$\int_{t_0}^{t} \mathrm{d}q_{i_k} \mathrm{d}q_{i_{k-1}} ... \mathrm{d}q_{i_0} = \int_{t_0}^{t} \mathrm{d}q_{i_k} dt' \int_{t_0}^{t'} \mathrm{d}q_{i_{k-1}} ... \mathrm{d}q_{i_0}. \quad (4.2.7)$$

The integral corresponding to the multi-index 0 is 1.

To a formal series in $(m+1)$ non-commutative indeterminates, one can associate a functional by adding all the products of the form

$$c(i_k, i_{k-1}, ..., i_0) \int_{t_0}^{t} \mathrm{d}q_{i_k} \mathrm{d}q_{i_{k-1}} ... \mathrm{d}q_{i_0}, \quad (4.2.8)$$

for all the elements of J^*.

Such a series is convergent, provided its coefficients $c(i_k, i_{k-1}, ..., i_0)$ satisfy certain conditions. More precisely, one can prove

Lemma 4.1 [33, 39]. *If there exists two positive constants* $K>0, M>0$ *such that*

$$\left|c\left(i_k, i_{k-1}, \ldots, i_0\right)\right| \leq K(k+1) M^{k+1}, \quad k \geq 0, \left(i_k, i_{k-1}, \ldots, i_0\right) \in J^*, \tag{4.2.9}$$

then one can find a real interval I such that for any $x \in I$ *and any set of real functions piecewise continuous on I, also satisfying the inequalities*

$$\sup_{t \in I}\left|p_i(t)\right|<1, \tag{4.2.10}$$

the series

$$y(t)=c(\varnothing)+\sum_{k=0}^{\infty} \sum_{i_0, i_0, \ldots, i_k=0}^{m} c\left(i_k, i_{k-1}, \ldots, i_0\right) \int_{t_0}^{t} \mathrm{d} q_{i_k} \ldots \mathrm{d} q_{i_1} \mathrm{d} q_{i_0} \tag{4.2.11}$$

be absolutely and uniformly convergent.

Proof. It is immediate [33, 39], observing that the inequalities (4.2.10) imply

$$\left|\int_{t_0}^{t} \mathrm{d} q_{i_k} \mathrm{d} q_{i_{k-1}} \ldots \mathrm{d} q_{i_0}\right| \leq \frac{\left|t-t_0\right|^{k+1}}{(k+1)!}. \tag{4.2.12}$$

Formula (4.2.11) defines a certain functional relatively to $p_1, p_2, \ldots, p_m$, called by M. Fliess *causal functional*, in the sense that it depends only on the restrictions imposed on the functions p_i. The uniqueness of such a representation is easily proved

Actually, Fliess series do not directly aim at the solutions of the above control system, but at the outputs φ_j, which are, in fact, the target of all control problems. These outputs are determined simultaneously with the solutions.

Let now λ be an analytic on $\Omega \subset \Re^n$ scalar function and consider, for $\mathbf{y}_0 \in \Omega$, the formal series defined by

$$\begin{aligned} & c(\varnothing)=\lambda\left(\mathbf{y}_0\right), \\ & c\left(i_k, i_{k-1}, \ldots, i_0\right)=\mathrm{L}_{\mathbf{f}^{i_0} \mathbf{f}^{I_1} \ldots \mathbf{f}^{i_k}} \lambda\left(\mathbf{y}_0\right). \end{aligned} \tag{4.2.13}$$

In [33] it was shown that one can find two positive constants *K* and *M* such that the inequality (4.2.9) be satisfied. Also taking into account lemma 4.1, it results that one can associate the functional

$$u(\mathbf{y}_0) = \lambda(\mathbf{y}_0) + \sum_{k=0}^{\infty} \sum_{i_0, i_1, \ldots, i_k = 0}^{m} \mathsf{L}_{\mathbf{f}^{i_0}} \mathsf{L}_{\mathbf{f}^{i_1}} \ldots \mathsf{L}_{\mathbf{f}^{i_k}} \lambda(\mathbf{y}_0) \int_{t_0}^{t} \mathrm{d}q_{i_k} \ldots \mathrm{d}q_{i_1} \mathrm{d}q_{i_0} \,. \tag{4.2.14}$$

to the vector functions $\mathbf{f}^0, \mathbf{f}^1, \ldots, \mathbf{f}^m$.

The next result is particularly useful in handling such functionals.

Lemma 4.2 [33, 39]. *Let* $\mathbf{f}^0, \mathbf{f}^1, \ldots, \mathbf{f}^m$ *be analytic in* $\mathbf{y}$, $\lambda_1, \lambda_2, \ldots, \lambda_l$ *l analytic functions defined on* Ω *and* Ψ_0 *a real analytic function, defined on* $\Re^l$. *Let* $u_1, u_2, \ldots, u_l$ *be functionals of type* (4.2.14), *associated to* $\lambda_1, \lambda_2, \ldots, \lambda_l$ *accordingly. Then the composition* $\Psi(u_1(x), u_2(x), \ldots, u_l(x))$ *is also a functional of type* (4.2.14), *with* $\lambda = \Psi(\lambda_1, \lambda_2, \ldots, \lambda_l)$.

A schedule of the proof would be also useful for the comparison with the LEM representations. We observe that the formal power series, defined by putting $\lambda = \alpha_1 \lambda_1 + \alpha_2 \lambda_2$, for $\alpha_1, \alpha_2 \in \Re$ in (4.2.14), is, in fact, $\alpha_1 u_1(x) + \alpha_2 u_2(x)$. Moreover, the functional corresponding to the product $\lambda_1 \lambda_2$ is precisely $u_1 u_2$. Indeed,

$$\begin{aligned} u_1(x) u_2(x) = & \left(\lambda_1 + \mathsf{L}_{\mathbf{f}^0} \lambda_1 \int_{t_0}^{t} \mathrm{d}p_1 + \mathsf{L}_{\mathbf{f}^1} \lambda_1 \int_{t_0}^{t} \mathrm{d}p_1 + \mathsf{L}_{\mathbf{f}^0} \mathsf{L}_{\mathbf{f}^0} \lambda_1 \int_{t_0}^{t} \mathrm{d}p_0 \mathrm{d}p_0 + \ldots \right) \\ & \cdot \left(\lambda_2 + \mathsf{L}_{\mathbf{f}^0} \lambda_2 \int_{t_0}^{t} \mathrm{d}p_0 + \mathsf{L}_{\mathbf{f}^1} \lambda_2 \int_{t_0}^{t} \mathrm{d}p_1 + \mathsf{L}_{\mathbf{f}^0} \mathsf{L}_{\mathbf{f}^0} \lambda_2 \int_{t_0}^{t} \mathrm{d}p_0 \mathrm{d}p_0 + \ldots \right); \end{aligned} \tag{4.2.15}$$

In the above relation, the values of all functions depending on $\mathbf{y}$ are taken at $\mathbf{y} = \mathbf{y}_0$. Multiplying term by term, it is obtained

$$\begin{aligned} u_1(x) u_2(x) = & \lambda_1 \lambda_2 + \left(\lambda_1 \mathsf{L}_{\mathbf{f}^0} \lambda_2 + \lambda_2 \mathsf{L}_{\mathbf{f}^0} \lambda_1 \right) \int_{t_0}^{t} \mathrm{d}p_0 \\ & + \left(\lambda_1 \mathsf{L}_{\mathbf{f}^1} \lambda_2 + \lambda_2 \mathsf{L}_{\mathbf{f}^1} \lambda_1 \right) \int_{t_0}^{t} \mathrm{d}p_1 \\ & + \left(\lambda_1 \mathsf{L}_{\mathbf{f}^0} \mathsf{L}_{\mathbf{f}^0} \lambda_2 + \lambda_2 \mathsf{L}_{\mathbf{f}^0} \mathsf{L}_{\mathbf{f}^0} \lambda_1 \right) \int_{t_0}^{t} \mathrm{d}p_0 \mathrm{d}p_0 + \left(\mathsf{L}_{\mathbf{f}^0} \lambda_1 \right) \left(\mathsf{L}_{\mathbf{f}^0} \lambda_2 \right) \left(\int_{t_0}^{t} \mathrm{d}p_0 \right)^2 + \ldots \end{aligned} \tag{4.2.16}$$

The factors multiplying $\int_{t_0}^{t} \mathrm{d}p_0$ and $\int_{t_0}^{t} \mathrm{d}p_1$ are, obviously, $\mathsf{L}_{\mathbf{f}^0}(\lambda_1\lambda_2)$ and $\mathsf{L}_{\mathbf{f}^1}(\lambda_1\lambda_2)$ accordingly. Further, for the next three terms we deduce

$$\mathsf{L}_{\mathbf{f}^0}\mathsf{L}_{\mathbf{f}^0}(\lambda_1\lambda_2)=\lambda_1\mathsf{L}_{\mathbf{f}^0}\mathsf{L}_{\mathbf{f}^0}\lambda_2+\lambda_2\mathsf{L}_{\mathbf{f}^0}\mathsf{L}_{\mathbf{f}^0}\lambda_1+2\left(\mathsf{L}_{\mathbf{f}^0}\lambda_1\right)\left(\mathsf{L}_{\mathbf{f}^0}\lambda_2\right), \quad (4.2.17)$$

and as, on the other hand,

$$\left(\int_{t_0}^{t} \mathrm{d}p_0\right)^2 = 2\int_{t_0}^{t} \mathrm{d}p_0\mathrm{d}p_0 , \quad (4.2.18)$$

it follows that the sum of the three terms is

$$\mathsf{L}_{\mathbf{f}^0}\mathsf{L}_{\mathbf{f}^0}(\lambda_1\lambda_2)\int_{t_0}^{t} \mathrm{d}p_0\mathrm{d}p_0 . \quad (4.2.19)$$

The above described procedure must be repeated for the remaining terms.

With these preparations, one can deduce the following representation, which will be used in parallel with the LEM solutions

Theorem 4.3 [33, 39]. *If p_j are at least piecewise continuous and satisfy the inequalities*

$$\sup_{t\in \mathrm{I}}\left|p_j(t)\right|<1,\ j=\overline{1,m}, \quad (4.2.20)$$

then the solution of the nonlinear initial problem(4.2.1), (4.2.2) *allows around t_0 the representation*

$$y_j(t)=y_{j0}+\sum_{k=0}^{\infty}\ \sum_{i_0,i_1,\dots,i_k=0}^{m}\mathsf{L}_{\mathbf{f}^{i_0}}\mathsf{L}_{\mathbf{f}^{i_1}}\dots\mathsf{L}_{\mathbf{f}^{i_k}}(y_{j0})\int_{t_0}^{t}\mathrm{d}p_{i_k}\dots\mathrm{d}p_{i_1}\mathrm{d}p_{i_0} . \quad (4.2.21)$$

Proof. Indeed, from the definition of the iterated integral one deduces

$$\frac{\mathrm{d}}{\mathrm{d}t}\int_{t_0}^{t}\mathrm{d}p_0\mathrm{d}p_{i_{k-1}}\dots\mathrm{d}p_{i_0}=\int_{t_0}^{t}\mathrm{d}p_{i_{k-1}}\dots\mathrm{d}p_{i_0} , \quad (4.2.22)$$

as well as

$$\frac{\mathrm{d}}{\mathrm{d}t}\int_{t_0}^{t}\mathrm{d}p_i\mathrm{d}p_{i_{k-1}}...\mathrm{d}p_{i_0}=p_i(t)\int_{t_0}^{t}\mathrm{d}p_{i_{k-1}}...\mathrm{d}p_{i_0},\quad i=\overline{1,m}. \tag{4.2.23}$$

Consequently, differentiating (4.2.21) with respect to t, it is obtained

$$\begin{aligned}\frac{\mathrm{d}y_j}{\mathrm{d}t}&=\mathsf{L}_{\mathbf{f}^0}(y_{j0})+\sum_{k=0}^{\infty}\sum_{i_0,i_1,...,i_k=0}^{m}\mathsf{L}_{\mathbf{f}^{i_0}}\mathsf{L}_{\mathbf{f}^{i_1}}...\mathsf{L}_{\mathbf{f}^{i_k}}\mathsf{L}_{\mathbf{f}^0}(y_{j0})\int_{t_0}^{t}\mathrm{d}p_{i_k}...\mathrm{d}p_{i_1}\mathrm{d}p_{i_0}\\&+\sum_{i=1}^{m}\left[\mathsf{L}_{\mathbf{f}^i}(y_{j0})+\sum_{k=0}^{\infty}\sum_{i_0,i_1,...,i_k=0}^{m}\mathsf{L}_{\mathbf{f}^{i_0}}\mathsf{L}_{\mathbf{f}^{i_1}}..\mathsf{L}_{\mathbf{f}^{i_k}}\mathsf{L}_{\mathbf{f}^i}(y_{j0})\int_{t_0}^{t}\mathrm{d}p_{i_k}...\mathrm{d}p_{i_1}\mathrm{d}p_{i_0}\right]p_i(t).\end{aligned} \tag{4.2.24}$$

By lemma 4.2, it follows that

$$\begin{aligned}&\mathsf{L}_{\mathbf{f}^0}(y_{j0})+\sum_{k=0}^{\infty}\sum_{i_0,i_1,...,i_k=0}^{m}\mathsf{L}_{\mathbf{f}^{i_0}}\mathsf{L}_{\mathbf{f}^{i_1}}...\mathsf{L}_{\mathbf{f}^{i_k}}\mathsf{L}_{\mathbf{f}^0}(y_{j0})\int_{t_0}^{t}\mathrm{d}p_{i_k}...\mathrm{d}p_{i_1}\mathrm{d}p_{i_0}=f_j^0(\mathbf{y}_0)\\&+\sum_{k=0}^{\infty}\sum_{i_0,i_1,...,i_k=0}^{m}\mathsf{L}_{\mathbf{f}^{i_0}}\mathsf{L}_{\mathbf{f}^{i_1}}..\mathsf{L}_{\mathbf{f}^{i_k}}f_j^0(\mathbf{y}_0)\int_{t_0}^{t}\mathrm{d}p_{i_k}...\mathrm{d}p_{i_1}\mathrm{d}p_{i_0}=f_j^0(y_1,y_2,...,y_n).\end{aligned} \tag{4.2.25}$$

Proceeding analogously with the remaining terms, one finally gets

$$\frac{\mathrm{d}y_j}{\mathrm{d}t}=f_j^0(y_1,y_2,...,y_n)+\sum_{i=1}^{m}f_j^i(y_1,y_2,...,y_n)p_i(t),\quad j=\overline{1,n}. \tag{4.2.26}$$

This is precisely the system (4.2.1), written component-wise. Moreover, the functions y_j also satisfy $y_j(t_0)=y_{j0}$.

It then results that the representation (4.2.21) formally satisfies the nonlinear problem (4.2.1), (4.2.2). Its consistency is an immediate consequence of lemma 4.1.

Based on this representation, a similar series, called the *Fliess series* was associated to the outputs of the control system (4.2.1). We do not present here the particularities of this series, because it is not directly connected with LEM.

Theorem 4.3 may be considered as a starting point for the comparison between the Fliess representations and the LEM solutions.

4.2.2. A PARALLEL BETWEEN FLIESS EXPANSIONS AND LEM SOLUTIONS

Clearly, this parallel can be realized only for control systems of the particular form (4.2.2). For such systems, the first LEM equivalent reads

$$\mathfrak{L}v \equiv \frac{\partial v}{\partial t} - \left\langle \xi, \mathbf{f}^0(\mathrm{D}_\xi) + \sum_{i=1}^{m} p_i(t)\mathbf{f}^i(\mathrm{D}_\xi) \right\rangle v = 0. \tag{4.2.27}$$

If, moreover, $\mathbf{y}$ satisfies the condition (4.2.2), then v satisfies

$$v(t_0, \xi) = \mathrm{e}^{\langle \xi, \mathbf{y}_0 \rangle}. \tag{4.2.28}$$

We can now apply the results of chapter 2. Indeed, let us define the operator

$$\begin{aligned} \left\langle \xi, \mathfrak{F}_{t_0}\mathbf{f}(t, \mathrm{D}_\xi) \right\rangle v \equiv \sum_{j=1}^{n} \xi_j f_j^0(\mathrm{D}_\xi) \int_{t_0}^{t} v(t', \xi)\,\mathrm{d}t' \\ + \sum_{j=1}^{n} \sum_{k=1}^{m} \xi_j f_j^k(\mathrm{D}_\xi) \left(\int_{t_0}^{t} p_k(t') v(t', \xi)\,\mathrm{d}t' \right). \end{aligned} \tag{4.2.29}$$

With this definition, the LEM solution of the LEM equivalent (4.2.27) can be expressed as

$$v(t, \xi) = \sum_{k \geq 0} \left\langle \xi, \mathfrak{F}_{t_0}\mathbf{f}(t, \mathrm{D}_\xi) \right\rangle^k \left(\mathrm{e}^{\langle \xi, \mathbf{y}_0 \rangle} \right), \tag{4.2.30}$$

and, according to theorem 4.2, we obtain for the solution of the nonlinear initial problem (4.2.1), (4.2.2) the LEM representation

$$\mathbf{y}(t) = \mathrm{grad}_\xi \left[\sum_{k \geq 0} \left\langle \xi, \mathfrak{F}_{t_0}\mathbf{f}(t, \mathrm{D}_\xi) \right\rangle^k \left(\mathrm{e}^{\langle \xi, \mathbf{y}_0 \rangle} \right) \right]_{\xi = \mathbf{0}}. \tag{4.2.31}$$

The above expansions are locally consistent.

We observe that, by comparing (4.2.31) and (4.2.21), we can easier establish a link between the Lie derivative and the LEM equivalent. This parallel is also extended to the associated linear equivalent system that reads

$$\frac{\mathrm{d}\mathbf{V}}{\mathrm{d}t} = \left[\mathbf{F}_0 + \sum_{i=1}^{m} p_i(t)\mathbf{F}_i \right] \mathbf{V}, \tag{4.2.32}$$

where $\mathbf{V}$ is the infinite vector having as components the coefficients $v_\mu(t)$ from the expansion

$$v(t, \xi) = 1 + \sum_{|\mu| > 0} v_\mu(t) \frac{\xi^\mu}{\mu!}, \tag{4.2.33}$$

and $\mathbf{F}_0, \mathbf{F}_i, i=\overline{1,m}$ are the LEM matrices corresponding to the operators $\mathbf{f}^0(\mathrm{D}_\xi)\mathbf{f}^i(\mathrm{D}_\xi)\, i=\overline{1,m}$, respectively.

The formal solution of the LEM system (4.2.32) can be represented as

$$\mathbf{V}(t)=\mathbf{V}(t_0)+\int_{t_0}^{t}\mathbf{A}(t_1)\int_{t_0}^{t_1}\mathbf{A}(t_2)..\int_{t_0}^{t_{k-1}}\mathbf{A}(t_{k-2})\mathrm{d}t_{k-1}\,\mathrm{d}t_{k-2}...\mathrm{d}t_1, \tag{4.2.34}$$

where

$$\mathbf{A}(t)=\mathbf{F}_0+\sum_{i=1}^{m}p_i(t)\mathbf{F}_i \tag{4.2.35}$$

and

$$\mathbf{V}(t_0)=\left[\mathbf{y}_0^\mu\right]_{|\mu|\in\mathfrak{N}}. \tag{4.2.36}$$

The first n components of the vector $\mathbf{V}(t)$ from (4.2.34) are locally consistent and coincide with $y_j(t)$, which are the components of the nonlinear control system (4.2.1); they also satisfy the initial conditions (4.2.2).

4.2.3. HIGHER LIE DERIVATIVES AND LEM MATRICES

The above considerations emphasize two methods of getting a solution of a nonlinear problem: by using Lie derivatives and by using LEM.

We firstly take the case of constant coefficients. Theorem 4.2 leads in fact to two representations of the same (unique) solution of the nonlinear initial problem (4.1.2), (4.1.3). Hence we can compare formula (4.1.11) with the first n components of the vector

$$\mathbf{V}(x)=\mathrm{e}^{\mathbf{A}(x-x_0)}\mathbf{V}_0, \tag{4.2.37}$$

thus getting

Theorem 4.4 [140, 141, 153]. *The k-th order Lie derivatives* $\mathsf{L}_\mathbf{f}^k(y_{j0})$ *can be determined by the formula*

$$\left[\mathsf{L}_\mathbf{f}^k(y_{j0})\right]_{j=\overline{1,n}}\equiv\boldsymbol{\pi}_n\left(\mathbf{A}^k\right)\left[\mathbf{y}_0^\mu\right]_{|\mu|\in\mathfrak{N}}, \tag{4.2.38}$$

where **A** *is the LEM matrix corresponding to* $\mathbf{f}$ *and* $\boldsymbol{\pi}_n$ *is the projector associating to an infinite matrix its first* n *rows.*

Proof. By corollary 2.1, the solution of the nonlinear initial problem (4.1.2), (4.1.3) can be written as

$$\mathbf{y}(x)=\boldsymbol{\pi}_n\left(\mathrm{e}^{\mathbf{A}(x-x_0)}\right)\left[\mathbf{y}_0^{\mu}\right]_{|\mu\in\mathfrak{N}|}, \tag{4.2.39}$$

or else

$$\mathbf{y}(x)=\sum_{k=0}^{\infty}\frac{(x-x_0)^k}{k!}\boldsymbol{\pi}_n\left(\mathbf{A}^k\right)\left[\mathbf{y}_0^{\mu}\right]_{|\mu|\in\mathfrak{N}}. \tag{4.2.40}$$

On the other hand, taking (4.1.11) into account, the same (unique) solution can also be written in the form

$$y_j(x)=\sum_{m=0}^{\infty}\mathsf{L}_{\mathbf{f}}^m\left(y_{j0}\right)\frac{(x-x_0)^m}{m!},\quad j=\overline{1,n}. \tag{4.2.41}$$

Identifying the coefficients, we obtain (4.2.38). ◘

Let us note that the formal development (4.2.34) can be written by using the iterated integrals (4.2.5), (4.2.7). On this way, we can deduce another LEM representation of the control system (4.2.1).

Theorem 4.5 [140, 141, 153]. *The solution of the Cauchy problem* (4.2.1), (4.2.2) *can be represented in the form*

$$\begin{bmatrix} y_1(t)\\ y_2(t)\\ \vdots\\ y_n(t)\end{bmatrix}=\sum_{k=0}^{\infty}\ \sum_{i_0,i_1,\ldots,i_k=0}^{m}\boldsymbol{\pi}_n\left(\mathbf{F}_{i_0}\mathbf{F}_{i_1}\ldots\mathbf{F}_{i_k}\right)\left[\mathbf{y}_0^{\mu}\right]_{|\mu\in\mathfrak{N}|}\int_{t_0}^{t}\mathrm{d}q_{i_k}\ldots\mathrm{d}q_{i_1}\mathrm{d}q_{i_0}, \tag{4.2.42}$$

where $\mathbf{F}_{i_s}$, $s=\overline{1,m}$, *are the LEM matrices corresponding to the analytic vector functions* $\mathbf{f}^{i_s}$ *and*

$$q_0(t)=t,\ q_i(t)=\int_{t_0}^{t}p_i(t')\mathrm{d}t',\quad i=\overline{1,m}. \tag{4.2.43}$$

Proof. From formulae (4.2.5), (4.2.7), defining the iterated integrals, we see that the vector

$$\mathbf{V}(t)=\sum_{k=0}^{\infty}\sum_{i_0,i_1,\ldots,i_k=0}^{m}\mathbf{F}_{i_0}\mathbf{F}_{i_1}\ldots\mathbf{F}_{i_k}\left[\mathbf{y}_0^{\mu}\right]_{\mu\in\mathfrak{N}|}\int_{t_0}^{t}\mathrm{d}q_{i_k}\ldots\mathrm{d}q_{i_1}\mathrm{d}q_{i_0} \tag{4.2.44}$$

formally satisfies the linear ODS

$$\frac{\mathrm{d}\mathbf{V}}{\mathrm{d}t}=\sum_{i=0}^{m}p_i(t)\mathbf{F}_i\mathbf{V},\quad p_0(t)=1, \tag{4.2.45}$$

which is precisely the second linear equivalent of (4.2.1).

It is easily seen that $\mathbf{V}$ also satisfies the initial conditions

$$\mathbf{V}(t_0)=\left[\mathbf{y}_0^{\gamma}\right]_{\gamma|\in\mathfrak{N}}. \tag{4.2.46}$$

According to LEM, the first n components of $\mathbf{V}$ are locally consistent and coincide with the solution of the nonlinear problem (4.2.1), (4.2.2).

As $\mathbf{F}_i, i=\overline{0,m}$, are column-finite, the matrix products from (4.2.44) are computed by blocks, involving only finite sums. This completes the proof. ◘

In this case too, one can compare Fliess' series with the LEM representations by

Corollary 4.1 [140, 141, 153]. *The higher Lie derivatives* $\mathsf{L}_{\mathbf{f}_0}^{k_0}\mathsf{L}_{\mathbf{f}_1}^{k_1}\ldots\mathsf{L}_{\mathbf{f}_s}^{k_s}(y_{j0})$ *can be computed in terms of LEM matrices by the formula*

$$\begin{bmatrix}\mathsf{L}_{\mathbf{f}^{i_0}}\mathsf{L}_{\mathbf{f}^{i_1}}\ldots\mathsf{L}_{\mathbf{f}^{i_k}}(y_{10})\\ \mathsf{L}_{\mathbf{f}^{i_0}}\mathsf{L}_{\mathbf{f}^{i_1}}\ldots\mathsf{L}_{\mathbf{f}^{i_k}}(y_{20})\\ \vdots\\ \mathsf{L}_{\mathbf{f}^{i_0}}\mathsf{L}_{\mathbf{f}^{i_1}}\ldots\mathsf{L}_{\mathbf{f}^{i_k}}(y_{n0})\end{bmatrix}=\boldsymbol{\pi}_n\left(\mathbf{F}_{i_0}\mathbf{F}_{i_1}\ldots\mathbf{F}_{i_k}\right)\left[\mathbf{y}_0^{\mu}\right]_{\mu|\in\mathfrak{N}}. \tag{4.2.47}$$

Formula (4.2.47) generalizes (4.2.38).

Another consequence of theorem 4.5 regards the outputs of the control system; in fact, this is a collateral result.

Let $\mathbf{H}_j$ be a vector of components

$$\mathbf{H}_j=\left[h_{j\mu}\right]_{\mu|\in\mathfrak{N}}, \tag{4.2.48}$$

and suppose that the analytic functions h_j from (4.2.3) allow the expansions

$$h_j(\mathbf{y}) = \sum_{|\mu|\geq 0} h_{j\mu}\mathbf{y}^{\mu}. \tag{4.2.49}$$

Then (4.2.3) may be written as a formal scalar product

$$\varphi_j = h_j(\mathbf{y}_0) + \left\langle \mathbf{H}_j, \sum_{k=0}^{\infty} \sum_{i_0,i_1,\ldots,i_k=0}^{m} \mathbf{F}_{i_0}\mathbf{F}_{i_1}\ldots\mathbf{F}_{i_k}\left[\mathbf{y}_0^{\mu}\right]_{\mu\in\mathfrak{N}|} \int_{t_0}^{t} \mathrm{d}q_{i_k}\ldots\mathrm{d}q_{i_1}\mathrm{d}q_{i_0} \right\rangle. \tag{4.2.50}$$

Taking into account (4.2.46), the outputs φ_j can be straightforwardly expressed in LEM terms

$$\varphi_j = h_j(\mathbf{y}_0) + \langle \mathbf{H}_j, \mathbf{V}\rangle, \quad j = \overline{1,l}. \tag{4.2.51}$$

In the above relation, the scalar product is also computed by block partitioning, each involved block and matrix sum being finite. The convergence of the LEM representations and the analyticity of h_j yield the convergence of the representation (4.2.51) around t_0.

4.3. TWO-POINT PROBLEMS TREATED BY LEM

In this section we show how to apply LEM to nonlinear two-point problems. The obtained results concern only polynomial ODEs, but, as it was previously seen, they can be extended to more general classes of nonlinear operators.

Treating such problems by LEM emphasizes another advantage of this method: it can be applied to two-point problems even if their solution is not unique. At the end of this section, it is presented such an example, for which all the solutions were identified and obtained by LEM.

Consider the second order polynomial ODE

$$y'' = \sum_{j+k=0}^{p} a_{jk}(x) y^j (y')^k, \quad a_{jk} \in \mathrm{C}^{\infty}(I), I \equiv [x_0, x_1], \tag{4.3.1}$$

to which we associate the two-point (Picard) conditions

$$y(x_0) = y_0,\ y(x_1) = y_1. \tag{4.3.2}$$

Let us apply LEM to this equation, taking for y the initial conditions

$$y(x_0) = y_0, \quad y'(x_0) = \beta, \tag{4.3.3}$$

β being a parameter. By theorem 1.7, denoting by $\mathbf{\Pi}$ the inverse of the polynomial operator defined by (4.3.1), it follows that the solution of the polynomial initial problem (4.3.1), (4.3.3) is given by the first component $v_{10}(x,\beta)$ of the vector

$$\mathbf{\Pi}(x - x_0)\mathbf{V}_\beta(x_0) + \mathbf{\Theta}(x - x_0), \quad \mathbf{V}_\beta(x_0) = \left[y_0^j \beta^k\right]_{j+k \in \mathfrak{N}}. \tag{4.3.4}$$

Denote by $f(\beta)$ the value of this component at the point x_1, if it is consistent

$$f(\beta) = v_{10}(x_1, \beta). \tag{4.3.5}$$

A direct application of theorem 1.9 is

Theorem 4.6 [134]. *Let* a_{jk} *be infinitely differentiable on* $I = [x_0, x_1]$ *and suppose that* $\|a_{jk}\|_m \leq C, m \in \mathfrak{N}^*$. *Then, if* $f(\beta)$ *is consistent,*

i) the set of the analytic on I solutions of the two-point problem (4.3.1), (4.3.2), *has the same cardinal as the set* Γ *of the solutions of the functional equation*

$$f(\beta) = y_1; \tag{4.3.6}$$

ii) the analytic on I solutions of the two-point problem (4.3.1), (4.3.2) *coincide with the first components of the vectors*

$$\mathbf{\Pi}(x_1 - x_0)\mathbf{V}_{\bar{\beta}}(x_0) + \mathbf{\Theta}(x_1 - x_0), \quad \bar{\beta} \in \Gamma. \tag{4.3.7}$$

Proof. According to theorem 1.7, the solution of the Cauchy problem (4.3.1), (4.3.3) is the first component of the vector (4.3.4). Hence, an analytic on *I* solution of the above two-point problem must satisfy the functional equation (4.3.6). Replacing now β by a solution $\bar{\beta}$ of the functional equation in the LEM expression of $v_{10}(x,\beta)$, it follows that $v_{10}(x,\bar{\beta})$ satisfied the two-point problem. ◘

Theorem 4.6 emphasizes some of LEM's advantages when applied to a two-point problem:

- LEM works, even if the solution is not unique;

- numerically, it leads to non-standard algorithms, completely distinct from the usual ones (e.g., collocation, shooting, etc.);
- if uniqueness does not hold, then, by these algorithms, one can separate and effectively determinate the solutions.

These advantages were put into evidence in the study by LEM of the simply supported bar, as well as in the case of the hyperstatic bar, as it will be seen in chapter 5.

In the case of constant coefficients, the calculation of **Π** and **Θ** are much simpler. If, moreover, the equation is homogeneous, then we can prove a useful particular case of theorem 4.6, also taking into account corollary 1.3.

Corollary 4.2. [134]. *Consider the ODE* (4.3.1), *in which the coefficients* a_{jk} *are constant and* $a_{j\theta} = 0,\ j = \overline{1,n}$. *Then the solutions of Picard's problem* (4.3.1), (4.3.2) *are represented by the first component* $v_{10}(x,\beta)$ *of the vector*

$$\mathbf{V}(x,\beta) = \mathrm{e}^{\mathbf{A}(x-x_0)}\,\mathbf{V}_\beta(x_0),\quad \mathrm{e}^{\mathbf{A}(x-x_0)} = \mathbf{E} + \sum_{m=1}^{\infty} \mathbf{A}^m \frac{(x-x_0)^m}{m!}, \tag{4.3.8}$$

where β *satisfies the functional equation*

$$v_{10}(x_1,\beta) = y_1 . \tag{4.3.9}$$

A step forward on the way of getting LEM algorithms for two-point problem is

Theorem 4.7 [134]. *Consider the sequence* $\{y_k(x)\}_{k\in\mathfrak{N}}$, *where* $y_k(x,\beta) = v_{10}^k(x,\beta)$ *is the first component of the solution of the linear finite ODS*

$$\frac{\mathrm{d}v_{ij}^k}{\mathrm{d}x} = i\sum_{m+l\le p} a_{ml} v_{i+m-1,j+l}^k + j\sum_{m+l\le p} a_{ml} v_{i+m,j+l-1}^k ,\quad i+j\le k, \tag{4.3.10}$$

also satisfying the Cauchy conditions

$$v_{ij}^k(x_0,\beta) = y_0^i\,\beta^j,\quad i+j\le k . \tag{4.3.11}$$

Suppose that β *satisfies*

$$v_{ij}^{ki}(x_1,\beta)=y_1, \quad i+j\le k. \tag{4.3.12}$$

If $\{y_k(x)\}_{k\in\mathfrak{N}}$ *converges uniformly on I, then its limit is a solution of the polynomial two-point problem* (4.3.1), (4.3.2).

Proof. The double inclusion (2.2.3) results in the convergence of the sequence $\{y_k(x)\}_{k\in\mathfrak{N}}$ to the solution of the nonlinear Cauchy problem (4.3.1), (4.3.3); the equation (4.3.12) introduces the second bilocal condition. ◘

Remark. For the calculation of $f(\beta)\equiv v_{10}(x_1,\beta)$, we can also use other LEM representations.

Let us give an example that emphasizes the advantage of LEM to separate and identify the solutions of a two-point problem for which the uniqueness does not hold.

Example 4.1. Consider the two-point problem

$$\mathfrak{S}y\equiv y''+yy'^2+y^3=0, \tag{4.3.13}$$

$$y(0)=0, \quad y(\pi)=0. \tag{4.3.14}$$

The equation (4.3.13) is polynomial and odd.

We easily observe that this problem allows three distinct analytic solutions on $[0,\pi]$:

$$y_1(x)=0, \quad y_2(x)=\sin x, \quad y_3(x)=-\sin x. \tag{4.3.15}$$

Let us apply LEM to (4.3.13), for the initial conditions depending on the parameter β

$$y(0)=0, \quad y'(0)=\beta. \tag{4.3.16}$$

The LEM mapping depends on two parameters

$$v(x,\sigma,\xi)=\mathrm{e}^{\sigma y+\xi y'}. \tag{4.3.17}$$

The first linear equivalent reads

$$\mathfrak{L}v\equiv\frac{\partial v}{\partial x}-\sigma\frac{\partial v}{\partial \xi}+\xi\left(\frac{\partial^3 v}{\partial\sigma\partial\xi^2}+\frac{\partial^3 v}{\partial\sigma^3}\right)=0. \tag{4.3.18}$$

Considering for $v(x,\sigma,\xi)$ the expansion

$$v(x,\sigma,\xi)=1+\sum_{j+k=1}^{\infty} v_{jk}(x)\frac{\sigma^j}{j!}\frac{\xi^k}{k!}, \tag{4.3.19}$$

we explicitly obtain the LEM system for the coefficients $v_{jk}(x)$

$$\frac{\mathrm{d}v_{jk}}{\mathrm{d}x}=jv_{j-1,k+1}+k\left(-v_{j+1,k+1}-v_{j+3,k-1}\right), \qquad j+k\in\mathfrak{N}. \tag{4.3.20}$$

This system can be written in matrix form

$$\mathfrak{S}\mathbf{V}(x)\equiv\frac{\mathrm{d}\mathbf{V}}{\mathrm{d}x}-\mathbf{AV}=\mathbf{0}; \tag{4.3.21}$$

it defines a "contracted" LEM matrix, as the considered differential operator is odd:

$$\mathbf{A}=\begin{bmatrix} \mathbf{A}_{11} & \mathbf{A}_{13} & \mathbf{0} & \dots & \mathbf{0} & \mathbf{0} & \dots \\ \mathbf{0} & \mathbf{A}_{33} & \mathbf{A}_{35} & \dots & \mathbf{0} & \mathbf{0} & \dots \\ \mathbf{0} & \mathbf{0} & \mathbf{A}_{55} & \dots & \mathbf{0} & \mathbf{0} & \dots \\ \mathbf{0} & \mathbf{0} & \mathbf{0} & \dots & \mathbf{0} & \mathbf{0} & \dots \\ \dots & \dots & \dots & \dots & \dots & \dots & \dots \\ \mathbf{0} & \mathbf{0} & \mathbf{0} & \dots & \mathbf{A}_{2p-1,2p-1} & \mathbf{A}_{2p-1,2p+1} & \dots \\ \dots & \dots & \dots & \dots & \dots & \dots & \dots \end{bmatrix}. \tag{4.3.22}$$

The vector $\mathbf{V}$ of the unknown functions will be divided, as previously, by finite vectors, also "contracted":

$$\mathbf{V}(x)=\left[\mathbf{V}_{2m-1}(x)\right]_{m\in\mathfrak{N}}, \quad \mathbf{V}_{2m-1}(x)=\left\lfloor v_{jk}(x)\right\rfloor_{j+k=2m-1}. \tag{4.3.23}$$

The initial conditions (4.3.16) are transferred to $\mathbf{V}$ by the LEM mapping

$$\mathbf{V}(0)=\left[\mathbf{V}_{2m-1}(0)\right]_{m\in\mathfrak{N}}, \quad \mathbf{V}_{2m-1}(0)=\begin{bmatrix}0 & 0 & \dots & \beta^{2j-1}\end{bmatrix}^{\mathrm{T}}. \tag{4.3.24}$$

As the LEM matrix is constant, the local inverse of the polynomial differential operator defined by (4.3.13) will be of the exponential form (1.4.49) from corollary 1.3. By virtue of corollary 4.2, one must compute the first row of the matrix

$$\mathrm{e}^{\mathbf{A}\pi}\equiv\sum_{k=0}^{\infty}\mathbf{A}^k\frac{\pi^k}{k!}. \tag{4.3.25}$$

The square diagonal cells with $2j$ rows and columns read

$$\mathbf{A}_{2j-1,2j-1} = \begin{bmatrix} 0 & 2j-1 & 0 & \dots & 0 & 0 \\ 0 & 0 & 2j-2 & \dots & 0 & 0 \\ \dots & \dots & \dots & \dots & \dots & \dots \\ 0 & 0 & 0 & \dots & 0 & 1 \\ 0 & 0 & 0 & \dots & 0 & 0 \end{bmatrix}; \tag{4.3.26}$$

the rectangular cells with $2j$ rows and $2j+2$ columns are

$$\mathbf{A}_{2j-1,2j+1} = -\begin{bmatrix} 0 & 0 & 0 & 0 & \dots & 0 & 0 & 0 & 0 \\ 1 & 0 & 1 & 0 & \dots & 0 & 0 & 0 & 0 \\ 0 & 2 & 0 & 2 & \dots & 0 & 0 & 0 & 0 \\ \dots & \dots & \dots & \dots & \dots & \dots & \dots & \dots & \dots \\ 0 & 0 & 0 & 0 & \dots & 2j-1 & 0 & 2j-1 & 0 \end{bmatrix}. \tag{4.3.27}$$

We can now compute step by step the approximates $f_m(\beta) = v_{10}^m(\pi, \beta)$ of the first component of the vector

$$\mathrm{e}^{\mathbf{A}\pi}\,\mathbf{V}(0). \tag{4.3.28}$$

Table 4.1 contains the expressions of these approximates up to order 7 included. Besides, it also contains the approximate solutions of the functional equation

$$v_{10}^m(\pi, \beta) = 0, \tag{4.3.29}$$

as well as the corresponding LEM approximations of these solutions.

The table clearly points out three distinct numerical sequences $\{\beta_j^m\}$ $j=\overline{1,3}$, converging with m to $0, 1, -1$ accordingly. These are, in fact, the initial values $y_j'(0)$, $j=\overline{1,3}$. The other roots of the polynomials $f_m(\beta)$ are not significant, as they do not converge with m. The sequences of functions $\{y_j^m(x)\}$, corresponding to β_j^m converge to $y_j(x)$, $j=\overline{1,3}$ accordingly, approximating them better and better on the whole interval $[0, \pi]$.

Table 4.1

m	$f_m(\beta)=v_{10}^m(\pi,\beta)$	β_j^m	$v_{10}^m(x,\beta_j^m)=y_j^m(x)$
2	$\pi\beta$	$\beta_1^2=0$	$y_1^2=0$
3	$\pi\beta\left(1-\dfrac{\pi^2\beta^2}{6}\right)$	$\beta_1^3=0$ $\beta_2^3=0.79$ $\beta_3^3=-0.79$	$y_1^3=0$ $y_2^3=0.79x-0.0808x^3$ $y_2^3=-0.79x+0.0808x^3$
4	IDEM		
5	$\pi\beta\left[1-\dfrac{\beta^2}{6}\left(\pi^2+\dfrac{3\pi^4}{10}\right)+\right.$ $\left.+\dfrac{7\pi^4\beta^4}{120}\right]$	$\beta_1^5=0$ $\beta_2^5=0.981$ $\beta_3^5=-0.981$ $\beta_{4,5}^5=\pm 0.435$	$y_1^5=0$ $y_2^5=0.981x-0.155x^3$ $+0.0039x^5$ $y_3^5=-0.981x+0.155x^3$ $-0.0039x^5$
6	IDEM		
7	$\pi\beta\left[1-\dfrac{\beta^2}{6}\left(\pi^2+\dfrac{3\pi^4}{10}\right)+\right.$ $+\beta^4\left(\dfrac{7\pi^4}{120}+\dfrac{\pi^6}{42}\right)-$ $\left.-\dfrac{127\pi^6\beta^6}{5040}\right]$	$\beta_1^7=0$ $\beta_2^7=0.9899$ $\beta_3^7=-0.9899$ β_{4-7}^7 complexe	$y_1^7=0$ $y_2^7=0.9899x-0.16166x^3+$ $+0.0069x^5-0.00056x^7$ $y_3^7=-0.9899x+0.16166x^3-$ $-0.0069x^5+0.00056x^7$

Let us mention that $f_m(\beta)$ are also odd functions, just like the solutions $y_j(x),\ j=\overline{1,3}$; this is a consequence of the oddness with respect to y of the ODE (4.3.13). We thus obtain a new evidence of LEM's fidelity to the operator to which it is applied.

4.4. REAL ZEROS OF POLYNOMIALS AND ANALYTIC FUNCTIONS

This problem will be treated starting from the idea that the solutions of a nonlinear algebraic system can be interpreted as stationary solutions of a nonlinear ODS with constant coefficients.

4.4.1. STATIONARY SOLUTIONS FOR POLYNOMIAL ODS

In [136] it was considered the algebraic system

$$\mathbf{P}(\mathbf{y}) = \mathbf{0}, \quad \mathbf{P}(\mathbf{y}) \equiv \left[P_j(\mathbf{y})\right]_{j=\overline{1,n}}, \tag{4.4.1}$$

where the polynomials $P_j(\mathbf{y})$ of degree p_j are given by

$$P_j(\mathbf{y}) = \sum_{|\mu| \le p_j} a_{j\mu} \mathbf{y}^{\mu}, \quad j = \overline{1,n}. \tag{4.4.2}$$

If $\mathbf{y} = \mathbf{c}, \mathbf{c} = \left[c_j\right]_{j=\overline{1,n}} \in \Re^n$ is a solution of (4.4.1), then $\mathbf{c}$ satisfies the polynomial initial problem

$$\frac{d\mathbf{y}}{dx} = \mathbf{P}(\mathbf{y}), \quad \mathbf{y}(x_0) = \mathbf{c}, \tag{4.4.3}$$

for an arbitrary $x_0 \in \Re$.

Applying the LEM mapping, it results that the vector

$$\mathbf{V} = \left[\mathbf{V}_m\right]_{m \in \mathfrak{N}}, \quad \mathbf{V}_m = \left[v_\gamma\right]_{|\gamma| \in \mathfrak{N}} \tag{4.4.4}$$

satisfies the infinite linear algebraic system written component-wise

$$\sum_{j=1}^{n} \gamma_j \sum_{|\mu| \le p_j} a_{j\mu} v_{\gamma+\mu-\mathbf{e}_j} = 0, \quad |\gamma| \in \mathfrak{N}, \tag{4.4.5}$$

where

$$v_\theta = 1, \quad v_\gamma = \mathbf{c}^\gamma, |\gamma| \in \mathfrak{N}. \tag{4.4.6}$$

Therefore, considering the roots of (4.4.1) as stationary solutions of a polynomial ODS, we can apply LEM and thus the whole problem is moved in a linear frame.

Let us take the case $n = 1$. Consider the algebraic equation

$$\sum_{k=0}^{p} a_k y^k = 0. \tag{4.4.7}$$

According to the above remarks, we search for its roots as stationary solutions of the polynomial ODE

$$\frac{\mathrm{d}y}{\mathrm{d}x} = \sum_{k=0}^{p} a_k y^k , \tag{4.4.8}$$

with y satisfying

$$y(x_0) = \alpha , \tag{4.4.9}$$

for an arbitrary but fixed up point $x_0 \in \Re$.

The LEM mapping depends in this case only on one parameter σ, and the first linear equivalent of (4.4.8) reads

$$\sum_{k=0}^{p} a_k \frac{\mathrm{d}^k v}{\mathrm{d}\sigma^k} = 0, \quad v(0) = 1 . \tag{4.4.10}$$

The linear equivalent ODS, a particular case of (4.4.5) for $n = 1$, is obtained admitting for v a development in the form

$$v(x, \sigma) = 1 + \sum_{j=1}^{\infty} v_j \frac{\sigma^j}{j!} , \tag{4.4.11}$$

which, introduced in (4.4.10), leads to the infinite linear algebraic system

$$\sum_{k=0}^{p} a_k v_{j+k} = 0, \quad j \in \mathfrak{N}, \quad v_0 = 1 . \tag{4.4.12}$$

The system (4.4.12) can be written in matrix form

$$\mathbf{AV} = -a_0 \mathbf{B} ; \tag{4.4.13}$$

the matrix **A** reads

$$\mathbf{A} = \begin{bmatrix} a_1 & a_2 & a_3 & \ldots & a_p & 0 & \ldots & 0 & 0 & 0 & \ldots \\ a_0 & a_1 & a_2 & \ldots & a_{p-1} & a_p & \ldots & 0 & 0 & 0 & \ldots \\ 0 & a_0 & a_1 & \ldots & a_{p-2} & a_{p-1} & \ldots & 0 & 0 & 0 & \ldots \\ \ldots & \ldots & \ldots & \ldots & \ldots & \ldots & \ldots & \ldots & \ldots & \ldots & \ldots \\ 0 & 0 & 0 & \ldots & 0 & 0 & \ldots & 0 & a_0 & a_1 & \ldots \\ \ldots & \ldots & \ldots & \ldots & \ldots & \ldots & \ldots & \ldots & \ldots & \ldots & \ldots \end{bmatrix} , \tag{4.4.14}$$

and the vector **B** is defined by using Kronecker's delta

$$\mathbf{B} = \left\lfloor \delta_1^j \right\rfloor_{j \in \mathfrak{N}} . \tag{4.4.15}$$

Let us note that the matrix **A** of (4.4.13) is band diagonal, made of the main diagonal and other p parallels to it, one of them being sub-diagonal.

We shall obtain a method of determining of a real root of (4.4.7) by using the linear equivalent system (4.4.12).

An example will make things clearer.

Example 4.2 [136]. Consider the equation

$$y^2 - 3y + 2 = 0\,, \tag{4.4.16}$$

of real roots $y_1 = 1, y_2 = 2$.

The linear equivalent system is

$$\mathbf{AV} = -2\mathbf{B}\,; \tag{4.4.17}$$

the associated LEM matrix is tridiagonal and reads

$$\mathbf{A} = \begin{bmatrix} -3 & 1 & 0 & 0 & 0 & \ldots & 0 & 0 & 0 & 0 & \ldots \\ 2 & -3 & 1 & 0 & 0 & \ldots & 0 & 0 & 0 & 0 & \ldots \\ 0 & 2 & -3 & 1 & 0 & \ldots & 0 & 0 & 0 & 0 & \ldots \\ \ldots & \ldots & \ldots & \ldots & \ldots & \ldots & \ldots & \ldots & \ldots & \ldots & \ldots \\ 0 & 0 & 0 & 0 & 0 & \ldots & 2 & -3 & 1 & 0 & \ldots \\ \ldots & \ldots & \ldots & \ldots & \ldots & \ldots & \ldots & \ldots & \ldots & \ldots & \ldots \end{bmatrix}. \tag{4.4.18}$$

The vector **B** was given in (4.4.15)

Let us truncate now the system (4.4.17). We have

$$\mathbf{A}_m \mathbf{V}^{(m)} = -2\mathbf{B}_m\,, \tag{4.4.19}$$

where the square matrices $\mathbf{A}_m$, with m rows and columns, are

$$\mathbf{A}_m = \begin{bmatrix} -3 & 1 & 0 & \ldots & 0 & 0 & 0 \\ 2 & -3 & 1 & \ldots & 0 & 0 & 0 \\ 0 & 2 & -3 & \ldots & 0 & 0 & 0 \\ \ldots & \ldots & \ldots & \ldots & \ldots & \ldots & \ldots \\ 0 & 0 & 0 & \ldots & -3 & 1 & 0 \\ 0 & 0 & 0 & \ldots & 2 & -3 & 1 \\ 0 & 0 & 0 & \ldots & 0 & 2 & -3 \end{bmatrix}, \tag{4.4.20}$$

$$\mathbf{B}_m = \left\lfloor \delta_1^j \right\rfloor_{j=\overline{1,m}}. \tag{4.4.21}$$

Denote by Δ_m the determinant of $\mathbf{A}_m$. We see that the minors Δ_m are all non-zero in this case. Also denote by $\Delta_m^{(1)}$ the determinant obtained from Δ_m by replacing its first column with $\mathbf{B}_m$

According to LEM representation theorems and by virtue of Cramer's rule, we expect that the ratio

$$r_m = -2\frac{\Delta_m^{(1)}}{\Delta_m} \tag{4.4.22}$$

approximate one of the roots of the polynomial (4.4.16). Computing these ratios for the first indices m, we obtain the following table

Table 4.2

m	3	4	5	6	7	8	9	10
$\Delta_m^{(1)}$	– 14	30	– 62	126	– 254	510	– 1022	2046
Δ_m	– 15	31	– 63	127	– 255	511	– 1023	2047
r_m	0.933	0.967	0.984	0.992	0.996	0.998	0.999	0.999

We see that the numerical sequence $\left\{\frac{\Delta_m^{(1)}}{\Delta_m}\right\}_{m\in\mathfrak{N}}$ tends with m to 1, which is the least of the two positive roots of (4.4.16).

The above sequence does not exist if at least one of the minors Δ_m vanishes. This obviously happens when $a_0 = 0$. In this case, the equation allows the root $y = 0$; after simplification with y, we obtain a $p-1$ degree equation, to which we can apply LEM.

The sequence $\{\Delta_m\}_{m\in\mathfrak{N}}$ might contain infinitely many zeros also if the equation (4.4.7) is even with respect to y, which means that $a_{2k+1} = 0$ for any k such that $2k \le p$, p being also even. This difficulty can be overcome by considering the equation

$$\sum_{k=1}^{\left[\frac{p}{2}\right]} a_{2k} z^k = 0 . \tag{4.4.23}$$

The case of complex roots is not as easily treated, as the sequence $\{\Delta_m\}_{m\in\mathfrak{N}}$ periodically contains zeros.

The linear algebraic systems whose associated matrix has a band diagonal structure are treated in various books and articles, in which there are presented different methods of getting the solutions. But the LEM matrices also have other properties allowing an effective calculation of the determinants implied in the root approximation.

Indeed, the sequences $\{\Delta_m\}_{m\in\mathfrak{N}}$, $\{\Delta_m^{(1)}\}_{m\in\mathfrak{N}}$ can be determined by recurrence formulae established by the following theorem.

Theorem 4.8 [136]. *Let* **A** *be the infinite matrix* (4.4.14). *Then*

i) the minors Δ_m *and the determinants* $\Delta_m^{(1)}$ *obtained by replacing the first column of* Δ_m *with the vector* $\mathbf{B}_m$ *from* (4.4.21) *can be computed by using the recurrence formulae*

$$\begin{aligned} &\Delta_m = \sum_{q=1}^{p} \prod_{s=1}^{q} (m-s+1)(-1)^{q-1} a_q \Delta_{m-q}, \\ &\Delta_m^{(1)} - m\Delta_{m-1}; \end{aligned} \tag{4.4.24}$$

ii) if $\{\Delta_m\}_{m\in\mathfrak{N}}$ *is bounded below by a non-zero constant, then the sequence of the ratios*

$$r_m = -a_0 \frac{\Delta_m^{(1)}}{\Delta_m} \tag{4.4.25}$$

converges to a root of (4.4.7).

Proof. Point *i)* is immediately proved by direct calculation, taking into account some well-known properties of determinants. Point *ii)* results from the fact that the roots are regarded as stationary solutions of polynomial ODS with constant coefficients, therefore all the LEM theorems and representations hold true in this case, corollary 1.3 included. ◘

The preceding theorem emphasizes the sequence $\{r_m\}_{m\in\mathfrak{N}}$ as playing an important part in the determination of the roots of equation

(4.4.7). Starting from the recurrence formulae (4.4.24), the calculation of r_m can still be simplified, as it is seen in the next corollary.

Corollary 4.3 [136]. *The sequence* $\{r_m\}_{m\in\mathfrak{N}}$ *given by* (4.4.25) *can be computed according to the recurrence formula*

$$r_m = -\frac{a_0}{a_1 + a_2 r_{m-1} + a_3 r_{m-1} r_{m-2} + \ldots + a_p r_{m-1} r_{m-2} \ldots r_{m-p+1}},$$
$$r_0 = -\frac{a_0}{a_1}. \tag{4.4.26}$$

Example 4.3 [136]. **LEM approximations for the golden ratio.** Consider the equation

$$y^2 + y - 1 = 0, \tag{4.4.27}$$

allowing as a root the inverse of the *golden number (ratio)*, i.e.

$$\alpha = \frac{-1+\sqrt{5}}{2}. \tag{4.4.28}$$

Let us also consider Fibonacci's sequence $\{c_m\}_{m\in\mathfrak{N}}$, whose elements are determined following the rule

$$c_{k+1} = c_k + c_{k-1}, \quad c_1 = c_2 = 1. \tag{4.4.29}$$

A notorious fact is that that the sequence of ratios formed with the consecutive elements of Fibonacci sequence tends to the golden number

$$\lim_{k\to\infty} \frac{c_{k+1}}{c_k} = \frac{1+\sqrt{5}}{2} \cong 1.6180339... \tag{4.4.30}$$

Applying corollary 4.3 to the equation (4.4.27), we obtain

Consequence 4.1 [136]. *The LEM sequence* $\{r_k\}_{k\in\mathfrak{N}}$ *associated to equation* (4.4.27) *is precisely*

$$r_k = \frac{c_k}{c_{k+1}}. \tag{4.4.31}$$

Proof. Indeed, by virtue of corollary 4.3, the LEM sequence corresponding to (4.4.27) can be computed by using formula (4.4.26), which, in this case, reads

$$r_k = \frac{1}{1 + r_{k-1}}, \quad r_1 = 1; \tag{4.4.32}$$

this completes the proof. ◘

4.4.2. ZEROS OF ANALYTIC FUNCTIONS

Let us take now the nonlinear system

$$\mathbf{f}(\mathbf{y}) = \mathbf{0}, \quad \mathbf{f}(\mathbf{y}) = \lfloor f_j(\mathbf{y}) \rfloor_{j=\overline{1,n}}, \tag{4.4.33}$$

where $f_j(\mathbf{y})$ are analytic with respect to $\mathbf{y}$, allowing the expansions

$$f_j(\mathbf{y}) \equiv \sum_{|\nu| \geq 0} a_{j\nu} \mathbf{y}^{\nu}, \quad j = \overline{1.n}. \tag{4.4.34}$$

Let $\mathbf{y} = \mathbf{c}, \mathbf{c} = \left[c_j\right]_{j=\overline{1,n}} \in \Re^n$ be a solution of (4.4.33). Then it follows, as in the case of polynomial systems, that $\mathbf{c}$ satisfies the nonlinear initial problem

$$\frac{\mathrm{d}\mathbf{y}}{\mathrm{d}x} = \mathbf{f}(\mathbf{y}), \quad \mathbf{y}(x_0) = \mathbf{c}, \tag{4.4.35}$$

for an arbitrary $x_0 \in \Re$.

Applying the LEM mapping, it results that the vector

$$\mathbf{V} = [\mathbf{V}_m]_{m \in \mathfrak{N}}, \quad \mathbf{V}_m = \lfloor v_\gamma \rfloor_{|\gamma| \in \mathfrak{N}} \tag{4.4.36}$$

satisfies the linear infinite algebraic system, written component-wise

$$\sum_{j=1}^{n} \gamma_j \sum_{|\mu| \geq 0} a_{j\mu} v_{\gamma+\mu-\mathbf{e}_j} = 0, \quad |\gamma| \in \mathfrak{N}, \tag{4.4.37}$$

where the relations

$$v_\theta = 1, \quad v_\gamma = \mathbf{c}^\gamma, |\gamma| \in \mathfrak{N}. \tag{4.4.38}$$

are true again.

This system can be written in matrix form

$$\mathbf{AV} = -\mathbf{B}, \quad \mathbf{B} = \left[a_{1\theta} \quad a_{2\theta} \ldots a_{n\theta} \quad 0 \quad 0 \ldots\right]^{\mathrm{T}}. \tag{4.4.39}$$

Let us take $n = 1$, corresponding to the functional equation

$$\sum_{k=0}^{\infty} a_k y^k = 0 . \tag{4.4.40}$$

Its roots will be at the same time stationary solutions of the nonlinear ODE

$$\frac{dy}{dx} = \sum_{k=0}^{\infty} a_k y^k , \tag{4.4.41}$$

with y satisfying

$$y(x_0) = \alpha , \tag{4.4.42}$$

for an arbitrary $x_0 \in \Re$.

We now apply LEM, exactly as previously, finally getting the following linear equivalent system of (4.4.41)

$$\sum_{k=0}^{\infty} a_k v_{j+k} = 0, \quad j \in \mathfrak{N}, \quad v_0 = 1 , \tag{4.4.43}$$

or, in matrix form

$$\mathbf{AV} = -\, a_0\, \mathbf{B} ; \tag{4.4.44}$$

the matrix $\mathbf{A}$ is, however, different from that obtained in the polynomial case

$$\mathbf{A} = \begin{bmatrix} a_1 & a_2 & a_3 & a_4 & a_5 & \dots \\ a_0 & a_1 & a_2 & a_3 & a_4 & \dots \\ 0 & a_0 & a_1 & a_2 & a_3 & \dots \\ \dots & \dots & \dots & \dots & \dots & \dots \end{bmatrix} . \tag{4.4.45}$$

The vector $\mathbf{B}$ has all components null, except for the first, which is 1, similar to the polynomial case. The matrix (4.4.45) is no more row finite, but it is still column-finite.

Truncating the system (4.4.44), we obtain finite linear algebraic systems

$$\mathbf{A}_m \mathbf{V}^{(m)} = -\, a_0\, \mathbf{B}_m , \tag{4.4.46}$$

In this case too, it should be natural that the first components $y_m = v_1^{(m)}$ of the solutions of these system converge to a solution of the functional equation, if these systems are compatible. Let us take the sequence of ratios computed by Cramer's rule

$$r_m = -a_0 \frac{\Delta_m^{(1)}}{\Delta_m}, \quad m \in \mathfrak{N}, \tag{4.4.47}$$

where we used the same notations as in section 4.4.1. Using the same pattern, we can prove a result similar to that of theorem 4.8..

Theorem 4.9 [136, 143]. *Consider the LEM system* (4.4.44) *and its truncations* (4.4.46). *Then*

i) the determinants Δ_m *and* $\Delta_m^{(1)}$ *can be computed by the recurrence formulae*

$$\Delta_m = \sum_{q=1}^{m} \prod_{s=0}^{q-1} (m-s)(-a_0)^{q-1} a_q \, \Delta_{m-q}, \quad \Delta_0 = 1, \qquad \Delta_m^{(1)} = m\Delta_{m-1}; \tag{4.4.48}$$

ii) if $\{\Delta_m\}_{m\in\mathfrak{N}}$ *is bounded below by a non-zero constant, then the sequence* $\{r_m\}_{m\in\mathfrak{N}}$ *given by formula* (4.4.47) *converges to a zero of* $f(y)$.

We can simplify the calculation of r_m in this case too.

Corollary 4.4 [136]. *The sequence* $\{r_m\}_{m\in\mathfrak{N}}$ *can be computed by recurrence by the formula*

$$r_m = -\frac{a_0}{a_1 + a_2 r_{m-1} + a_3 r_{m-1} r_{m-2} + \ldots + a_m r_{m-1} r_{m-2} \ldots r_1}, \qquad r_0 = -\frac{a_0}{a_1}. \tag{4.4.49}$$

If $a_m = 0$, $m > p$, then the function f becomes a polynomial of degree p; we easily see that formula (4.4.26) is a particular case of (4.4.49).

Remark. In the case of rational a_k, the sequence $\{r_m\}_{m\in\mathfrak{N}}$ is rational too. Consequently, by using LEM, one can set up rational sequences converging to irrational or transcendental numbers, viewed as zeros of certain analytic functional.

Example 4.4 [136]. **A rational LEM sequence, convergent to** π^2.

Let us take the equation

$$\sin y = 0 . \tag{4.4.50}$$

Replacing the sine by its Taylor series, simplifying by y and applying the substitution $z = y^2$ in (4.4.50), we obtain the functional equation

$$\sum_{k=0}^{\infty} \frac{(-1)^k}{(2k+1)!} z^k = 0 . \tag{4.4.51}$$

By virtue of theorem 4.9 and corollary 4.4, we deduce the rational sequence $\{r_m\}_{m \in \mathfrak{N}}$, determined by the following formula

$$r_m = \frac{1}{\frac{1}{3!} - \frac{1}{5!} r_{m-1} + \frac{1}{7!} r_{m-1} r_{m-2} + \ldots + \frac{(-1)^{m+1}}{(2m+1)!} r_{m-1} r_{m-2} \ldots r_1}, \tag{4.4.52}$$

$$r_0 = 3!;$$

this sequence converges to π^2.

4.5. POLYNOMIAL AND OPERATORIAL GRAPHS GENERATED BY LEM

In [30], Oana Firică introduced *the polynomial graph* by using LEM; the study of this graph as a representation of the associated algebraic equation emphasizes a series of properties, specific to each of them.

The polynomial graph has not only a theoretical, but also a practical importance; by using it, Oana Firică established an algorithm which serves to get,. by graphs, the real roots of a polynomial.

The notion of polynomial graph was then extended to that of *operatorial graph* [31], applied to an interesting problem of chemistry.

Some spectral properties of the *s*-operatorial graphs were put into evidence by Oana Firică in [32].

4.5.1. LEM POLYNOMIAL GRAPHS

In the previous paragraph, we considered the pth-degree algebraic equation

$$\sum_{k=0}^{p} c_k x^k = 0, \qquad c_p = 1, c_0 \neq 0, \tag{4.5.1}$$

whose solutions coincide with the stationary solutions of the first order ODE

$$\frac{\mathrm{d}x}{\mathrm{d}t} = \sum_{k=0}^{p} c_k x^k, \quad x \in \mathrm{C}^1([0,1]), \tag{4.5.2}$$

solved in the Cauchy conditions

$$x(0) = \alpha, \quad \alpha \in \Re. \tag{4.5.3}$$

The LEM transform v depends on a unique parameter

$$y(x,\sigma) = \mathrm{e}^{\sigma x} \tag{4.5.4}$$

and leads to the first linear equivalent (4.4.10). Using now the development (4.4.11), one gets the second LEM equiivalent

$$y_{p+j} = -\sum_{k=0}^{p-1} c_k y_{j+k}, \quad j \in \mathcal{N}^*, \quad v_0 = 1. \tag{4.5.5}$$

To clarify the presentation, Oana Firică introduced some notations and definitions from the general theory of graphs.

Let V be a finite or countable set, composed of the vertices (points) v_j and $|V|$ – its cardinal. A *directed* graph G is defined by the couple (V,Γ), where Γ maps V to itself, i.e. $\Gamma v_j \in V$. If $|V| < \infty$, then the graph G is finite. If $v_j \in \Gamma v_i$, or, else, if $v_i \in \Gamma^{-1} v_j$, then the pair of vertices $(v_i, v_j) \in V \times V$ reprezsents an *arc* of G; if $v_i = v_j$, it is a *loop*. The in-degree of v_i is $|\Gamma^{-1} v_i|$ and the out-degree is $|\Gamma v_i|$. *A path* of G, formed of j arcs, from v_{i_0} to v_{i_j}, $j \geq 1$, is a sequence of $j+1$ vertices $v_{i_0}, v_{i_1}, \ldots, v_{i_j}$, in which $v_{i_{k+1}} \in \Gamma v_{i_k}$, $k = \overline{0, j-1}$. $\Gamma^2(v_i) = \Gamma(\Gamma v_i)$ is the set of vertices attained by two-arcs paths starting from v_i; $\Gamma^k(v_i)$ este is the set of vertices attaines by k-arcs paths, starting from v_i. If $v_j \in \Gamma v_i$ or $v_i \in \Gamma v_j$, then we say that v_i and v_j are joined by an *edge* of endpoints v_i and v_j. A *chain* of j edges, $j \geq 1$, joining the vertices v_{i_0}

and v_{i_j}, is a sequence of vertices $v_{i_0}, v_{i_1}, ..., v_{i_j}$ consecutively joined by edges. A graph G is called *connected* if any two vertices can be joined by a chain completely contained in G, otherwise the graph is *unconnected.* A *subgraph* G_1 of G, spanned by a subset of vertices $V_1 \subset V$, is defined by the couple (V_1, Γ_1), where $\Gamma_1 = \Gamma|_{V_1}$, $\Gamma V_1 \subset V_1$. A graph is called *progressively bounded* with respect to a vertex v if there is a natural number m such that any path starting from v has at most m arcs. A graph G is called *valuated* if to any arc (v_i, v_j) one associates a certain real value $e(v_i, v_j)$; the notation for a valuated graph is $G_e = (V, \Gamma, e)$.

Definition 4.1 [9]. Let $G_e = (V, \Gamma, e)$ be a valuated graph. The *P*-length of a path of j arcs, $j \geq 1$, is the product

$$\prod_{k=0}^{j-1} e\left(v_{i_k}, v_{i_{k+1}}\right). \tag{4.5.6}$$

If the valuated graph has no loops and the value on each arc is nonzero, then one can associate a *value-matrix* $\mathbf{A} = \left[a_{ij}\right]_{i,j \in \mathfrak{N}}$, by setting

$$a_{ij} = \begin{cases} e\left(v_i, v_j\right) & \text{for } v_j \in \Gamma v_i, \\ 0 & \text{for } v_j \notin \Gamma v_i. \end{cases} \tag{4.5.7}$$

If $\mathbf{A}$ is subdiagonal, i.e. $a_{ij} = 0$ for $j \geq i$, then its square $\mathbf{A}^2$ is well defined and its entries are computed by the formula

$$a_{ij}^{(2)} = \sum_{s \in \mathfrak{N}} a_{is} a_{sj}. \tag{4.5.8}$$

Obvioulsy, if $\mathbf{A}$ is subdiagonal, then any other positive power of $\mathbf{A}$ is also subdiagonal.

Theorem 4.10 [9]. *Let be the valuated* $G_e = (V, \Gamma, e)$. *If its associated matrix* $\mathbf{A}$ *is subdiagonal, then the entry* $a_{ij}^{(k)}$ *of* $\mathbf{A}^k$, $k \geq 1$, *is the sum of the P-lengths of all paths composed of k arcs that join the vertices* v_i *and* v_j.

With these preparations, let us go back to the algebraic equation. Let $H = \left\{h \middle| c_h \neq 0, h = \overline{0, p-1}\right\}$, c_h being the coefficients of the equation

(4.5.1), $i_1, i_2, ... i_{|H|}$ – the subscripts in *H*, in increasing order, starting from $i_1 = 0$, $i_1 < i_2 < ... < i_{|H|}$, and $|H| \geq 2$. Using LEM, Oana Firică defined in a natural way the polynomial graph.

Definition 4.2 [30]. A *polynomial graph* associated to equation (4.5.1) is the graph $\widetilde{G} = (V, \Gamma)$, where $V = \{v_0, v_1, v_2, ...\}$, $\Gamma v_j = \emptyset$, $j = \overline{0, p-1}$, and

$$\Gamma v_{p+j} = \left\{ v_{j+i_1}, v_{j+i_2}, ..., v_{j+i_{|H|}} \right\}, \quad j \in \mathfrak{N}^*. \tag{4.5.9}$$

In [30] there are proved some properties of the polynomial graphs. The calculation of vertex degrees leads to

Theorem 4.11 [30]. *In a polynomial graph, the following relations hold true:*

$$\begin{gathered} |\Gamma v_j| = 0, \quad j = \overline{0, p-1}, \quad |\Gamma v_{p+j}| = |H|, \quad j \in \mathfrak{N}^*, \\ |\Gamma^{-1} v_{p+j}| = |H|, \; j \in \mathfrak{N}^*, \\ |\Gamma^{-1} v_0| = 1, \quad |\Gamma^{-1} v_k| = \begin{cases} |\Gamma^{-1} v_{k-1}| & \text{for } c_k = 0, \\ |\Gamma^{-1} v_{k-1}| + 1 & \text{for } c_k \neq 0, \end{cases} \quad k = \overline{1, p-1}. \end{gathered} \tag{4.5.10}$$

Definition 4.2 and theorem 4.11 show that any polynomial graph is progressively bounded.

Let $C(v)$ be the connected component corresponding to the vertex *v* and $m \equiv r (\mathrm{mod}\, p)$, $1 \leq r \leq p-1$. Then $v_m \in C(v_r)$, hence it suffices to prove connectivity properties only for the components $v_0, v_1, ..., v_{p-1}$. Let *d* be the greatest common divisor of the numbers $p, i_2, ..., i_{|H|}$. Then,

Theorem 4.12 [30]. *If* $d \neq 1$, *then the polynomial graph* $\widetilde{G} = (V, \Gamma)$ *is disconnected.*

From here, it immediately follows

Corollary 4.5 [30]. *The number of the connected components of the polynomial graph is d.*

4.5.1.1. The solvent of a polynomial graph

One can naturally associate to the polynomial graph the valuation $a_{ij} = e(v_i, v_j) = -c_{i_h}$, if $v_j \in \Gamma v_i$ ($j = i + i_h - p$, $h = \overline{1, |H|}$, $i \geq p$), as well as the value matrix $\mathbf{A} = \lfloor a_{ij} \rfloor_{i,j \in \mathfrak{N}}$. The matrix $\mathbf{A}$ is infinite, but subdiagonal, even row- and column-finite, as the graph is locally finite. Let $\widetilde{G}_e = (V, \Gamma, e)$ be the polynomial graph endowed with this valuation and $\widetilde{G}_{ne}$ – the subgraph of $\widetilde{G}_e$, spanned by $v_0, v_1, ..., v_n$. Let us associate to each vertex v of the polynomial graph the variable y_n from (4.5.5). We use the notations $V_0 = \{v_0, v_1, ..., v_{p-1}\}$ and $\gamma_k^n = V_0 \cup \Gamma^k v_n$.

Oana Firică proved the following representation theorem

Theorem 4.13 [30]. *If* $n \geq p$, *then*

$$y_n = \sum_{k=s}^{L_n} \sum_{v_i \in \gamma_k^n} a_{ni}^{(k)} y_i . \tag{4.5.11}$$

The number L_n can be determined as follows: if $n = (k + t_0)(p - i_{|H|}) + r$, $0 \leq r < p - i_{|H|}$, where $t_0 = \max\{t \mid 0 \leq r + t(p - i_{|H|}) \leq p - 1\}$, then $L_n = k$. It should be noted that if $i_{|H|} \neq p - 1$, then $L_n < n - p + 1$ and the calculation of the coefficients $y_j, j = \overline{0, p-1}$ is substantially simplified. If $n \geq kp$, then $y_n = \sum_{i=1}^{p-1} \sum_{v_i \in \gamma_k^n} a_{ni}^{(k)} y_i$. By induction, it is proved that in this case $a_{ni}^{(k)}$ are the multinomial coefficients of $\left(\sum_{r=0}^{p-1} c_r x^r \right)^k$ [30].

Oana Firică called formula (4.5.11) *the solvent of the polynomial graph* $\widetilde{G}$, a denomination perfectly justified, as it will be seen further.

4.5.1.2. Applications of the solvent of a polynomial graph

The solvent of a polynomial graph leads to the practical determination of the roots of the corresponding polynomial.

Suppose that (4.5.1) allows only the real and distinct roots $\lambda_j, j=\overline{1,p}$, ordered as follows

$$0<|\lambda_1|<|\lambda_2|<\ldots<|\lambda_p| . \tag{4.5.12}$$

We shall use the notations from [30]:

$$\lambda=(\lambda_j)_{j=\overline{1,p}}, \quad \lambda'=(\lambda_j)_{j=\overline{1,p-1}}, \quad P_0(\lambda)=1, \quad P_0(\lambda')=1,$$
$$P_n(\lambda)=\sum_{|i|=n}\lambda^i, \quad i\in(\mathfrak{N}^*)^p, \quad P_n(\lambda')=\sum_{|i|=n}\lambda'^{i'}, \quad i'\in(\mathfrak{N}^*)^{p-1}. \tag{4.5.13}$$

By $\tau_j(\lambda)$, $\tau_j(\lambda')$ we mean the fundamental symmetric polynomials of degree j with respect to the variables λ, λ' accordingly; we also have $\tau_0(\lambda)=1$, $\tau_0(\lambda')=1$.

The Vandermonde determinant of the roots is

$$W=W(\lambda)=\prod_{n\geq i>j\geq 1}(\lambda_i-\lambda_j)=\begin{vmatrix} 1 & 1 & \ldots & 1 \\ \lambda_1 & \lambda_2 & \ldots & \lambda_p \\ \ldots & \ldots & \ldots & \ldots \\ \lambda_1^{p-1} & \lambda_2^{p-1} & \ldots & \lambda_p^{p-1} \end{vmatrix}, \tag{4.5.14}$$

and $W_i=W_i(\lambda)$ is the Vandermonde determinant obtained from W by cutting off the i-th column and the last row, therefore a $(p-1)$-determinant in which the terms containing λ_i are missing.

By using some well known formulae from the polynomial algebra, Oana Firică succeded to write the solvent of a polynomial graph in terms of symmetric polynomials

$$y_n=\sum_{k=0}^{p-1}\sum_{j=0}^{p-k-1}(-1)^j\tau_j(\lambda)P_{n-k-j}(\lambda)y_k . \tag{4.5.15}$$

Taking into account that, under the above hypotheses on equation (4.5.1), one can find $n_0\in\mathfrak{N}$ such that $P_n(\lambda)\neq 0$ for $n>n_0$, another form for the solvent was obtained

$$y_n=P_{n-p+1}(\lambda)\sum_{k=0}^{p-1}\alpha_k^{(n)}y_k , \tag{4.5.16}$$

which is true for $n>n_0$; here,

$$\alpha_k^{(n)} = \sum_{j=0}^{p-k-1} (-1)^j \tau_j(\lambda) \frac{P_{n-k-j}(\lambda)}{P_{n-p+1}(\lambda)}, \quad \alpha_{p-1}^{(n)} = 1. \tag{4.5.17}$$

As one can easily be seen, the solvent of a polynomial graph provides a numerical method of getting the roots of a polynomial, if these roots are all real and of distinct moduli, more precisely, the following

ALGORITHM [30]
INPUT:

- the value matrix of the polynomial graph, built up by the coefficients $-c_j = c_j^{(1)}$;
- the relative error ε.

OUTPUT:

- the root of maximum modulus λ_p;
- the coefficients of the $(p-1)$-th degree polynomial of roots $\lambda_j, j = \overline{1, p-1}$.

Step 1. Choose $n > p$ and compute the coefficients

$$d_{ni} = \sum_{k=s}^{L_n} a_{ni}^{(k)}, \quad i = \overline{0, p-1} \tag{4.5.18}$$

by means of the solvent (4.5.11), making use of the powers of the value matrix.

Let n_0 be the index for which $d_{ni} \neq 0$, $n > n_0$ and compute

$$\delta_{in} = \left\| \frac{d_{ni}}{d_{n,p-1}} \right| - \left| \frac{d_{n+1,i}}{d_{n+1,p-1}} \right\|, \quad i = \overline{0, p-2}, \tag{4.5.19}$$

as well as

$$ER_n = \frac{1}{p-1} \sum_{i=0}^{p-1} \delta_{in} . \tag{4.5.20}$$

If $ER_n < \varepsilon$, then go to step 2, n being the number of iterations; otherwise, compute $d_{n+2,i}$, $i = \overline{0, p-1}$ and repeat step 1.

Step 2. Compute approximately the coefficients of the $(p-1)$-th degree equation allowing the roots $\lambda_j, j = \overline{1, p-1}$:

$$c_j^{(2)} = \frac{d_{n+1,j}}{d_{n+1,p-1}}, \quad j = \overline{0, p-2}, \tag{4.5.21}$$

and also the approximating root

$$\lambda_p = \frac{c_0^{(1)}}{c_0^{(2)}}. \tag{4.5.22}$$

The algorithm is repeated for all $(p-k)$-th degree equations, $k = \overline{1, p-2}$, by using the coefficients computed at each previous step and also the corresponding solvent representations.◘

Compared with other algorithms built up for the same purpose, this one is explicit and purely algebraic; moreover, it requires only matrix computation, saving up a lot of time and memory. The corresponding programming code involves $p-1$ applications of the algorithm, each one requiring several iterations. The number of iterations per step at step k depends on $l_k = |\lambda_{p-k+1}| - |\lambda_{p-k}|$ and increases as l_k is decreasing.

Running the programming code emphasized some surprising aspects of the proposed method. Namely, if at a step k the coefficients of the intermediate equations were not obtained with a satisfactory precision, this fact has no important consequences on the next steps, because the next roots are obtained more and more precisely, unlike expected.

Table 4.3

Nr.	Exact coefficients	Exact roots	Computed coefficients	Roots obtained by the solvent	Nr. of iterations
1	0.1 7.1 – 6.6	4	0.09994 7.10023 – 6.60027	3.99984	32
2	3.1 – 2.2	– 3	3.10009 2.20012	– 2.99995	28
3	– 1.1	2	–1.09981	2.00046	12
		1.1		1.09981	

Example 4.5 [30]. Table 4.3 shows the results obtained by applying the above algorithm to the equation

$$x^4 = 4.1x^3 + 6.7x^2 - 35x + 26.6 . \tag{4.5.23}$$

The case of complex and/or multiple roots must be treated separately. Let us note, however, that even if the roots are not real and distinct, the method introduced by Oana Firică can still provide qualitative informations on the nature of the roots.

4.5.2. LEM OPERATORIAL GRAPHS

Using LEM, Oana Firică also introduced the operatorial graph [31. Consider the polynomial ODS with constant coefficients

$$\mathcal{P}\mathbf{y} \equiv \frac{\mathrm{d}\mathbf{y}}{\mathrm{d}x} - \mathbf{P}(\mathbf{y}) = \mathbf{0}, , \tag{4.5.24}$$

where

$$\mathbf{P} = \left[P_j(\mathbf{y})\right]_{j=\overline{1,n}}, \quad P_j(\mathbf{y}) = \sum_{|\gamma| \le p_j} a_{j\gamma}\, \mathbf{y}^{\gamma},\ j = \overline{1,n}, \tag{4.5.25}$$

that must be solved under the Cauchy conditions

$$\mathbf{y}(a) = \mathbf{y}_0, \quad a \in \mathrm{I}, \mathbf{y}_0 \equiv \left[y_{j0}\right]_{j=\overline{1,n}} \in \Re^n . \tag{4.5.26}$$

One can apply LEM to this system. As $a_{j\gamma}$ are constant, it follows by corollary 1.3 that the first n components of the vector

$$\mathbf{V}(x) = \mathrm{e}^{\mathbf{T}(x-x_0)}\, \mathbf{V}(a), \quad \mathrm{e}^{\mathbf{T}(x-x_0)} = \mathbf{E} + \sum_{m=1}^{\infty} \mathbf{T}^m \frac{(x-a)^m}{m!}, \tag{4.5.27}$$

i.e.

$$\left[v_{\mathbf{e}_j}(x)\right]_{j=\overline{1,n}}, \tag{4.5.28}$$

coincide with the Taylor series expansion around a of the solution of the polynomial Cauchy problem (4.5.24), (4.5.26). In (4.5.27), $\mathbf{T}$ is the LEM matrix associated to the polynomial system and it is row-and column-finite. In other words, the LEM representation (4.5.27) leads to a specific way of determining the coefficients of the Taylor series

$$v_{\mathbf{e}_j}(x) = \sum_{m \ge 0} v_{\mathbf{e}_j}^{(m)}(a) \frac{(x-a)^m}{m!}, \quad \mathbf{e}_j = \left(\delta_k^j\right)_{k=\overline{1,n}} \tag{4.5.29}$$

Denote by

$$S_j = \left\{ \gamma \in \left(\mathfrak{N}^*\right)^n ,\ a_{j\gamma} \neq 0 \right\}, \quad 1 \leq \left|S_j\right| \leq p_j + 1 , \quad j = \overline{1,n} . \tag{4.5.30}$$

Each set S_j can be lexicografically ordered. Let us define by recurrence, for every $s = \overline{1,n}$ and $m \in \mathfrak{N}$, the finite set of multiindices

$$V_0^s = \{\mathbf{e}_s\},\ V_m^s = \bigcup_{\mu \in V_{m-1}^s} \bigcup_{j=1}^{n} \left\{ \left(\gamma + \mu - \mathbf{e}_j\right) \mid \mu_j \neq 0, \gamma \in S_j \right\}, \tag{4.5.31}$$

V_m^s having distinct elements.

Conditions $V_{m-1}^s \neq \emptyset$, $V_m^s \neq \emptyset$ are both satisfied if and only if $P_s(\mathbf{y}) = \text{const.}$, $m = 1$; this is the trivial case, to be rejected in what follows. The sets V_m^s are not disjoint; but if an index μ belongs to the intersection $V_m^s \cap V_{m+r}^s$, it will be denoted by μ^m, μ^{m+r} accordingly, whenever it must be considered as an element of V_m^s, V_{m+r}^s respectively. This convention induces a new order on these sets, indicated by vinculi (bars) in order to put it into evidence. Consequently,

$$\overline{V}_{m_1}^s \cap \overline{V}_{m_2}^s = \emptyset, \qquad m_1 \neq m_2, \quad m_1, m_2 \in \mathfrak{N} . \tag{4.5.32}$$

Starting from this and from (4.5.27), Oana Firică introduced the operatorial graph.

Definition 4.3 [31]. A *s-operatorial graph generated by LEM* is the graph $GO^s = \left(V^s, \Gamma\right)$, whose vertex set is

$$V^s = \bigcup_{m \in \mathfrak{N}^*} \overline{V}_m^s , \tag{4.5.33}$$

and whose multi-valued transform Γ associates to each multiindice μ belonging to $\overline{V}_m^s$ the set of multiindices

$$\Gamma\mu = \bigcup_{j=1}^{n} \left\{ \left(\gamma + \mu - \mathbf{e}_j\right) \mid \mu_j \neq 0, \gamma \in S_j \right\} \subset \overline{V}_{m+1}^s . \tag{4.5.34}$$

Remark. If the system (4.5.24) has a non-zero constant free term, then the operatorial graph contains the vertex $\theta = (0, 0, \ldots, 0)$ and $\Gamma\theta = \emptyset$. If the polynomial system has a non-zero linear part, then the sets V_m^s are not disjoint.

The set of arcs is

$$A^s = \bigcup_{m \in \mathfrak{N}} A^s_m, \quad A^s_m = \left\{ (\gamma, \mu) \mid \gamma \in \overline{V}^s_{m-1}, \mu \in \Gamma\gamma \right\}$$
$$A^s_{m_1} \cap A^s_{m_2} = \emptyset \ \text{ for } m_1 \neq m_2 . \tag{4.5.35}$$

Certainly, one can always associate to the system (4.5.24) n distinct s-operatorial graphs, but there is no loss of generality by fixing up an arbitrary s. As a consequence, in what follows we drop the index s.

Oana Firică emphasized and proved one of the characteristic properties of the operatorial graph, very useful in applications.

Theorem 4.14 [31]. *The operatorial graph generated by LEM is locally finite.*

Let us point out that the operatorial graphs are connected and each of the unoriented subgraphs spanned by $\overline{V}^s_{m-1}, \overline{V}^s_m$ is bipartite; the unoriented graph deduced from GO^s is itself bipartite.

4.5.2.1. The valuation of the operatorial graph and its solvent

The valuation function $\mathrm{d} : A \to \mathfrak{R}$ for GO was defined similarly to that introduced in [30] for the LEM polynomial graph.

More precisely, if $(\eta, \mu) \in A$, then it exists $\eta_j \neq 0$ and $\gamma \in S$ such that $\mu = \gamma + \eta - \mathbf{e}_j$ and we put

$$\mathrm{d}(\eta, \mu) = \eta_j a_{j\gamma} . \tag{4.5.36}$$

Let us use the notation $G\widetilde{O}$ for the valuated operatorial graph. Its value matrix $\mathbf{D}$, similar to that defined in [30], will be built up by blocks $\mathbf{B}_{ij}, i, j \in \mathfrak{N}$, each of them of dimensions $\left|\overline{V}_i\right| \times \left|\overline{V}_j\right|$. If $\eta \in \overline{V}_m$ and $(\eta, \mu) \in A$, then $\mu \in \overline{V}_{m+1}$ and $\mathrm{d}(\eta, \mu) \neq 0$, therefore $\mathbf{B}_{m,m+1} \neq \mathbf{0}$. We denote by $b_{\eta\mu} \equiv b^{(1)}_{\eta\mu}$ the entries of the block $\mathbf{B}_{m,m+1}$. From (4.5.36) it also follows that $\mathbf{B}_{ij} = \mathbf{0}$ for $i \neq m, j \neq m+1$.

To conclude, the value matrix $\mathbf{D}$ of the operatorial graph is superdiagonal, being row- and column-finite; thus, any natural power of $\mathbf{D}$ can be computed by block partitioning..

Theorem 4.15 [31]. *One can partition the matrix* $\mathbf{D}^m$ *off by the blocks* $\mathbf{B}^{(m)}_{ij}$ *of dimensions* $\left|\overline{V}_i\right| \times \left|\overline{V}_j\right|$. *Moreover,*

$$\mathbf{B}_{ij}^{(m)} = \begin{cases} \prod_{l=1}^{m} \mathbf{B}_{i+l-1,i+1}, & \text{for } j = i + m, \\ \mathbf{0}, & \text{for } j \neq i + m, \end{cases} \tag{4.5.37}$$

and the entry $b_{\theta\mu}^{(m)}$ *of the block* $\mathbf{B}_{0m}^{(m)}$ *is the sum of the P-lengths of all m-arcs paths in* $G\widetilde{O}^s$ *joining* $\mathbf{e}_s$ *with* $\mu \in \overline{V}_m^s$.

We saw that the solution of the polynomial Cauchy problem (4.5.24), (4.5.26) is given by the first n components of the vector $\mathbf{V}$, computed by the LEM formula (4.5.27).

The operatorial graph provides a new tool to get this solution, by using P-lengths.

Theorem 4.16 [31]. *The m-th order derivative of the solution of the polynomial Cauchy problem* (4.5.24), (4.5.26) *allows the local representation*

$$v_{\mathbf{e}_s}^{(m)}(x) = \sum_{\mu \in \overline{V}_m^s} b_{\theta\mu}^{(m)} \, v_\mu(x). \tag{4.5.38}$$

The coeficients $v_{\mathbf{e}_s}^{(m)}(a)$ *from the expansion* (4.5.29) *are computed by the formula*

$$v_{\mathbf{e}_s}^{(m)}(a) = \sum_{\mu \in \overline{V}_m^s} b_{\theta\mu}^{(m)} \, \mathbf{y}_0^{\mu}. \tag{4.5.39}$$

Oana Firică called formula (4.5.39) *the solvent of the operatorial graph*; we see that the solvent conserves the specific LEM property of separating the contribution of the initial data from that of the coefficients of the polynomial system.

4.5.2.2. An application in chemistry

The above mentioned ideas were applied to a chemistry problem [31]. The application concerns the design of a reactor for the catalytic dehydrogenation of ethylbenzene to produce styrene. The material balances of the chemical reactions can be modelled by a system of polynomial differential equations with parametric coefficients, in view of determining the size of the industrial scale dehydrogenation reactor.

Let t be the length of the reactor tube and let

$$x(t),\quad y(t),\quad z(t),\qquad x,y,z:\left[0,t^*\right]\mapsto[0,1] \tag{4.5.40}$$

be three functions representing the degree of conversion of specifical chemical reaction: x for styrene, y for benzene and z for toluene; the last trwo are by-products. The material balances, as well as the stoichiometric relations, lead to the following polynomial ODS:

$$\begin{aligned} x'(t)&=a\left(1-x-y-z\right)-b\left(x^2-xz\right),\\ y'(t)&=c\left(1-x-y-z\right),\\ z'(t)&=d\left(1-x-y-z\right)(x-z), \end{aligned} \tag{4.5.41}$$

which must be solved under null Cauchy conditions

$$x(0)=0,\quad y(0)=0,\quad z(0)=0\,. \tag{4.5.42}$$

The parameters a,b,c,d depend on another parameter p, involving the pressure. They are of the form

$$a=\alpha p,\quad b=\beta p^2,\quad c=\gamma p,\quad d=\delta p^2, \tag{4.5.43}$$

where $p\in(2.5,6.5)$ and $\alpha,\beta,\gamma,\delta$ are given constants.

The usual methods to solve such a system is to approximate it by finite difference equations. The major shortcoming of a numerical approach is that, in order to start the algorithm, one must give numerical values to the coefficients; this does not provide a qualitative study of the solution. Moreover, one must check the algorithm stability at each choice of the coefficients, otherwise undesired errors may occur.

Making use of the solvent (4.5.39), one overcomes these difficulties. Applying the change of functions

$$x_1=x,\qquad x_2=1-x-y-z,\qquad x_3=x-z, \tag{4.5.44}$$

the system (4.5.41) is simplified, remainig still polynomial

$$\begin{aligned} x_1'(t)&=ax_2-bx_1x_3,\\ x_2'(t)&=-\left(a+c\right)x_2+bx_1x_3-dx_2x_3,\\ x_3'(t)&=ax_2-bx_1x_3-dx_2x_3, \end{aligned} \tag{4.5.45}$$

and the initial conditions become

$$x_1(0)=0,\qquad x_2(0)=1,\qquad x_3(0)=0\,. \tag{4.5.46}$$

The LEM mapping depends here on three parameters

$$v(t,\sigma_1,\sigma_2,\sigma_3)=\mathrm{e}^{\sigma_1 x_1+\sigma_2 x_2+\sigma_3 x_3}, \qquad \sigma_1,\sigma_2,\sigma_3 \in \Re ; \qquad (4.5.47)$$

Applying it to the polynomial ODS (4.5.45), one gets its linear equivalent system

$$\begin{aligned} v'_{ijk} &= i\left(av_{i-1,j+1,k} - bv_{ij,k+1}\right) + j\left[-(a+c)v_{ijk} - dv_{ij,k+1} + bv_{i+1,j-1,k+1}\right] \\ &+ k\left(av_{i,j+1,k-1} - dv_{i,j+1,k} - bv_{i+1,jk}\right), \qquad i+j+k \in \mathfrak{N}. \end{aligned} \qquad (4.5.48)$$

Firstly, the three associated operatorial graphs were built up, based on this system. The set of their vertices was considered up to $\overline{V}_4^s$, $s = 1,2,3$. For each of them the valuations of the arcs and the first $\left|\bigcup_{i=0}^{4} \overline{V}_i^s\right|$ rows and columns of the corresponding matrices $\mathbf{D}_s$ were set up.

Then, the powers of the $\mathbf{D}_s$, up to the fourth order included, were computed according to theorem 4.14 and the parametric representations of the solution were obtained by using theorem 4.16.

The results were listed in table 4.4. In this table, x_1 represents the styrene fraction and x_2 the unconverted ethylbenzene fraction.

The tube length is $t = 3.048\,\mathrm{m}$, its diameter has 10.16 cm. and the inlet temperature is $467.32°$ C.

Table 4.4

p	3.5	4.0	4.5	5.0	5.5	6.0	6.5
x_1	0.0684	0.0752	0.0810	0.0854	0.0889	0.0921	0.0927
x_2	0.9393	0.9230	0.9167	0.9115	0.9073	0.9040	0.0916
x_3	0.0874	0.0738	0.0790	0.0829	0.0855	0.0871	0.0877

These values are in conformity whith those obtained in [11] by means of finite differences. While the finite differences induce only a purely numeric study, the method proposed by Oana Firică led to parametric representations of x_1, x_2, x_3, which are appropriate for a qualitative study and from the computing point of view, saved a lot of time and memory.

Chapter 5

THE STRAIGHT BAR

In this chapter, some problems referring to the nonlinear bending of an elastic straight bar are solved by LEM. The bar is considered in classical hypotheses: cantilever, simply supported and hyperstatic. The corresponding models are mathematically distinct and they require distinct numerical methods. We shall see that, by using LEM, all these models could be treated in the same frame of a unique programming code.

Another problem studied here by LEM is the stability of an elastic bar. In this case too, there were obtained LEM formulae for the critical and postcritical behaviour of the bar that turned out better than the standard ones.

In Section 5.3, there are described the wobble solitons obtained by Veturia Chiroiu and Ligia Munteanu for heavy elastica. The last section is dedicated to the critical behaviour of the nonlinear elastic frame.

5.1. THE NONLINEAR BENDING OF A STRAIGHT BAR

Far off, in XVIIth and XVIIIth centuries, Jacob Bernoulli, his younger brother, Johann Bernoulli, and Leonhard Euler elaborated "elastica", that is, the theory of the elastic displacement of the central axis of a bar subjected to bending, a problem that became classical in the frame of the strength of materials. The problem can be considered solved for infinitesimal displacements in the most complex cases of material, loading, as well as geometry of the cross section. For finite displacements, the problem was studied only in particular cases, using methods of calculation more or less accurate – polynomial approximations, finite difference method, etc.

In what follows, we consider a straight bar, elastic and isotropic, of variable rigidity, for various diagrams of the bending moment. To this problem one associates the nonlinear Bernoulli-Euler (B.-E.) equation, and to the B.-E. equation, immaterial the boundary conditions, one can apply LEM.

LEM allows the treatment of the straight bar problems in a common mathematical frame. Due to LEM, we can study, basically using a unique LEM representation, various bar problems, which are not otherwise mathematically connected than by the B.-E. equation: the cantilever bar – a Cauchy problem, the simply supported bar – a two-point problem and the hyperstatic bar – a mixed problem, depending on a parameter.

All these mathematically distinct problems were treated in the literature, in the case of small displacements, by distinct numerical methods. Except for the benefit of a common mathematical tool for all these problems, LEM also produced numerical results which proved to be better that the standard ones. This fact is mainly due to one of its advantages: by LEM, one can set up the local inverse of a nonlinear operator in matrix form. In turn, this inverse allows the building of the transport matrix of the solution; this is why the LEM algorithms separate the contribution of the data from that of the coefficients. The involved errors are thus controlled by using specific LEM conditions of convergence.

5.1.1. STATEMENT OF THE PROBLEM

Consider a straight bar of length l, reported to a reference system xOy. In the B.-E. hypothesis on plane sections, the bar axis $y = y(x)$ satisfies the B.-E. equation

$$\mathcal{Y}(y) \equiv y'' - f(x)\left(1 + y'^2\right)^{\frac{3}{2}} = 0, \tag{5.1.1}$$

which represents a nonlinear mathematical model for the straight bar, also called the B.-E. bar. The function

$$f(x) = \frac{1}{\rho(x)} = -\frac{M(x)}{E(x)I(x)} \tag{5.1.2}$$

is the curvature, of sign contrary to that of the bending moment $M(x)$. The rigidity $E(x)I(x)$ may vary following the mechanical properties of the material (e.g., the longitudinal elasticity modulus E, a continuous or discontinuous inhomogeneity along the bar axis) and the geometry of the cross section (the moment of inertia I with respect to the axis, which may also vary).

The equation (5.1.1) must be completed with boundary value conditions whose nature depends on the nature of the physical problem.

Thus, to the cantilever bar one associates Cauchy conditions, while to the simply supported bar – two-point conditions. On the other hand, the hyperstatic bar case supposes both Cauchy and two-point conditions and, moreover, the obtained nonlinear mixed problem depends on a parameter that must be determined.

All these distinct problems could be treated in the same frame, starting from a basic element: the Cauchy problem with arbitrary initial data associated to the B.-E. equation and solved by LEM. The specific parametric LEM representation obtained for this general Cauchy problem easily allowed a differentiate treatment of each of the above considered cases.

5.1.2. THE CAUCHY PROBLEM FOR THE B.-E. BAR TREATED BY LEM

LEM is easier applied in the case of polynomial differential operators, but the B.-E. equation is not polynomial. However, by using a convenient change of function, we will find a polynomial equivalent for this equation, in the sense of the definitions introduced in chapter 1.

Consider therefore the transformation

$$T(z) \equiv [w, z] \equiv \left[\sqrt{1+y'^2}, y'\right], \quad y' = z, \tag{5.1.3}$$

by means of which the B.-E. equation becomes

$$\mathcal{P}([w, z]) \equiv \begin{bmatrix} w' - f(x)w^2 z \\ z' - f(x)w^3 \end{bmatrix} = \begin{bmatrix} 0 \\ 0 \end{bmatrix}. \tag{5.1.4}$$

The operator $\mathcal{P}$, defined on $\left(C^1([0,l])\right)^2$ with values in $\left(C^0([0,l])\right)^2$, is polynomial. By l we mean the bar length (figure 5.1).

Let $x_0 \in [0,l]$ be an arbitrary point and let us associate to the B.-E. equation the Cauchy conditions

$$y(x_0) = \alpha, \quad y'(x_0) = \lambda. \tag{5.1.5}$$

By the above mentioned transformation T, the second condition (5.1.5) yields

$$T(z(x_0)) = \left[\sqrt{1+\lambda^2}, \lambda\right], \tag{5.1.6}$$

which means

$$w(x_0)=\sqrt{1+\lambda^2}, \qquad z(x_0)=\lambda . \tag{5.1.7}$$

Thus, we set up the initial polynomial problem with arbitrary Cauchy data (5.1.4), (5.1.7), to which we shall apply LEM.

The exponential LEM mapping depends on two parameters

$$v(x,\sigma,\xi)=\mathrm{e}^{\sigma w+\xi z}; \tag{5.1.8}$$

it leads to the first linear equivalent of the polynomial system

$$\mathfrak{L}v \equiv \frac{\partial v}{\partial x} - f(x)\left(\sigma\frac{\partial^3 v}{\partial\sigma^2\xi} + \xi\frac{\partial^3 v}{\partial\sigma^3}\right) = 0 . \tag{5.1.9}$$

Conditions (5.1.7) become

$$v(x_0,\sigma,\xi)=\mathrm{e}^{\sigma\sqrt{1+\lambda^2}+\xi\lambda} . \tag{5.1.10}$$

We are looking for the analytic LEM solutions of the linear problem (5.1.9), (5.1.10) in the form of series (4.3.19); the coefficients $v_{jk}(x)$ of this development will satisfy the second linear equivalent – the LEM system

$$\mathfrak{S}\mathbf{V} \equiv \frac{\mathrm{d}\mathbf{V}}{\mathrm{d}x} - f(x)\mathbf{A}\mathbf{V} = \mathbf{0} . \tag{5.1.11}$$

As the polynomial system is odd, the matrix **A** will "contract" to cells with even numbers of rows and columns

$$\mathbf{A} = \begin{bmatrix} \mathbf{0} & \mathbf{B}_1 & \mathbf{0} & \mathbf{0} & \dots & \mathbf{0} & \dots \\ \mathbf{0} & \mathbf{0} & \mathbf{B}_3 & \mathbf{0} & \dots & \mathbf{0} & \dots \\ \dots & \dots & \dots & \dots & \dots & \dots & \dots \\ \mathbf{0} & \mathbf{0} & \mathbf{0} & \mathbf{B}_{2j-1} & \dots & \mathbf{0} & \dots \\ \dots & \dots & \dots & \dots & \dots & \dots & \dots \end{bmatrix}; \tag{5.1.12}$$

so will the vector **V**

$$\mathbf{V}(x)=\left[\mathbf{V}_{2j-1}(x)\right]_{j\in\mathfrak{N}}, \quad \mathbf{V}_{2j-1}(x)=\left[v_{2j-1-k,k}(x)\right]_{k=\overline{0,2j-1}} . \tag{5.1.13}$$

The matrices **B** can be written in compact form by using Kronecker's δ

$$\mathbf{B}_{2j-1} = \left[(2j-1)\delta_{k+1}^{s} + (k-1)\delta_{k}^{s+1}\right]_{\substack{k=\overline{1,2j},\\ s=\overline{1,2j+2}}} . \tag{5.1.14}$$

Conditions (5.1.10) become for **V**

$$\mathbf{V}(x_0)=\left[\mathbf{V}_{2j-1}(x_0)\right]_{j\in\mathcal{N}},\quad \mathbf{V}_{2j-1}(x_0)=\left[\lambda^k\left(1+\lambda^2\right)^{\frac{2j-1-k}{2}}\right]_{k=\overline{0,2j-1}}. \tag{5.1.15}$$

The appropriate functional frame for the above LEM mappings is given by the Exp-type spaces, introduced in chapter 1 by formulae (1.3.34) and (1.4.21) accordingly. We claim that $\mathcal{L}:\mathcal{A}_2^1([0,l])\to\mathcal{A}_2^0([0,l])$ and $\mathcal{S}:\mathcal{B}_2^1([0,l])\to\mathcal{B}_2^0([0,l])$.

Let us index by x_0 the restrictions of $\mathcal{Y},\mathcal{P},\mathcal{L},\mathcal{S}$ to sets of admissible functions, satisfying the conditions (5.1.5), (5.1.7), (5.1.10) and (5.1.13) respectively.

Definitions 1.3 and 1.4 involve that $\mathcal{Y}$ is *T*-immersible in $\mathcal{P}$, and that $\mathcal{Y}_{x_0}$ is the *T*-equivalent of $\mathcal{P}_{x_0}$. With this remark, also taking into account Theorem 1.9, we deduce

Theorem 5.1. [97, 100]. *The chain $\mathcal{Y},\mathcal{P},\mathcal{L},\mathcal{S}$ is resonant and the chain of their restrictions $\mathcal{Y}_{x_0},\mathcal{P}_{x_0},\mathcal{L}_{x_0},\mathcal{S}_{x_0}$ is resolvent.*

The terms in the middle of the chain are representative for LEM. Considering the chain extremities, we deduce that the function $z=y'$ is the second component of the solution of the linear initial problem (5.1.11), (5.1.13).

Remark. The form of the LEM matrix allows two modes of using the linear equivalent system: by recurrence formulae of differential type, applying Theorem 1.7, or of integral type, according to Theorem 3.5 and, particularly, to corollary 3.2. We shall analyse each of them separately.

5.1.2.1. LEM solutions of differential type

As previously mentioned, theorem 1.7 ensures the constructive local existence of the solution of the Cauchy problem with arbitrary data associated to the B.-E. equation.

Theorem 5.2. [94, 95]. *Let $f\in C^\infty([0,l])$. Then the solution of the Cauchy problem* (5.1.1), (5.1.5), *locally allows the LEM representation*

$$\begin{aligned} y(x)=&\ \alpha+(x-x_0)\langle\mathbf{V}_1(x_0),\mathbf{e}\rangle\\ &+\sum_{s\geq 1}(x-x_0)[f]_s(x)\langle\mathbf{B}_1\mathbf{B}_1\ldots\mathbf{B}_{2s-1}\mathbf{V}_{2s+1}(x_0),\mathbf{e}\rangle; \end{aligned} \tag{5.1.16}$$

its derivative $z = y'$ *is locally represented as follows*

$$y'(x) = \langle \mathbf{V}_1(x_0), \mathbf{e} \rangle + \sum_{s \geq 0} \widetilde{f}_s(x) \langle \mathbf{B}_1 \mathbf{B}_1 \ldots \mathbf{B}_{2s-1} \mathbf{V}_{2s+1}(x_0), \mathbf{e} \rangle, \qquad (5.1.17)$$

where $\mathbf{e} = [0,1]^{\mathrm{T}}$, *and*

$$\widetilde{f}_s(x) = \sum_{k \geq 0} f_{s,k+s}(x_0) \frac{(x - x_0)^{k+s}}{(k+s)!}. \qquad (5.1.18)$$

The coefficients $f_{s,k+s}$ *are determined by recurrence*

$$f_{j,k} = f'_{j,k-1} + f_{j-1,k-1} f, \quad f_{0,k} = \delta_k^0, \quad f_{k,0} = 0, \qquad (5.1.19)$$

and $(x - x_0)[f]_s(x)$ *is a primitive of* $\widetilde{f}_s(x)$

$$[f]_s(x) = \sum_{k \geq 0} f_{s,k+s}(x_0) \frac{(x - \widetilde{x})^{k+s}}{(k+s+1)!}, \quad s, k \in \mathfrak{N}. \qquad (5.1.20)$$

Proof. By virtue of Theorem 1.6, we shall compute the Taylor series of $z = y'$ around $\widetilde{x}$. The practical problem is to write explicitly the matrices $\mathbf{A}^{(m)}$ from the recurrence formulae (1.4.38). In this case, the LEM matrix has the form $f(x)\mathbf{A}$, where $\mathbf{A}$ is the constant matrix from (5.1.12). Thus,

$$\mathbf{A}^{(1)} = f(x)\mathbf{A}; \qquad (5.1.21)$$

then

$$\mathbf{A}^{(2)} = \frac{\mathrm{d}\mathbf{A}^{(1)}}{\mathrm{d}x} + f^2(x)\mathbf{A}^2. \qquad (5.1.22)$$

Consequently, for $\mathbf{A}^{(m)}$ we get

$$\mathbf{A}^{(m)}(x) = \sum_{j=1}^{m} f_{j,m}(x)\mathbf{A}^j, \qquad (5.1.23)$$

$f_{j,m}$ being determined by the recurrence formulae (5.1.19). By Theorem 1.6, $z = y'$ can be represented as the second component of the vector

$$\mathbf{V}(x_0)+\sum_{k\geq 1}\left(\sum_{j=1}^{k} f_{j,k}(x_0)\mathbf{A}^j\right)\frac{(x-x_0)^k}{k!}\mathbf{V}(x_0). \tag{5.1.24}$$

We must now compute the successive powers of **A**. Due to its cell-diagonal form, we immediately find

$$\mathbf{A}^j=\begin{bmatrix} \mathbf{0} & \mathbf{0} & \dots & \mathbf{0} & \mathbf{B}_1\mathbf{B}_3\cdots\mathbf{B}_{2j-1} & \mathbf{0} & \mathbf{0} & \dots \\ \mathbf{0} & \mathbf{0} & \dots & \mathbf{0} & \mathbf{0} & \mathbf{B}_3\mathbf{B}_5\cdots\mathbf{B}_{2j+1} & \mathbf{0} & \dots \\ \dots & \dots & \dots & \dots & \dots & \dots & \dots & \dots \end{bmatrix}, \tag{5.1.25}$$

where the finite matrix product $\mathbf{B}_1\mathbf{B}_3\cdots\mathbf{B}_{2j-1}$ is the $j+1$-th entry on the first row of cells.

Let now

$$\mathbf{e}_j^i=\left[\delta_k^i\right]_{k=\overline{1,j}},\quad i=\overline{1,j} \tag{5.1.26}$$

be a basis in $\Re^j$ and denote by

$$\mathbf{e}=\mathbf{e}_2^2=[0\quad 1]^{\mathrm{T}}. \tag{5.1.27}$$

The formulae (5.1.24) and (5.1.25) yield the LEM solution for $z=y'$

$$\begin{aligned} z(x)=&\left\langle \mathbf{V}_1(x_0),\mathbf{e}_2^2\right\rangle+\sum_{k\geq 1}\Big[f_{1,k}(x)\left\langle \mathbf{B}_1\mathbf{V}_3(\widetilde{x}),\mathbf{e}_4^2\right\rangle+\\ &+.f_{2,k}(x_0)\left\langle \mathbf{B}_1\mathbf{B}_3\mathbf{V}_3(x_0),\mathbf{e}_6^2\right\rangle+\dots\\ &\dots+f_{k,k}(x_0)\left\langle \mathbf{B}_1\mathbf{B}_3\cdots\mathbf{B}_{2k-1}\mathbf{V}_3(x_0),\mathbf{e}_{2k+2}^2\right\rangle\Big]\frac{(x-x_0)^k}{k!}. \end{aligned} \tag{5.1.28}$$

By virtue of series convergence, as well as by the analyticity of the LEM solution, the terms in this formula can be re-arranged, to give

$$z(x)=\left\langle \mathbf{V}_1(x_0),\mathbf{e}\right\rangle+\sum_{s\geq 0}\widetilde{f}_s\left\langle \mathbf{B}_1\mathbf{B}_3\cdots\mathbf{B}_{2s-1}\mathbf{V}_{2s+1}(x_0),\mathbf{e}\right\rangle\frac{(x-x_0)^k}{k!}; \tag{5.1.29}$$

obviously, $\widetilde{f}_s$ is given by (5.1.18). Integrating once the expansion (5.1.29) term by term and taking into account the first condition (5.1.5), we immediately obtain the LEM representation (5.1.16) for y. ◘

In the particular case $x_0=0, \alpha=0, \lambda=0$, the above LEM solution, as well as its derivative, are brought to the simplified form

$$y(x) = x[f]_1 + \sum_{k\geq 2} x[f]_{2k-1}\left(\prod_{s=1}^{k-1}(2s-1)\right)^2 (2k-1),$$
$$y'(x) = x\tilde{f}_1 + \sum_{k\geq 2} \tilde{f}_{2k-1}\left(\prod_{s=1}^{k-1}(2s-1)\right)^2 (2k-1). \tag{5.1.30}$$

Indeed, in this case, the scalar products $\langle \mathbf{B}_1\mathbf{B}_3 \cdots \mathbf{B}_{2s-1}\mathbf{V}_{2s+1}(0), \mathbf{e}\rangle$ are easily computed, as conditions (5.1.15) read

$$\mathbf{V}(0) = \left[\mathbf{V}_{2j-1}(0)\right]_{j\in\mathfrak{N}}, \quad \mathbf{V}_{2j-1}(0) = \left[\delta_k^1\right]_{k=\overline{0,2j-1}}. \tag{5.1.31}$$

5.1.2.2. LEM solutions of integral type

We notice that the form (5.1.4), obtained by transforming the B.-E. equation, allows the application of corollary 3.2. We get at once

Consequence 5.1 [116, 153]. *The second component of the vector*

$$\mathbf{V}(x) = \mathrm{e}^{\mathbf{A}\varphi(x)}\,\mathbf{V}(x_0), \quad \varphi(x) = \int_{x_0}^{x} f(t)\,\mathrm{d}t, \tag{5.1.32}$$

where $\mathbf{A}$ *is the LEM matrix* (5.1.12) *and* $\mathbf{V}(x_0)$ *satisfies* (5.1.15), *locally coincides with the derivative* $y'(x)$ *of the solution of the nonlinear initial problem* (5.1.1),(5.1.5).

Proof. $\mathcal{Y}_{x_0}, \mathcal{P}_{x_0}$ are equivalent by Theorem 5.1; they are also locally equivalent with their corresponding LEM operators, $\mathfrak{L}_{x_0}, \mathfrak{S}_{x_0}$. According to Theorem 1.6 and Corollary 3.2, the solution of the initial linear problem $\mathfrak{S}_{x_0}\mathbf{V} = \mathbf{0}$ is given by the representation (5.1.32). By virtue of the transformation T from (5.1.3), the second component of $\mathbf{V}$ is precisely $z = y'(x)$, the derivative of the solution of the Cauchy problem (5.1.1), (5.1.5). ◘

Using the Consequence 5.1, we obtain the derivative of the LEM solution

$$
\begin{aligned}
y'(x) \cong \sqrt{1+\lambda^2} \cdot & \left[\omega(x) + \frac{\omega^3(x)}{3!}\left(3 + 2^2.3\mu^2\right) + \right. \\
& + \frac{\omega^5(x)}{5!}\left(3^2 \cdot 5 + 2^2 \cdot 3^3 \cdot 5\mu^2 + 2^3 \cdot 3^2 \cdot 5\mu^4\right) + \\
& + \frac{\omega^7(x)}{7!}\left(3^2 \cdot 5^2 \cdot 7 + 2^3 \cdot 3^3 \cdot 5^2 \cdot 7\mu^2 + 2^4 \cdot 3^3 \cdot 5^2 \cdot 7\mu^4 + \frac{8!}{2}\mu^6\right) + \\
& + \frac{\omega^9(x)}{9!}\left(3^4 \cdot 5^2 \cdot 7^2 + 2^3 \cdot 3^4 \cdot 5^3 \cdot 7^2\mu^2 + 2^5 \cdot 3^4 \cdot 5^3 \cdot 7^2\mu^4 \right. \\
& \left.\left. + 2^7 \cdot 3^4 \cdot 5^2 \cdot 7^2\mu^6 + \frac{10!}{2}\mu^8\right)\right] \qquad (5.1.33) \\
& + \lambda\left[1 + \frac{3\omega^2(x)}{2}\right) + \frac{\omega^4(x)}{4!}\left(3^2 \cdot 5 + 2^2 \cdot 3 \cdot 5\mu^2\right) \\
& + \frac{\omega^6(x)}{6!}\left(3^2 \cdot 5^2 \cdot 7 + 2^2 \cdot 3^2 \cdot 5^2 \cdot 7\mu^2 + \frac{7!}{2}\mu^4\right) \\
& \left. + \frac{\omega^8(x)}{8!}\left(3^4 \cdot 5^2 \cdot 7^2 + 2^3 \cdot 3^4 \cdot 5^2 \cdot 7\mu^2 + 2^4 \cdot 3^5 \cdot 5 \cdot 7^2\mu^4 + \frac{9!}{2}\mu^6\right)\right].
\end{aligned}
$$

In this LEM representation for y', we used the notations

$$
\varphi(x) = \int_{x_0}^{x} f(t)\mathrm{d}t, \quad \omega(x) = \left(1 + \lambda^2\right)\varphi(x), \quad \mu = \frac{\lambda}{\sqrt{1+\lambda^2}}. \qquad (5.1.34)
$$

The LEM formula (5.1.33) for the 9th order approximation of this derivative was written also putting into evidence the numerical coefficients, as they have some particularities and symmetries that could lead to a future guess in advance of the higher order approximating terms.

The following representation holds true for any Cauchy problem associated to the B.-E. equation, immaterial the choice of x_0.

The displacement y is obtained by direct integration from the above formula for y'

$$
y(x) = \alpha + \int_{x_0}^{x} y'(t)\mathrm{d}t. \qquad (5.1.35)
$$

In the particular case $x_0 = 0$, the two previous formulae do not change their form; we must only replace φ defined by (5.1.34) with

$$\varphi(x) = \int_0^x f(t)\,\mathrm{d}t\,. \tag{5.1.36}$$

5.1.3. SPECIFIC FEATURES OF SOME TYPICAL CASES IN THE NONLINEAR BENDING OF A STRAIGHT BAR

The general solution of the B.-E. equation may be put in integral form

$$y(x) = \int_0^x \frac{\varphi(t) + c}{\sqrt{1 - [\varphi(t) + c]^2}}\,\mathrm{d}t\,, \tag{5.1.37}$$

with φ given by (5.1.36) and

$$c = \frac{y'(0)}{\sqrt{1 + [y'(0)]^2}}\,. \tag{5.1.38}$$

This is an immediate consequence of the equation

$$\frac{\mathrm{d}z}{\mathrm{d}\varphi} = \left(1 + z^2\right)^{\frac{3}{2}}, \qquad z = y'\,, \tag{5.1.39}$$

obtained by performing the following change of variable in the B.-E. equation:

$$\varphi(x) = \int_0^x f(t)\,\mathrm{d}t\,. \tag{5.1.40}$$

The equation (5.1.39) is invariant on the class of the B.-E. -type equations.

Let us note that formula (5.1.37) is not always efficiently applied, as it generally leads to elliptic integrals. This shortcoming makes it practically inadequate for the simply supported bar problem, except for problems in which $f \in \mathcal{G}$.

Definition 5.1 [93, 94]. f is of class $\mathcal{G}$ if

i) f allows a primitive $Pf(x)$ and $Pf(0) = 0$;

ii) for any $k \in \mathcal{N}$, $(Pf)^k$ also allows a primitive in closed form and $P\left(\left(Pf\right)^k\right)(0) = 0$.

We introduced this class of functions to avoid cumbersome computation and this is also the reason for the supplementary requirement ii), according to which the primitives are expressed in terms of analytic functions, easily dealt with.

Following these ideas, in [94] it was studied the cantilever bar problem for various practical cases in which f belongs to the $\mathcal{G}$-class.

But LEM allows the study of straight bar problems for much more general than $\mathcal{G}$ classes of functions. Also, LEM allows the study of mathematically distinct problems in the general frame offered by the Cauchy problem for the B.-E. equation with arbitrary data, as previously specified.

We present here three hypostases, typical for the straight bar and frequently met in problems of the statics of constructions.

5.1.3.1. The cantilever bar

The mathematical model for the cantilever (or built-in) bar is represented by the B.-E. equation to which one adds null Cauchy conditions, because the displacement *y*, as well as the rotation of the cross section y', must be null at the built-in end.

Along with the equation (5.1.1), the displacement *y* must also satisfy the conditions

$$y(0) = 0, \quad y'(0) = 0. \qquad (5.1.41)$$

The physical problem requires the bar displacement and rotation of the cross section at the point $x = l - u$, where *u* is the displacement along the bar axis at the bar free end, determined by

$$u = \int_0^{l-u} \left(\sqrt{1 + y'^2(x)} - 1 \right) \mathrm{d}x. \qquad (5.1.42)$$

So, the cantilever bar problem is not only a standard nonlinear problem, because its solution *y*, as it is seen from the formula of *u*, must be computed at a point depending on *y*. Even if we know y', formula (5.1.42) has to be considered as an integral equation rather than a simple computation formula.

For small or moderate displacements, (5.1.42) can be relaxed to formula

$$u = \frac{1}{2}\int_0^l y'^2(x)\,\mathrm{d}x\,. \tag{5.1.43}$$

The physical magnitude u is called *shortening*.

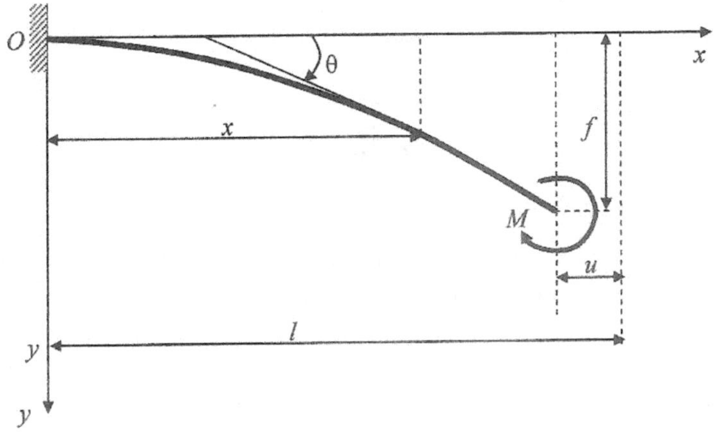

Figure 5.1. The cantilever bar

Obviously, in a complete nonlinear study, l must be replaced with $l-u$ in the expression of $f(x)$; but solving this problem is far too difficult and, actually, in this case, u can be neglected with respect to l.

Consequently, for moderate displacements of the cantilever bar, the maximum bar deflection is obtained for $x = l - u$, with u given by formula (5.1.43). By using LEM, one can determine both the bending and the shortening.

As a consequence of the theorem 5.2, one can prove

Corollary 5. 1 [95, 100]. *Suppose that f satisfies the conditions*

$$\left|f_{s,j+s}(0)\right| \le K^{j+s}C^j\,\frac{(j+s)!}{s!}, \qquad K = \|f\|_0,\ j \in \mathfrak{N},\ s \in \mathfrak{N}^*, \tag{5.1.44}$$

where $\|\cdot\|_0$ *is the "sup" norm on* $\mathrm{C}^0([0,l])$ *and K simultaneously satisfies the inequalities*

$$2Kl < 1, \quad KlC < 1\,. \tag{5.1.45}$$

Then the formulae (5.1.16), (5.1.17) *are valid on the interval* $[0,l]$.

Proof. The hypothesis, as well as (5.1.18), (5.1.19), yield

$$\begin{aligned}\| y' \|_0 &\le \sum_{s,k\ge 1} \frac{l^{s+k}}{(s+k)!} K^{s+k} C^k \frac{(s+k)!}{s!}(2s-1)!! \\ &\le \sum_{s>1} (2KL)^s \sum_{k>0} (KlC)^k \le \frac{1}{(1-2Kl)(1-Klc)}.\end{aligned} \tag{5.1.46}$$

Similarly, one can prove that

$$\| y \|_0 \le \frac{1}{(1-2Kl)(1-Klc)}. \tag{5.1.47}$$

This completes the proof. ▫

Let us note that, as we have the benefit of an analytic representation for y', the calculation of u from formula (5.1.43) is easily performed and, moreover, within the limits of a desired precision. There is more: by Corollary 5.1, the LEM representations become global, extending their consistency over the whole interval of definition of the unknown functions y, y'.

Thus, Corollary 5.1 creates a practical frame for applications, as it will be seen in what follows.

5.1.3.2. The simply supported bar

In this case, the B.-E. equation (5.1.1) must be solved with the two-point conditions

$$y(0)=0, \quad y(l-u)=0, \tag{5.1.48}$$

u being the shortening, previously defined.

Thus, for the simply supported bar, we obtained a two-point nonlinear problem, which, as in the case of the cantilever bar, is yet non-standard, because it depends, through the shortening u, on the solution – more precisely, on its derivative y'.

We shall try to separate this problem in other two nonlinear problems, (P_n^1) and (P_n^2), by using a two-step iterative method:

$$(\mathsf{P}_n^1) \quad \begin{cases} y_n'' = f(x)(1+y_n')^{\frac{3}{2}} \\ y_n(0)=0, y_n(l_n)=0 \end{cases}, \qquad n \in \mathfrak{N}, \quad l_1 = l, \tag{5.1.49}$$

and

$$\left(\mathsf{P}_n^2\right) \quad l_{n+1} = l - \frac{1}{2}\int_0^{l_n} y_n'^2(x)\,\mathrm{d}x\,. \tag{5.1.50}$$

$\left(\mathsf{P}_n^1\right)$ is a nonlinear two-point problem; to solve $\left(\mathsf{P}_n^2\right)$, one must know y', either explicitly, or by a semi-analytic approach.

For a convenient approximation of the solution of $\left(\mathsf{P}_n^1\right)$, we can apply Theorems 4.6 and 4.7,which lead to

Corollary 5.2. [95, 100, 126] *Let f satisfy* (5.1.44) *for K and* l *such that*

$$Kl < \frac{1}{\sqrt{5}}, \tag{5.1.51}$$

and let $[f]_s \geq 0$. *Consider the two-point conditions*

$$y(0) = 0, \quad y(l) = 0\,. \tag{5.1.52}$$

associated to the B.-E. equation. Then

i) the solution of the two-point problem (5.1.1), (5.1.52) *exists, is unique and allows the representation* (5.1.16) *for* $\alpha = 0$, λ *being determined from the second condition of* (5.1.52);

ii) for $y'(0) = \lambda$ *one can establish the following approximating formula*

$$y'(0) \cong -\frac{\sum_{s\geq 1} [f]_{2s-1}(l)\, a_s}{1 + \sum_{s\geq 1} [f]_{2s}(l)\, a_{s+1}}, \tag{5.1.53}$$

where

$$a_{s+1} = \left\langle \mathbf{B}_1 \mathbf{B}_3 \cdots \mathbf{B}_{4s-1} \mathbf{e}_{4s+1}^2, \mathbf{e} \right\rangle, \quad a_1 = 1\,. \tag{5.1.54}$$

Proof. The solution of the B.-E. equation can be expressed, as previously mentioned, in the integral form (5.1.37). Naturally, $|c| < 1$, and

$$\| \varphi + c \|_0 < 1\,. \tag{5.1.55}$$

As $\frac{\mathrm{d}}{\mathrm{d}c} y(l, c) \neq 0$ for any c, the condition $y(l) = 0$ is satisfied for at most one real c. Therefore $y'(0)$ is unique.

The constant c must satisfy

$$\int_0^l \frac{\varphi(t)}{\sqrt{1-[\varphi(t)+c]^2}}\,\mathrm{d}t + \int_0^l \frac{c}{\sqrt{1-[\varphi(t)+c]^2}}\,\mathrm{d}t = 0\,. \tag{5.1.56}$$

By hypothesis, $[f]_s \geq 0$ and therefore, according to (5.1.16), one has $y(x) > 0$ too, for $x > 0$. This yields $y'(0) < 0$, and thus $c < 0$. Taking now c such that

$$|c| > Kl\,, \tag{5.1.57}$$

the left member of (5.1.56) will be negative, hence $c \in [-Kl, 0]$. Consequently,

$$y'(0) \in I_0\,, \quad I_0 \equiv \left[-\frac{KL}{\sqrt{1-K^2l^2}}, 0\right]. \tag{5.1.58}$$

Using the same notations as in Corollary 5.1, we deduce that the LEM representation (5.1.16) is consistent for

$$2Kl\sqrt{1+\lambda^2} < 1\,, \tag{5.1.59}$$

which means

$$|\lambda| < \sqrt{\frac{1}{4K^2l^2} - 1}\,. \tag{5.1.60}$$

This inequality makes sense due to the relation (5.1.51) from the hypothesis. As $y'(0) \in I_0$, the interval of convergence of the LEM representation must include I_0; this is also ensured by the inequality (5.1.51) from the hypothesis. Thus *i)* is proved.

To prove *ii)*, let us note that, as $y'(0)$ belongs to I_0 , we have

$$\|y'\|_0 < \frac{KL}{\sqrt{1-K^2l^2}}\,. \tag{5.1.61}$$

The second bilocal condition (5.1.52(5.1.52) provides the following equation in λ, whose solution is $y'(0)$

$$\lambda + \sum_{s\geq 1} [f]_s(l)\langle \mathbf{B}_1\mathbf{B}_3 \cdots \mathbf{B}_{2s-1}\mathbf{V}_{2s+1}(0), \mathbf{e}\rangle = 0\,. \tag{5.1.62}$$

Neglecting higher order terms in λ, we get

$$\lambda = y'(0) \cong -\frac{\sum_{s\geq 1} [f]_{2s-1}(l)\left\langle \mathbf{B}_1\mathbf{B}_3\cdots\mathbf{B}_{4s-3}\mathbf{e}_{4s}^1, \mathbf{e}\right\rangle}{1+\sum_{s\geq 1} [f]_{2s}(l)\left\langle \mathbf{B}_1\mathbf{B}_3\cdots\mathbf{B}_{4s-1}\mathbf{e}_{4s+2}^2, \mathbf{e}\right\rangle}. \tag{5.1.63}$$

Taking into account the special form of the **B** matrices, we conclude that

$$a_s = \left\langle \mathbf{B}_1\mathbf{B}_3\cdots\mathbf{B}_{4s-1}\mathbf{e}_{4s+2}^2, \mathbf{e}\right\rangle = \left\langle \mathbf{B}_1\mathbf{B}_3\cdots\mathbf{B}_{4s+1}\mathbf{e}_{4s+4}^1, \mathbf{e}\right\rangle, \tag{5.1.64}$$

which immediately leads to formula (5.1.53). ◘

We must now establish conditions of convergence for the two-step procedure (P_n^1), (P_n^2).

Theorem 5.3. [95, 100,126]. *If the hypotheses of corollary* 5.2 *are fulfilled, then the algorithm* (P_n^1), (P_n^2) *converges with n to the solution of the nonlinear problem* (5.1.1), (5.1.48), *with u given by the approximating formula* (5.1.43).

Proof. Let

$$\tilde{l} = l - \frac{K^2 l^3}{2(1-K^2 l^2)}. \tag{5.1.65}$$

From the formula of the shortening u, we immediately deduce that l_n defined in (P_n^2) is such that

$$\tilde{l} < l_n < l; \tag{5.1.66}$$

this means that the limit of the sequence $\{l_n\}_{n\in\mathfrak{N}}$, if it exists, is finite and does not vanish. Denoting by c_n the constant corresponding to the condition $y(l_n)=0$, computed by formula (5.1.38), from (5.1.56) it follows that c_n satisfy

$$\int_0^{l_n} \frac{\varphi(x)+c_n}{\sqrt{1-[\varphi(x)+c_n]^2}}\,\mathrm{d}x - \int_0^{l_{n-1}} \frac{\varphi(x)+c_{n-1}}{\sqrt{1-[\varphi(x)+c_{n-1}]^2}}\,\mathrm{d}x = 0, \tag{5.1.67}$$

hence

$$\left|c_n - c_{n-1}\right| L \le \int_{l_{n-1}}^{l_n} \frac{\varphi(x) + c_n}{\sqrt{1 - [\varphi(x) + c_n]^2}} \mathrm{d}x \le \frac{Kl\left|l_n - l_{n-1}\right|}{\sqrt{1 - K^2 l^2}}, \tag{5.1.68}$$

where we used the notation

$$L \equiv \tilde{l} + \frac{1}{2}\int_0^{\tilde{l}} \left[\phi_n^2(x) + \phi_n(x)\phi_{n-1}(x) + \phi_{n-1}^2(x) + \ldots\right]\mathrm{d}x, \qquad \phi_j(x) = \varphi(x) + c_j, \quad j \in \mathfrak{N}. \tag{5.1.69}$$

Obviously, $L > \tilde{l}$. Also taking into account the bounds (5.1.66) of the sequence $\{l_n\}_{n\in\mathfrak{N}}$, it results

$$\left|c_n - c_{n-1}\right| \le \frac{Kl\left|l_n - l_{n-1}\right|}{\sqrt{1 - K^2 l^2}\left(1 - \frac{K^2 l^2}{1 - K^2 l^2}\right)}. \tag{5.1.70}$$

The difference between two successive elements of the sequence $\{l_n\}_{n\in\mathfrak{N}}$ is therefore estimated by

$$\begin{aligned} \left|l_n - l_{n-1}\right| &\le \frac{1}{2}\int_0^l \frac{[\varphi(x) + c_{n-1}]^2 - [\varphi(x) + c_{n-2}]^2}{\left(1 - K^2 l^2\right)^2}\mathrm{d}x \\ &\le \frac{Kl^2}{\left(1 - K^2 l^2\right)^2} \cdot \frac{K\left|l_{n-1} - l_{n-2}\right|}{\sqrt{1 - K^2 l^2}\left(1 - \frac{K^2 l^2}{1 - K^2 l^2}\right)}, \end{aligned} \tag{5.1.71}$$

so that, finally,

$$\left|l_n - l_{n-1}\right| \le \omega^2 \left|l_{n-1} - l_{n-2}\right|. \tag{5.1.72}$$

In the above inequality, we used the notation

$$\omega^2 \equiv \frac{K^2 l^2}{2\left(1 - K^2 l^2\right)^{\frac{3}{2}}\left(1 - 2K^2 l^2\right)}. \tag{5.1.73}$$

To ensure the convergence of $\{l_n\}_{n\in\mathfrak{N}}$, one should have $\omega^2 < 1$, but this inequality is obviously satisfied if $Kl < 1/\sqrt{5}$; it even holds true for bigger values, e.g. for $Kl = 0.52$.

From these inequalities, as well as from $y \in C^2([0,l])$, it is immediately seen that the recurrence algorithm converges and that its limit is the solution of the nonlinear problem associated to the simply supported bar.◘

We can now affirm that theorem 5.3 and corollary 5.2 completely solve the simply supported bar problem. Indeed, let $y_n \in C^2([0,l])$ be the solution of problem (P_n^1). By corollary 5.2, this solution exists, is unique and can be approximated by using the LEM formula (5.1.16). Also by virtue of corollary 5.2, we can affirm that, for any $\varepsilon > 0$, an approximate $\widetilde{y}_n$ of the LEM solution y_n exists such that

$$\left\| y_n - \widetilde{y}_n \right\|_0 < \varepsilon . \qquad (5.1.74)$$

5.1.3.3. The hyperstatic bar

In this case, the bending moment has the form

$$M(x) = M_p(x) + R(l - x), \qquad (5.1.75)$$

where $p = p(x)$ represents the loading and R – the reaction.

Consider the bar built-in at the left end (at $x = 0$) and simply supported at its right end ($x = l$). The statically indeterminate unknown is the reaction at the simply supported end (figure 5.2).

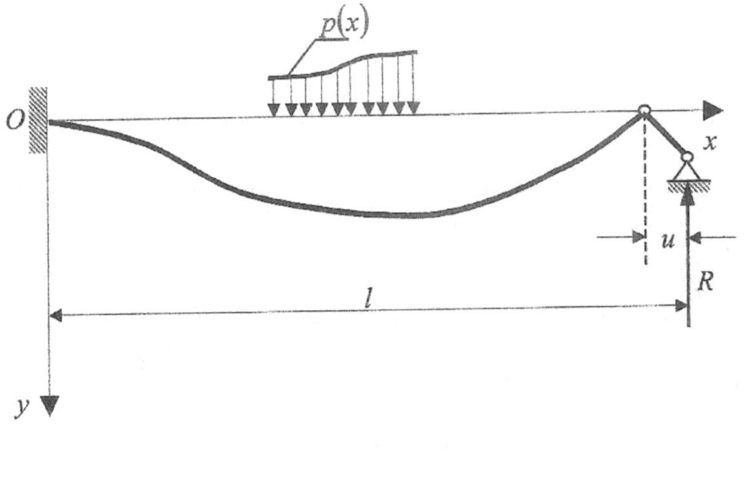

Figure 5.2. The bending of a straight bar built-in at the left and simply supported at the right end

Due to the nonlinearity, the superposition principle – a usual method in the linear case – is not working, therefore we shall consider the bar simultaneously subjected to an arbitrary loading and to a unknown concentrated force *R*.

At the left end, one must consider the null Cauchy conditions (5.1.41); at $x = l - u$, *u* being the shortening given by formula (5.1.43), we take

$$y(l - u) = 0 . \tag{5.1.76}$$

In fact, the reaction *R* introduces a parameter, requiring some additional results concerning the convergence of the LEM representations.

Corollary 5.3 [97, 100, 126]. *The LEM representations* (5.1.16), (5.1.17) *are consistent on the interval* $[x_0, x] \subseteq [0, l]$ *if there exists an interval* $[R_1, R_2]$ *and a constant K independent of* $R \in [R_1, R_2]$ *such that*

$$\left| f_{s,j+s}(x_0, R) \right| \le K^{j+s} C^j \frac{(j+s)!}{s!} , \tag{5.1.77}$$

where

$$R \in [R_1, R_2], \qquad \sup_{x \in [0,l]} \left| f(x, R) \right| \le K , \tag{5.1.78}$$

K and $[x_0, x]$ *satisfying*

$$K |x_0 - x| < \frac{1}{2} . \tag{5.1.79}$$

To show this, we use a proof similar to those of corollaries 5.1 and 5.2.

Obviously, the LEM representations associated to this problem depend in this case on *x*, as well as on the parameter *R*. Actually, corollary 5.3 fixes up – with respect to both *x* and *R* – the intervals of convergence of these representations.

Consider now the two-point problem composed of the B.-E. equation in which the bending moment *M* is expressed by (5.1.75), of the null Cauchy conditions (5.1.41) and of the condition

$$y(l) = 0 . \tag{5.1.80}$$

We introduce the notation

$$f(x) = f_1(x) - Rf_2(x), \tag{5.1.81}$$

where

$$f_1(x) = -\frac{M_p(x)}{E(x)I(x)}, \quad f_2(x) = -\frac{l-x}{E(x)I(x)}. \tag{5.1.82}$$

Let

$$\varphi(x) = \int_0^x f_1(t)\,\mathrm{d}t, \quad \psi(x) = \int_0^x f_2(t)\mathrm{d}t\,. \tag{5.1.83}$$

We shall establish sufficient conditions for the existence and uniqueness of the solution of the above problem; these conditions will respect the physically admitted limits for small and moderate displacements of a straight bar.

Theorem 5.4 [97, 100]. *Let* $f_1, f_2 : [0,l] \to \Re_+$ *be continuous and suppose that*

$$\|\psi\| < \|\varphi\| < 1, \quad \int_0^l \psi(x)\mathrm{d}x \neq 0 \tag{5.1.84}$$

and that

$$\int_0^l \varphi(x)\mathrm{d}x + l > \frac{1}{\sqrt{1-\|\varphi\|^2}} \int_0^l \varphi(x)\mathrm{d}x \tag{5.1.85}$$

is fulfilled. Then the problem of the hyperstatic bar, represented by the B.-E. equation (5.1.1), *where the bending moment M has the form* (5.1.75), *and by the conditions* (5.1.41), (5.1.80), *allows a unique solution.*

Proof. As mentioned before, the solution of the B.-E. equation with null Cauchy data can be written in the form

$$y(x) = \int_0^x \frac{\varphi(t) - R\psi(t)}{\sqrt{1-[\varphi(t) - R\psi(t)]^2}}\mathrm{d}t\,. \tag{5.1.86}$$

The parameter R must be determined such that y also satisfy the condition $y(l) = 0$, i.e.

$$R=\frac{\displaystyle\int_0^l \frac{\varphi(x)}{\sqrt{1-[\varphi(x)-R\psi(x)]^2}}\,\mathrm{d}x}{\displaystyle\int_0^l \frac{\psi(x)}{\sqrt{1-[\varphi(x)-R\psi(x)]^2}}\,\mathrm{d}x}. \tag{5.1.87}$$

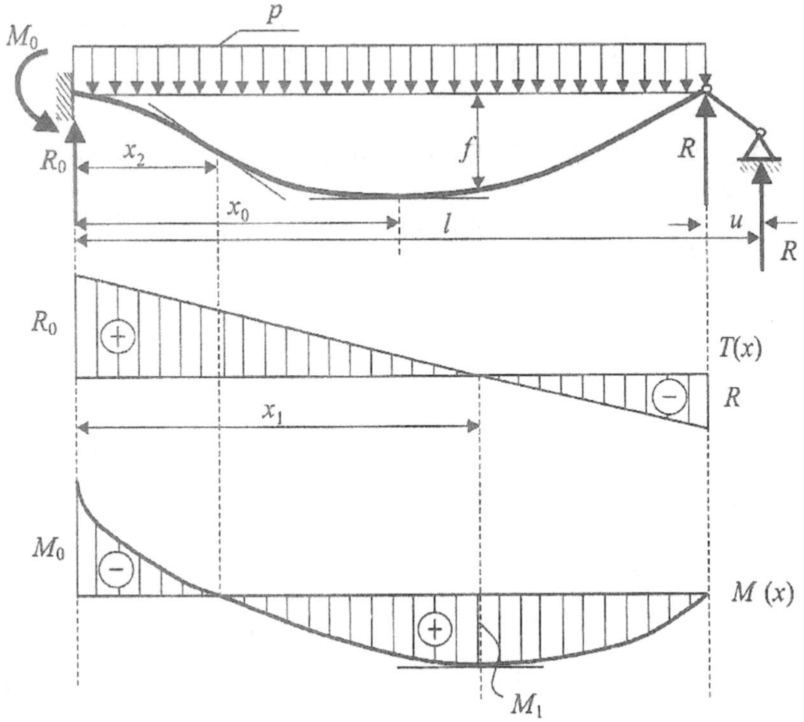

Figure 5.3. The displacement and stress diagrams in the case of a straight bar built-in at the left and simply supported at the right end, acted upon by a uniformly distributed loading

Naturally, to obtain real values for y and R, it is necessary that

$$|\varphi(x)-R\psi(x)|<1,\quad x\in[0,l]; \tag{5.1.88}$$

as R must be positive, this comes back to require that

$$R < q_1 \equiv \frac{\int_0^l \varphi(x)\mathrm{d}x + l}{\int_0^l \psi(x)\mathrm{d}x}. \tag{5.1.89}$$

From (5.1.87), it also follows that

$$R < q_2 \equiv \frac{1}{\sqrt{1-\|\varphi\|^2}} \frac{\int_0^l \varphi(x)\mathrm{d}x}{\int_0^l \psi(x)\mathrm{d}x}, \tag{5.1.90}$$

as a consequence of the assumptions on f_1, f_2.

Inequality (5.1.89) shows that the solution of the B.-E. equation with null Cauchy data may be written in the form (5.1.86) only if $R \in (0, q_1)$. The solution satisfies $y(l) = 0$ too, if

$$q_2 < q_1, \tag{5.1.91}$$

therefore if

$$\frac{1}{\sqrt{1-\|\varphi\|^2}} \int_0^l \varphi(x)\mathrm{d}x < \int_0^l \varphi(x)\mathrm{d}x, \tag{5.1.92}$$

which is precisely the inequality (5.1.85) from the hypothesis.

To prove the uniqueness of the solution, it suffices to show that R is unique. Indeed, we have

$$\frac{\mathrm{d}}{\mathrm{d}R} y(l, R) = \int_0^l \frac{-\psi(x)}{\left\{-[\varphi(x) - R\psi(x)]^2\right\}^{\frac{3}{2}}} \mathrm{d}x < -\int_0^l \psi(x)\mathrm{d}x \tag{5.1.93}$$

and thus

$$\frac{\mathrm{d}}{\mathrm{d}R} y(l, R) < 0, \tag{5.1.94}$$

due to the hypothesis. Theorem 5.4 is proved. ◘

Although we used the integral form of the solution to prove the preceding theorem, this form is not easily handled, as it leads to elliptic integrals and therefore is does not fit in concrete cases. Formula (5.1.37) has another disadvantage: it does not permit qualitative interpretations.

In applications, we shall use the constructive alternative offered by LEM, setting up a two-step algorithm, as in the case of the simply supported bar. This algorithm reads

$$\left(\mathrm{Q}_n^1\right) \quad \begin{cases} y_n'' = [f_1(x) - R_n f_2(x)]\left(1 + y_n'\right)^{\frac{3}{2}} \\ y_n(0) = 0,\, y_n'(0) = 0 \end{cases}, \quad n \in \mathfrak{N}, \quad l_1 = l, \tag{5.1.95}$$

$$\left(\mathrm{Q}_n^2\right) \quad l_{n+1} = l - \frac{1}{2}\int_0^{l_n} y_n'^2(x)\,\mathrm{d}x\,. \tag{5.1.96}$$

If we succeed to establish satisfactory conditions of convergence, then the problems fulfilling them will be completely solved numerically. As we dispose of LEM representations, both for the solution and for its derivative, one can easily set up a programming code for the two steps of the above algorithm.

Lemma 5.1. [97, 100, 126]. *The sequences* $\{R_n\}_{n\in\mathfrak{N}}$ *and* $\{l_n\}_{n\in\mathfrak{N}}$ *are included in independent of* n *intervals.*

Proof. From the definition of $\left(\mathrm{Q}_n^2\right)$, we deduce

$$l_n \geq l - \frac{1}{2}\int_0^l y_n'^2(x)\,\mathrm{d}x = l - \frac{1}{2}\int_0^l \frac{[\varphi(x) - R_n\psi(x)]^2}{1 - [\varphi(x) - R_n\psi(x)]^2}\,\mathrm{d}x, \tag{5.1.97}$$

hence

$$l_n \geq \frac{l}{2}\,. \tag{5.1.98}$$

Then (5.1.87) yields for R_n

$$R_n > \frac{1}{\sqrt{1 - \|\varphi\|^2}}\,\frac{\int_0^{l_n}\varphi(x)\mathrm{d}x}{\int_0^{l_n}\psi(x)\mathrm{d}x} > \frac{1}{\sqrt{1 - \|\varphi\|^2}}\,\frac{\int_0^{\frac{l}{2}}\varphi(x)\mathrm{d}x}{\int_0^{l}\psi(x)\mathrm{d}x} \equiv r\,. \tag{5.1.99}$$

As $R \in (0, q_1)$, it follows that

$$R_n \in (r, q_1), \quad l_n \in \left(\frac{l}{2}, l\right). \tag{5.1.100}$$

The lemma is thus proved. ◘

This partial result will be used to prove the convergence of the two-step procedure (5.1.95), (5.1.96).

Theorem 5.5. [97, 100]. *Let* f_1, f_2 *satisfy the hypotheses of Theorem* 5.4 *and suppose that the following inequality holds*

$$\Omega \equiv \frac{\|\varphi\|^2 l}{\left(1-\|\varphi\|^2\right)^{\frac{5}{2}}} \frac{\int_0^l \psi(x)\mathrm{d}x}{\int_0^{\frac{l}{2}} \psi(x)\mathrm{d}x} < 1. \tag{5.1.101}$$

Then the algorithm $\left(\mathrm{Q}_n^1\right), \left(\mathrm{Q}_n^2\right)$ *converges with n to the solution of the hyperstatic bar problem* (5.1.1), (5.1.41), (5.1.76).

Proof. According to Lemma 5.1, the sequences $\{R_n\}_{n\in\mathfrak{N}}$ and $\{l_n\}_{n\in\mathfrak{N}}$ are bounded. Let us estimate the differences $|R_n - R_{n-1}|$ and $|l_n - l_{n-1}|$. We have

$$\begin{aligned}|R_n - R_{n-1}| &\le \left| \frac{\int_0^{l_n} \frac{\varphi(x)}{\sqrt{1-[\varphi(x)-R_n\psi(x)]^2}}\mathrm{d}x}{\int_0^{l_n} \frac{\psi(x)}{\sqrt{1-[\varphi(x)-R_n\psi(x)]^2}}\mathrm{d}x} - \frac{\int_0^{l_{n-1}} \frac{\varphi(x)}{\sqrt{1-[\varphi(x)-R_{n-1}\psi(x)]^2}}\mathrm{d}x}{\int_0^{l_{n-1}} \frac{\psi(x)}{\sqrt{1-[\varphi(x)-R_{n-1}\psi(x)]^2}}\mathrm{d}x} \right| \\ &\le \frac{\int_{l_{n-1}}^{l_n} \varphi(x)\mathrm{d}x \int_0^{l_{n-1}} \psi(x)\mathrm{d}x}{\left[\int_0^{l/2} \psi(x)\mathrm{d}x\right]^2} \cdot \frac{\sqrt{1-\|\varphi\|^2}}{1-\|\varphi\|^2},\end{aligned} \tag{5.1.102}$$

i.e.

$$\left|R_n - R_{n-1}\right| \le \left|l_n - l_{n-1}\right| \frac{\|\varphi\|}{1-\|\varphi\|^2} \cdot \frac{\left|\int_0^l \psi(x)\mathrm{d}x\right|}{\left[\int_0^{l/2} \psi(x)\mathrm{d}x\right]^2}. \tag{5.1.103}$$

For $\left|l_n - l_{n-1}\right|$, we deduce

$$\left|l_n - l_{n-1}\right| \le \frac{1}{2}\left|\int_0^l \frac{[\varphi(x) - R_{n-1}\psi(x)]^2}{1-[\varphi(x) - R_{n-1}\psi(x)]^2}\mathrm{d}x - \int_0^l \frac{[\varphi(x) - R_{n-2}\psi(x)]^2}{1-[\varphi(x) - R_{n-2}\psi(x)]^2}\mathrm{d}x\right|, \tag{5.1.104}$$

which involves

$$\left|l_n - l_{n-1}\right| \le \frac{\|\varphi\|\, l}{\left[1-\|\varphi\|^2\right]^2} \cdot \left|R_n - R_{n-1}\right|. \tag{5.1.105}$$

The last inequality, combined with (5.1.103), implies

$$\left|l_n - l_{n-1}\right| \le \frac{\|\varphi\|^2 l}{\left[1-\|\varphi\|^2\right]^{\frac{5}{2}}} \cdot \frac{\left|\int_0^l \psi(x)\mathrm{d}x\right|}{\left[\int_0^{l/2} \psi(x)\mathrm{d}x\right]^2} \left|l_{n-1} - l_{n-2}\right|, \tag{5.1.106}$$

or, using the notation from (5.1.101),

$$\left|\, l_n - l_{n-1}\right| \le \Omega \left|\, l_{n-1} - l_{n-2}\right|. \tag{5.1.107}$$

As, by hypothesis, $\Omega < 1$, it follows that $\{l_n\}_{n\in\mathfrak{N}}$ converges and, implicitly, so does the sequence $\{R_n\}_{n\in\mathfrak{N}}$. ◘

Finally, theorems 5.4 and 5.5, as well as corollary 5.3, establish an approximating procedure for the solution of the hyperstatic bar problem. Indeed, the solution y_n of the nonlinear problem $\left(\mathrm{Q}_n^1\right)$ can be approximated on an interval $[x_0, x] \subseteq [0, l]$, satisfying the inequality (5.1.79). Suppose that $[R_1, R_n] = [r, q_1]$. If $[x_0, x]$ does not coincide with $[0, l]$ – and this is the case in the following examples – one can divide $[0, l]$ into subintervals such that (5.1.79) be satisfied on each of them. Due to the LEM representations, easily handled as functions of x_0 and R, the values R_n are obtained as zeros of some perfectly determined functions, depending on R. As mentioned in Theorem 5.4, these functions allow a unique real zero, fact which is also confirmed in applications. Moreover, one can easily compute l_n, due to the LEM solution (5.1.16). Lastly, theorem 5.5 ensures the convergence of the sequence $\{y_n\}_{n\in\mathfrak{N}}$ thus obtained to the solution of the hyperstatic bar problem.

5.1.4. APPLICATIONS

In the previous sections, we set up a general method based on LEM, enabling the study of the problems associated to the Bernoulli-Euler bar in a common theoretical frame. This method was then particularized to some typical bar problems. Let us mention that the cases considered here are not the only cases that could be treated by this method; a case in point is the LEM representation (5.1.16) or formula (5.1.17).

In subsections 5.1.3.1, 5.1.3.2 and 5.1.3.3, there were established convergent algorithms for the cantilever, simply supported and hyperstatic bar. We shall apply them to bars of rectangular cross sections, constant or of heights h or widths b varying linearly along the bar axis and of constant, or proportionally varying along the bar axis, bending moments.

We take $l = 100\,\text{cm}$ and $E = 21000\,\text{kN/cm}^2$ (steel). We admit the action of a constant bending moment $M = 40\,\text{kNcm}$ for the cantilever or simply supported bar (indexed by 1), or the action of a bending moment varying linearly along the bar axis (indexed by 2), of the form

$$M(x) = -Pl\left(1 - \frac{x}{l}\right), \quad Pl = M, \tag{5.1.108}$$

in the case of the cantilever bar (corresponding to a concentrated force $P = 0.4\,\text{kN}$ at its free end), or of the form

$$M(x) = M\left(1 - \frac{x}{l}\right), \tag{5.1.109}$$

in the case of a simply supported bar (given by a bending moment $M = 40\,\text{kNcm}$ concentrated at one of the bar ends).

In the case of a hyperstatic bar, we consider the action of a uniformly distributed loading $p = 3.2\,\text{daN/cm}$, hence

$$M_p(x) = -\frac{1}{2}p(l-x)^2, \tag{5.1.110}$$

so that f becomes

$$f(x) = f_1(x) - Rf_2(x),$$
$$f_1(x) = \frac{pl^2}{2EI(x)}\left(1-\frac{x}{l}\right),\ f_2(x) = \frac{l}{EI(x)}\left(1-\frac{x}{l}\right). \tag{5.1.111}$$

(a) Consider a bar of constant height $h = 1\,\text{cm}$, its width varying linearly along the bar axis, such that $b_0 = 9\,\text{cm}, b_l = b_0 / c = 6\,\text{cm}, c = 1.5$; hence

$$b = b_0 + \frac{b_l - b_0}{l}x = b_l\left[c + (1-c)\frac{x}{l}\right]. \tag{5.1.112}$$

The maximum normal stress (according to Navier's formula ([89]) is

$$|\sigma|_{\max} = 6|M|b_l h^2 = 40\,\text{kN/cm}^2. \tag{5.1.113}$$

Observing that $I = \frac{bh^3}{12}$, we easily get

$$f(x) = \frac{\gamma x + 1}{\alpha x + \beta},$$
$$k = \frac{12|M|}{Eh^3},\ \alpha = \frac{b_l}{k}(1-c),\ \beta = \frac{b_l}{k}c,\ \gamma = -\frac{1}{l}. \tag{5.1.114}$$

For both cantilever and simply supported bar, one has, for the above data, $\alpha = -1.3125, \beta = 393.75\,\text{cm}$ and $\gamma = 0$ in the case (a_1) or $\gamma = -0.01\,\text{cm}^{-1}$ in the case (a_2).

For the hyperstatic bar it is obtained

$$f_1(x) = \frac{(\gamma x + 1)^2}{\alpha x + \beta},\ f_2(x) = \frac{\gamma x + 1}{\delta(\alpha x + \beta)}, \tag{5.1.115}$$

where

$$k = \frac{6pl^2}{Eh^3},\ \alpha = \frac{b_l}{k}\frac{1-c}{l},\ \beta = \frac{b_l}{k}c,\ \gamma = -\frac{1}{l},\ \delta = \frac{pl}{2}. \tag{5.1.116}$$

According to the chosen data, it results that $\alpha = -0.328125$, $\beta = 98.4375\,\text{cm}$, $\gamma = -0.01\,\text{cm}^{-1}$ and $\delta = 1.6\,\text{kN} = 160\,\text{daN}$.

(b) Let us also consider a bar of constant width $b = 6\,\text{cm}$ and of height varying linearly along the bar axis, with a proportionality factor $c = 1.5$, that is, $h_0 = 1.5$ cm, $h_l = h_0 / c = 1\,\text{cm}$. Hence

$$h = h_0 + \frac{h_l - h_0}{l} x = h_l \left[c + (1 - c)\frac{x}{l} \right]. \tag{5.1.117}$$

The maximum normal stress is

$$|\sigma|_{\max} = \frac{6|M|}{bh_l^2} = 40\,\text{kN/cm}^2. \tag{5.1.118}$$

Similarly,

$$f(x) = \frac{\gamma x + 1}{(\alpha x + \beta)^3},$$
$$k = \frac{12|M|}{Eb},\ \alpha = \frac{h_l}{\sqrt[3]{k}}(1 - c),\ \beta = \frac{h_l}{\sqrt[3]{k}}c,\ \gamma = -\frac{1}{l}, \tag{5.1.119}$$

for the cantilever and simply supported bar.

Numerically, $\alpha = -0.038014\,\text{cm}^{-2/3}$, $\beta = 9.604\,\text{cm}^{-1/3}$ and $\gamma = 0$ in the case (b_1) or $\gamma = -0.01\,\text{cm}^{-1}$ in the case (b_2).

For the hyperstatic bar,

$$f_1(x) = \frac{(\gamma x + 1)^2}{(\alpha x + \beta)^3},\quad f_2(x) = \frac{\gamma x + 1}{\delta(\alpha x + \beta)^3}, \tag{5.1.120}$$

where

$$k = \frac{6pl^2}{Eb},\ \alpha = \frac{h_l}{\sqrt[3]{k}}\frac{1 - c}{l},\ \beta = \frac{h_l}{\sqrt[3]{k}}c,\ \delta = \frac{pl}{2},\ \gamma = -\frac{1}{l}. \tag{5.1.121}$$

For the chosen values, we get $\alpha = -0.020168\,\text{cm}^{-2/3}$, $\beta = 6.050141\,\text{cm}^{1/3}$, $\gamma = -0.01\,\text{cm}^{-1}$ and $\delta = 1.6\,\text{kN} = 160\,\text{daN}$.

(c) Admitting a constant bar rigidity, more precisely, $I = bh^3/12 = 6 \cdot 1^3/12 = 0.5\,\text{cm}^4$, the normal stress is

$$|\sigma|_{\max} = \frac{6|M|_{\max}}{bh_l^2} = 40\ \text{kN/cm}^2; \tag{5.1.122}$$

in the case of the cantilever or simply supported bar,

$$f(x) = k(\gamma x + 1), \qquad k = \frac{12|M|}{Ebh}, \quad \gamma = -\frac{1}{l}. \tag{5.1.123}$$

For the chosen data, we obtain $k = 0.0038095\ \text{cm}^{-1}$ and $\gamma = 0$ in the case (c_1) or $\gamma = -0.01\ \text{cm}^{-1}$ in the case (c_2).

For the hyperstatic bar, we have

$$f_1(x) = k(\gamma x + 1)^2, \quad f_2(x) = \frac{k}{\delta}(\gamma x + 1),$$
$$k = \frac{6pl^2}{Ebh^3}, \quad \delta = \frac{pl}{2}, \quad \gamma = -\frac{1}{l}; \tag{5.1.124}$$

numerically, $k = 0.015238\ \text{cm}^{-1}$, $\gamma = -0.01\ \text{cm}^{-1}$, $\delta = 1.6\ \text{kN} = 160\ \text{daN}$.

5.1.5. NUMERICAL RESULTS

Due to the generality of LEM, also emphasized by formulae (5.1.16), (5.1.17), we could set up a unique programming code, applicable to all the problems considered in the preceding section.

The numerical results for the cantilever bar are given in tables 5.1 and 5.2, containing the values of the shortening u, the maximum bar deflection f, attained in this case at the bar free end, and also of the derivative y' and of the rotation of the cross section at the point at which these magnitudes are maximum.

Table 5.1

Magnitude \ Problem	(a_1)	(b_1)	(c_1)
u cm	1.65	0.61	2.87
f cm	14.21	8.39	18.63
$y'_{\max}$	0.316863	0.211890	0.397949
$\theta_{\max}$	$17^0 34' 53''$	$11^0 57' 48''$	$21^0 42''$

Table 5.2

Magnitude \ Problem	(a_2)	(b_2)	(c_2)
u cm	0.56	0.17	1.04
f cm	9.26	5.06	12.60
y'_{max}	0.145299	0.086243	0.189701
θ_{max}	$8^0 16' 02''$	$4^0 55' 45''$	$10^0 44' 29''$

The results corresponding to the simply supported bar are given in tables 5.3 and 5.4, for the shortening u, the maximum bar deflection f, realized at the point x_m, as well as for y' and the rotation angle at both bar ends, where these magnitudes have maximum values.

Table 5.3

Magnitude \ Problem	(a_1)	(b_1)	(c_1)
u cm	0.28	0.08	0.62
x_m cm	51.53	54.92	49.73
f cm	3.84	2.56	4.75
$y'(0)$	0.154109	0.084690	0.192953
θ_0	$8^0 15' 23''$	$4^0 50' 23''$	$10^0 55' 16''$
$y'(l-u)$	-0.166191	-0.126134	-0.193720
$\theta(l-u)$	$-9^0 26' 09''$	$-7^0 11' 20''$	$-10^0 57' 43''$

Table 5.4

Magnitude \ Problem	(a_2)	(b_2)	(c_2)
u cm	0.34	0.06	0.85
x_m cm	43.43	46.41	41.52
f cm	1.88	1.15	2.38
$y'(0)$	0.093021	0.050713	0.126360
θ_0	$5^0 18' 52''$	$2^0 54' 13''$	$7^0 12' 06''$
$y'(l-u)$	-0.051273	-0.035318	-0.057348
$\theta(l-u)$	$-2^0 56' 06''$	$-2^0 01' 22''$	$-10^0 57' 43''$

For the hyperstatic bar, the values of the shortening u and the reaction R are given in table 5.5, corresponding to problems (a), (b) and (c). Hence the shearing force and the bending moment are

$$T(x) = p(l - u - x) - R,$$
$$M(x) = -\frac{1}{2}p(l - u - x)^2 + R(l - u - x). \qquad (5.1.125)$$

Let us note that each chosen function $f(x)$ satisfies the hypothesis of corollaries 5.1 and 5.2. For instance, $Kl = 0.382$ in all cases, which obviously involves that (5.1.45) and (5.1.51) are satisfied.

Table 5.5

Magnitude \ Problem	(a)	(b)	(c)
u cm	0.05	0.01	0.07
R kN	1.1767	1.1813	1.1999

Observing that $T(x) = M'(x)$ ($T(x_1) = 0$ implies $M_{max} = M(x_1)$) and that $M(x_2) = 0$ corresponds to the inflection point x_2 ($y''(x_2) = 0$), it results

$$R_0 = T(0) = p(l - u) - R, \quad M_0 = M(0) = \frac{1}{2}p(l - u)^2 + R(l - u),$$
$$x_2 = l - u - 2\frac{R}{p}, \quad x_1 = l - u - \frac{R}{p}, \quad M_1 = M(x_1) = \frac{R^2}{2p}. \qquad (5.1.126)$$

The corresponding numerical results are given in table 5.6.

Table 5.6

Magnitude \ Problem	(a)	(b)	(c)
R_0 cm	2.0218	2.0184	1.9977
$-M_0$ kN cm	42.2389	41.8492	39.8550
x_2 cm	26.41	26.16	24.93
x_1 cm	63.18	63.07	62.43
M_1 kN cm	21.6334	21.8054	22.4987

The maximum deflection, i.e. y_{max} at x_m, is given in table 5.7.

Table 5.7

Magnitude \ Problem	(a)	(b)	(c)
x_m cm	59.28	58.96	58.00
f cm	1.30	0.56	1.66

In this case too, the chosen functions $f(x)$ satisfy the hypotheses of corollary 5.3; otherwise, the interval $[0, l]$ was divided into subintervals on which (5.1.79) be true.

To conclude about the contribution of the nonlinearities, we also compared the above values of physical magnitudes with those computed in the linear case. This corresponds to the equation

$$\bar{y}'' = f(x), \tag{5.1.127}$$

that must be solved in null Cauchy conditions $\bar{y}(0) = 0, \bar{y}'(0) = 0$ for the cantilever and null two-point conditions $\bar{y}(0) = 0, \bar{y}(l) = 0$ for the simply supported bar. Obviously, the shortening will be no more considered, the values of various magnitudes being computed at $x = l = 100$ cm.

Tables 5.8 and 5.9 contain the results for the cantilever bar.

Table 5.8

Magnitude \ Problem	(a_1)	(b_1)	(c_1)
$\bar{f}$ cm	14.41	8.47	19.05
$\bar{y}'_{max}$	0.299779	0.198723	0.380952
$\bar{\theta}'_{max}$	$16^0 28' 37''$	$11^0 14' 22'''$	$20^0 51' 16''$

Table 5.9

Magnitude \ Problem	(a_2)	(b_2)	(c_2)
$\bar{f}$ cm	9.28	5.07	12.68
$\bar{y}'_{max}$	0.144500	0.085933	0.128400
$\bar{\theta}'_{max}$	$8^0 13' 20''$	$4^0 54' 15''$	$7^0 19' 02''$

Tables 5.10 and 5.11 were set up for the simply supported bar.

Table 5.10

Magnitude \ Problem	(a$_1$)	(b$_1$)	(c$_1$)
$\bar{x}_{\mathrm{m}}$ cm	51.60	55.05	50.00
$\bar{f}$ cm	3.84	2.57	4.78
$\bar{y}'(0)$	0.144053	0.084656	0.190476
$\bar{\theta}_0$	$8^0 11' 50''$	$4^0 50' 20''$	$10^0 47' 03''$
$\bar{y}'(l)$	-0.164873	-0.125984	-0.190476
$\bar{\theta}(l)$	$-9^0 14' 42''$	$-7^0 10' 50''$	$-10^0 47' 03''$

Table 5.11

Magnitude \ Problem	(a$_2$)	(b$_2$)	(c$_2$)
$\bar{x}_{\mathrm{m}}$ cm	43.55	46.50	42.20
$\bar{f}$ cm	1.88	1.15	2.44
$\bar{y}'(0)$	0.092846	0.050528	0.125984
$\bar{\theta}_0$	$5^0 18' 16''$	$2^0 53' 31'''$	$7^0 10' 50''$
$\bar{y}'(l)$	-0.051207	-0.034138	-0.053492
$\bar{\theta}(l)$	$-2^0 55' 30''$	$-1^0 57' 18''$	$-3^0 03' 43''$

The linear hyperstatic bar requires a more detailed examination. As one neglects y'^2 with respect to 1 and the shortening u with respect to l, the equation (5.1.127) must be solved with the conditions

$$\bar{y}(0) = 0, \quad \bar{y}'(0) = 0, \quad \bar{y}(l) = 0, \tag{5.1.128}$$

which also allow to determine the reaction R.

For the problem (a), considering the expressions (5.1.115) of f_1, f_2, we get

$$\begin{aligned} \bar{y}'(x) = {} & \frac{\gamma^2}{\alpha^2}\left[\left(\frac{\alpha x}{3} - \beta\right)x + \frac{\beta^2}{\alpha}\ln\left(\frac{\alpha}{\beta}x + 1\right)\right] \\ & + \left(2 - \frac{\bar{R}}{\delta}\right)\frac{\gamma}{\alpha}\left[x - \frac{\beta}{\alpha}\ln\left(\frac{\alpha}{\beta}x + 1\right)\right] + \left(1 - \frac{\bar{R}}{\delta}\right)\frac{1}{\alpha}\ln\left(\frac{\alpha}{\beta}x + 1\right), \end{aligned} \tag{5.1.129}$$

$$\bar{y}(x)=\frac{\gamma^2}{\alpha^2}\left\{\left(\frac{\alpha x}{3}-\beta\right)\frac{x^2}{2}+\frac{\beta^2}{\alpha^2}\left\{\left(\frac{\alpha}{\beta}x+1\right)\left[\ln\left(\frac{\alpha}{\beta}x+1\right)-1\right]+1\right\}\right\}$$
$$+\left(2-\frac{\overline{R}}{\delta}\right)\frac{\gamma}{\alpha}\left\{\frac{x^2}{2}-\frac{\beta^2}{\alpha^2}\left\{\left(\frac{\alpha}{\beta}x+1\right)\left[\ln\left(\frac{\alpha}{\beta}x+1\right)-1\right]+1\right\}\right\} \tag{5.1.130}$$
$$+\left(1-\frac{\overline{R}}{\delta}\right)\frac{\beta}{\alpha^2}\left\{\left(\frac{\alpha}{\beta}x+1\right)\left[\ln\left(\frac{\alpha}{\beta}x+1\right)-1\right]+1\right\}.$$

The last condition leads to

$$\overline{R}=\left[1+\frac{\beta}{\alpha l}+\frac{1}{3\left\{\frac{1}{2}+\left(1+\frac{\beta}{\alpha l}\right)\left[1-\left(1+\frac{\beta}{\alpha l}\right)\ln\left(1+\frac{\beta}{\alpha l}\right)\right]\right\}}\right]\delta \tag{5.1.131}$$
$$=\left\{\frac{1}{2(1-c)}+\frac{(1-c)^2}{3[(1-c)(3-c)+2\ln c]}\right\}pl.$$

For the problem (b), one uses the relations (5.1.120) and we have

$$\bar{y}'(x)=-\frac{\gamma^2}{2\alpha^3}\left[\frac{(3\alpha x+2\beta)x}{(\alpha x+\beta)^2}-2\ln\left(\frac{\alpha}{\beta}x+1\right)\right]$$
$$+\frac{\gamma}{2}\left(2-\frac{\overline{R}}{\delta}\right)\frac{x^2}{\beta(\alpha x+\beta)^2}+\frac{1}{2}\left(1-\frac{\overline{R}}{\delta}\right)\frac{(\alpha x+2\beta)x}{\beta^2(\alpha x+\beta)^2}, \tag{5.1.132}$$

$$\bar{y}(x)=-\frac{\gamma^2\beta}{\alpha^4}\left[\frac{(5\alpha x+6\beta)\alpha x}{2\beta(\alpha x+\beta)}-\left(\frac{\alpha}{\beta}x+3\right)\ln\left(\frac{\alpha}{\beta}x+1\right)\right]$$
$$+\frac{1}{2}\left(2-\frac{\overline{R}}{\delta}\right)\frac{\gamma}{\alpha^3}\left[\frac{(\alpha x+2\beta)\alpha x}{\beta(\alpha x+\beta)}-2\ln\left(\frac{\alpha}{\beta}x+1\right)\right]+ \tag{5.1.133}$$
$$+\frac{1}{2}\left(1-\frac{\overline{R}}{\delta}\right)\frac{x^2}{\beta^2(\alpha x+\beta)}.$$

Same way, we get

$$\overline{R} = 3\left\{1 + \frac{\beta}{\alpha l} - \frac{1}{\dfrac{1}{2} - \dfrac{\beta}{\alpha l}\left[1 - \dfrac{\beta}{\alpha l}\ln\left(\dfrac{\alpha l}{\beta} + 1\right)\right]}\right\}\delta$$
$$= \left\{\frac{3}{2(1-c)} - \frac{(1-c)^2}{(1-c)(1-3c) - 2c^2 \ln c}\right\} pl. \tag{5.1.134}$$

Problem (c) corresponds to formulae (5.1.124); we deduce

$$\overline{y}'(x) = \frac{1}{3}k\gamma^2 x^3 + \frac{1}{2}\left(2 - \frac{\overline{R}}{\delta}\right)k\gamma x^2 + \left(1 - \frac{\overline{R}}{\delta}\right)kx, \tag{5.1.135}$$

$$\overline{y}(x) = \frac{1}{12}k\gamma^2 x^4 + \frac{1}{6}\left(2 - \frac{\overline{R}}{\delta}\right)k\gamma x^3 + \frac{1}{2}\left(1 - \frac{\overline{R}}{\delta}\right)kx^2, \tag{5.1.136}$$

as well as

$$\overline{R} = \frac{3}{4}\delta = \frac{3}{8}pl. \tag{5.1.137}$$

According to the above considerations, we can write – for all the three problems – in the case of the hyperstatic bar

$$\overline{T}(x) = p(l - x) - \overline{R}, \quad \overline{M}(x) = -\frac{1}{2}p(l - x)^2 + \overline{R}(l - x). \tag{5.1.138}$$

It results

$$\overline{R}_0 = pl - \overline{R}, \quad \overline{M}_0 = -\frac{pl^2}{2} + \overline{R}l,$$
$$\overline{x}_2 = l - 2\frac{\overline{R}}{p}, \quad \overline{x}_1 = l - \frac{\overline{R}}{p}, \quad \overline{M}_1 = \frac{\overline{R}^2}{2p}. \tag{5.1.139}$$

Thus, for problem (c) we have

$$\overline{y}'(x) = \frac{kl}{24}\left[6 - 15\frac{x}{l} + 8\left(\frac{x}{l}\right)^2\right]\frac{x}{l},$$
$$\overline{y}(x) = \frac{kl^2}{24}\left(3 - 2\frac{x}{l}\right)\left(1 - \frac{x}{l}\right)\left(\frac{x}{l}\right)^2, \tag{5.1.140}$$

and

$$\overline{T}(x)=\frac{p}{8}(5l-8x),\quad \overline{M}(x)=-\frac{p}{8}(l-x)(l-4x),$$
$$\overline{R}_0=\frac{5}{8}pl,\quad \overline{M}_0=-\frac{pl^2}{8},\quad \overline{x}_2=\frac{l}{4},\quad \overline{x}_1=\frac{5}{8}l,\quad \overline{M}_1=\frac{9}{128}pl^2 \qquad (5.1.141)$$

We specify that $\overline{x}_{\text{m}}=0.578465\,l$ and $\overline{f}=\overline{y}(\overline{x}_{\text{m}})=0.010832\,kl^2$.

The corresponding numerical results are given in tables 5.12 and 5.13.

Table 5.12

Magnitude \ Problem	(a)	(b)	(c)
$\overline{R}$ cm	2.0234	2.0751	2.0000
$-\overline{M}_0$ kN cm	42.4086	47.5132	40.0000
$\overline{x}_2$ cm	26.46	29.70	25.00
$\overline{x}_1$ cm	63.23	64.85	62.50
$\overline{M}_1$ kN cm	21.6311	19.7719	22.5000

Table 5.13

Magnitude \ Problem	(a)	(b)	(c)
$\overline{R}$ cm	1.17666	2.0751	2.0000
$\overline{x}_{\text{m}}$ cm	42.4086	47.5132	40.0000
$\overline{f}$ cm	26.46	29.70	25.00

5.1.6. COMMENTS

The above examples were chosen in order to represent the most interesting situations in straight bar problems: the variation of the bending moment or variations of the rigidity along the bar axis. The programming code required 10.5 kB. The relative errors were almost all around 10^{-4} (absolute errors around 10^{-2} cm) for the maximum bar deflection f and for the shortening u.

For the numerical study of the simply supported bar, only two – maximum three – iterations were necessary to realize a precision of

10^{-2} cm (relative error around 10^{-4}), $y'(0)$ being determined with 10^{-4} accuracy. In the case of the hyperstatic bar, the relative errors were around 10^{-3} for the maximum bar deflection and around 10^{-4} for the statically indeterminate reaction R (absolute errors of 10^{-2} cm and 10^{-4} kN respectively).

Let us observe that the difference between the linear and the nonlinear approximation is not significant for the cantilever and the simply supported bar, both for problems indexed by 1 and 2.

The maximum deflection of a cantilever bar is smaller in the nonlinear than in the linear case ($f < \bar{f}$) and, conversely, the rotation of the cross section is smaller in the linear than in the nonlinear case ($\bar{\theta}_{max} < \theta_{max}$). This fact is easily put into evidence comparing tables 5.1 and 5.2 with tables 5.8 and 5.9 accordingly.

Also, the maximum bar deflection at the point x_m of the simply supported bar is smaller in the nonlinear case than that computed at $\bar{x}_m$, corresponding to the linear case ($f < \bar{f}$); we remark that the point x_m is on the left of $\bar{x}_m$ ($x_m < \bar{x}_m$), due to the shortening. On the contrary, the maximum values of the rotation of the cross section at the bar ends are smaller (in absolute value) in the linear than in the nonlinear case ($\bar{\theta}_0 < \theta_0$ and $\left|\bar{\theta}(l)\right| < \left|\theta(l-u)\right|$).

The problem (c_1) for the cantilever bar allows a solution in closed form. Indeed, $f(x) = \dfrac{1}{\rho} = \text{const}$, and the bar displacement is an arc of circle. The numerical results can be obtained directly with the formulae

$$\theta_{max} = \frac{1}{\rho}, \quad u = l - \rho \sin\theta_{max}, \quad f = \rho\left(1 - \cos\theta_{max}\right). \tag{5.1.142}$$

Table 5.14 presents the numerical comparison between the solution of the simplified model of the cantilever bar, expressed by formulae (5.1.30), and the LEM approximate solution.

For the hyperstatic bar too, we observe that there are no significant numerical differences, when comparing the nonlinear approximations with the solution of the linear models, except for case (b). A higher value of the reaction R is noticed in the nonlinear cases (a) and (c) than the values of R obtained in their linear correspondents.

Table 5.14

Magnitude	u cm	f cm	θ_{max}
Exact solution	2.40	18.82	$21^0 49' 37''$
LEM solution	2.39	18.62	$21^0 44' 01''$

On the contrary, problem (b) leads to a greater value of the maximum deflection in the linear case. For problems (a) and (c) we have similarly, $x_{\mathrm{m}} < \bar{x}_{\mathrm{m}}$. The bar deflection in the case (b) can be explained, from the mechanical point of view, by an increased rigidity, due to the variation of the height of the bar cross section.

All the given examples correspond to moderate displacements and rotations of the cross section of a straight bar. For greater values of these magnitudes, one must modify accordingly the intervals of approximation of the solution; however, this does not affect structurally the method based on LEM presented here.

One can also study other models of straight bars; this method represents a sufficiently general tool, able to treat any physical phenomenon associated to the B.-E. bar. In what follows, we shall see that LEM not only offers a basis for a numerical study of the straight bar problems, but it also provides instruments for their qualitative study.

5.1.7. COMPARISON BETWEEN DIFFERENTIAL AND INTEGRAL LEM FORMULAE

The LEM formula (5.1.33) of the derivative of the solution of the B.-E. problem (5.1.1), (5.1.5) is valid for any Cauchy data and at any point $x_0 \in [0, l]$.

The formula of y can be obtained by direct integration from (5.1.33)

$$y(x) = \alpha + \int_{x_0}^{x} y'(t)\mathrm{d}t\,. \tag{5.1.143}$$

Yet, in the applications presented here we shall use only the particular case $x_0 = 0$; hence we take $x_0 = 0$ in the formulae (5.1.34), (5.1.35), (5.1.143), φ being given by (5.1.36).

5.1.7.1. Applications

We shall compare the numerical values obtained by using algorithms based on differential LEM formulae with those obtained by direct pointwise application of the LEM formulae of integral type; this comparison is realized for the cantilever and simply supported bar, in the case of a constant bending moment $M = 40\,\mathrm{kN\,cm}$ [125, 126].

We considered, as before, elastic bars made of steel, of constant rectangular cross section, or of heights h or width b varying linearly along the bar axis.

We also keep the same basis of comparison: the bar length $l = 100\,\mathrm{cm}$ and the longitudinal elasticity modulus $E = 21000\,\mathrm{kN/cm}^2$ and we consider the three cases from the previous sections:

(a) a bar of constant height and of width varying linearly along the bar axis, such that $b_0 = 9\,\mathrm{cm}$, $b_1 = \dfrac{b_0}{c}$, $c = 1.5$;

(b) a bar of constant width $b = 6\,\mathrm{cm}$ and of height varying linearly across the bar axis, such that $h_0 = 1.5\,\mathrm{cm}$, $h_1 = \dfrac{h_0}{c}$, $c = 1.5$;

(c) a bar of constant cross section, of dimensions $h = 9\,\mathrm{cm}$, $b = 1\,\mathrm{cm}$.

The expressions of $f(x)$ were already calculated in section 5.1.4; their simplified forms are found in table 5.15.

Table 5.15

Case	(a)	(b)	(c)
$f(x)$	$-\dfrac{1}{\alpha(300-x)}$	$-\dfrac{1}{\alpha^3(300-x)^3}$	$-\alpha$
α	1.3125	$0.032014\,\mathrm{cm}^{-2/3}$	$0.0038095\,\mathrm{cm}^{-1}$

In all the above cases, the shortening was approximated by formula (5.1.43), which it true for small and moderate displacements.

The cantilever bar

From the LEM formula (5.1.33) written for $\lambda = 0$, we get the following approximating expression for the shortening

$$u \cong \frac{1}{2}\int_0^l \varphi^2(x)\left[1+\varphi^2(x)\right]dx, \qquad (5.1.144)$$

with φ defined in (5.1.36).

In all the above mentioned cases, the second term in the integrand is of order 10^{-5}. The maximum bar deflection was computed by using formula (5.1.143), taken at $l-u$, $y'(x)$ being determined from (5.1.33).

The comparison with the previous results is emphasized in table 5.16.

Table 5.16

Case	(a)	(b)	(c)	Method
u cm	1.65 1.43	0.61 0.54	2.87 2.40	LEM *diff.* LEM *int.*
f cm	14.21 14.28	8.39 8.43	18.63 18.82	LEM *diff.* LEM *int.*
y'_{max}	0.316863 0.309663	0.211890 0.211635	0.397949 0.371731	LEM *diff.* LEM *int.*
θ_{max}	17⁰34'53" 17⁰12'14"	11⁰57'48" 11⁰57'06"	21⁰42' 20⁰23'29"	LEM *diff.* LEM *int.*

On the first row there are written, for each physical magnitude, the previous results, corresponding to the considered case, denoted by LEM *diff.* The values obtained by LEM integral formulae are written on the second row and are denoted by LEM *int.*

The simply supported bar

A serious difficulty in this case is to determine $y'(0)$. Formula (5.1.53) gives $y'(0)$ by means of the LEM solution of differential type. Imposing to (5.1.33) the condition $y(l-u)=0$, we obtain another approximating formula for $y'(0)$, this time based on integral LEM formulae [125, 126]

$$y'(0) \cong -\frac{\displaystyle\int_0^l \left(\varphi+\frac{1}{2}\varphi^3+\frac{3}{8}\varphi^5+\frac{5}{16}\varphi^7+\frac{35}{256}\varphi^9\right)dx}{\displaystyle\int_0^l \left(1+\frac{3}{2}\varphi^2+\frac{15}{8}\varphi^4+\frac{35}{16}\varphi^6+\frac{315}{256}\varphi^8\right)dx}, \qquad (5.1.145)$$

that can be easily extended to the following expression

$$y'(0) \cong -\frac{\int_0^l \varphi\left(1-\varphi^2\right)^{-\frac{1}{2}} \mathrm{d}x}{\int_0^l \left(1-\varphi^2\right)^{-\frac{3}{2}} \mathrm{d}x}. \qquad (5.1.146)$$

For small and moderate displacements, one can also use the simplified formula [125]

$$y'(0) \cong -\frac{4p\left(\frac{l}{2}\right)+p(l)}{1+4q\left(\frac{l}{2}\right)+q(l)}, \qquad (5.1.147)$$

p and q being given by

$$\begin{aligned} p(x) &= \varphi(x)+\frac{3}{2}\varphi^3(x)+\frac{15}{8}\varphi^5(x), \\ q(x) &= 1+\frac{1}{2}\varphi^2(x)+\frac{3}{8}\varphi^4(x), \end{aligned} \qquad (5.1.148)$$

with φ defined by (5.1.36). For the shortening u, we deduce

$$u \cong \frac{1}{2}\int_0^l (\varphi(x)+\mu)^2\left[1+(\varphi(x)+\mu)^2\right]\mathrm{d}x, \qquad (5.1.149)$$

where φ and μ are given by (5.1.34).

Let us note that formula (5.1.146) immediately leads to the following inequality

$$|y'(0)| \leq \frac{\|\varphi\|}{\sqrt{1-\|\varphi\|^2}}, \qquad (5.1.150)$$

It should be also mentioned that in all the three cases the ratio in the right member approximately keeps the same value

$$\frac{\|\varphi\|}{\sqrt{1-\|\varphi\|^2}} \cong 0.412. \qquad (5.1.151)$$

Table 5.17

Case	(a)	(b)	(c)	Method
$y'(0)$	0.145109 0.140953	0.084690 0.084087	0.192953 0.183012	LEM *diff.* LEM *int.*
θ_0	$8^0 15'23''$ $8^0 01'23''$	$4^0 50'27''$ $4^0 48'23''$	$10^0 55'16''$ $10^0 22'16''$	LEM *diff.* LEM *int.*
u cm	0.28 0.40	0.08 0.18	0.62 0.62	LEM *diff.* LEM *int.*
$y'(l-u)$	-0.166191 -0.171229	-0.126134 -0.128212	-0.193720 -0.202784	LEM *diff.* LEM *int.*
$\theta(l-u)$	$-9^0 26'09''$ $-9^0 42'53''$	$-7^0 11'20''$ $-7^0 18'22''$	$-10^0 57'43''$ $-11^0 27'05''$	LEM *diff.* LEM *int.*
x_0 cm	51.53 50.22	54.92 54.62	49.73 47.26	LEM *diff.* LEM *int.*
f cm	3.84 3.63	2.56 2.50	4.75 4.27	LEM *diff.* LEM *int.*

Table 5.17 emphasizes a comparison between the numerical values obtained by LEM algorithms based on recurrence formulae of differential type and those obtained by directly computing the LEM integral formulae. The notations are the same as before.

5.1.7.2. Conclusions

The LEM representations of integral type have some important advantages, among which a larger interval of convergence and the release of regularity class of the involved physical magnitudes are of most interest in applications.

The numerical comparison put into evidence a perfect concordance with the previous results, based on differential LEM formulae. By using the integral LEM representations, some new approximation formulae were obtained for the shortening, in the case of the cantilever bar, and for $y'(0)$, in the case of the simply supported bar. Let us also note that this comparison emphasized coincident numerical results, up to 6th decimal place.

The polynomial system (5.1.4), which increased the efficiency of LEM, deserves a special attention. In fact, instead of (5.1.4), we might have considered the intrinsic system

$$\mathcal{P}(w,z)\equiv\begin{bmatrix}\dfrac{\mathrm{d}w}{\mathrm{d}\varphi}-zw^2\\ \dfrac{\mathrm{d}z}{\mathrm{d}\varphi}-w^3\end{bmatrix}=\begin{bmatrix}0\\0\end{bmatrix}, \tag{5.1.152}$$

with φ given by (5.1.36). This system is effectively independent of any physical magnitude, so that it can be considered as a "mathematical hardcore", governing from the abstract frame the mechanical phenomenon.

5.2. THE STABILITY OF A STRAIGHT BAR

The nonlinear study of the bifurcations of the elastic equilibrium of a straight bar involves, in a way, a change of the physical point of view; this change is mainly due to the difficulties of tackling nonlinear problems of any kind – and the B.-E. equation is strongly nonlinear. This aspect gave rise to various models, describing the same phenomenon, e.g. Kirchhoff's analogy between the nonlinear rigid pendulum and the B.-E. bar, as well as to various methods of determining the critical and postcritical values of the magnitudes of interest; among these methods, one can quote Thompson's [129], based on the minimization of some energetic functionals.

By using LEM, a systematic study concerning the critical and postcritical behaviour of the B.-E. bar in various hypostases was elaborated. The results of Chapter 3 can be used to enlighten several aspects of the bar stability problem.

The LEM solutions led to closed form critical and postcritical formulae. Firstly, there were obtained, by LEM too, criticity conditions for two fundamental cases: 1. the cantilever bar and 2. the simply supported bar [101]. Then we elaborated a similar study of criticity for two special cases of B.-E. bar: 3. the bar built-in at one end and simply supported at the other and 4. the cantilever bar, with slight geometric imperfections [96, 100, 102].

To get postcritical formulae, there were used the LEM representations deduced for the cantilever bar. Thus, a direct formula was obtained for f/l (f – the maximum bar deflection, l – the bar length) as a function of the supraunitary ratio P/P_{cr} (P – the axial force, P_{cr} –

Euler's critical loading, which remains unaltered in a nonlinear study [103-107, 126]).

Introducing the expansions of y and P with respect to f in the above mentioned formula, there were obtained two more approximating formulae for the same ratio f/l. The exact values of f/l from the tables, some standard approximating formulae, well known in the literature, on the one hand, and the three LEM formulae, on the other hand, were then compared. Same way, there were obtained postcritical formulae for other magnitudes: the displacement of the bar end along the straight axis and the rotation of the cross section.

It should be mentioned that the result of this comparison was in favour with the LEM formulae, which lead to numerical values very close to the exact ones and much better than other postcritical estimations [103-107].

5.2.1. STATEMENT OF THE PROBLEM. CRITICAL CONDITIONS

Consider a cantilever bar of length l, acted upon by an axial force P. The mathematical model of this problem may be, in general, expressed in nonlinear form

$$\frac{dy}{ds} = \sin\theta, \qquad \frac{d\theta}{ds} = \alpha^2 (f - y), \quad \alpha^2 = \frac{P}{EI}, \tag{5.2.1}$$

where $ds = \sqrt{dx^2 + dy^2}$, Ox is the direction along the central bar axis, O corresponds to the bar left end, Oy is the transversal axis, θ – the rotation of the cross section and EI is the bar rigidity, supposedly constant (as previously, E is the longitudinal elasticity modulus and I – the moment of inertia of the cross section with respect to the bar axis).

From (5.2.1) we immediately get the B.-E. equation

$$y'' - \alpha^2 (f - y)\left(1 + y'^2\right)^{\frac{3}{2}} = 0, \quad y' = \frac{dy}{dx}, y'' = \frac{d^2 y}{dx^2}, \tag{5.2.2}$$

valid for $\theta < \dfrac{\pi}{2}$.

We consider the following criticity problems:

1. *The cantilever bar.* In this case, the criticity problem can be solved by determining the solution of the B.-E. equation (5.2.2) that satisfy the Cauchy conditions

$$y(0) = 0, \quad y'(0) = 0. \tag{5.2.3}$$

2. *The simply supported bar.* This involves the equation

$$y'' + \alpha^2 y \left(1 + y'^2\right)^{\frac{3}{2}} = 0, \tag{5.2.4}$$

whose solution must satisfy the two-point conditions

$$y(0) = 0, \quad y(l) = 0. \tag{5.2.5}$$

3. *The bar built-in at one end and simply supported at the other.* In this case, the displacement $y(x)$ satisfies the B.-E. equation

$$y'' + \left[\alpha^2 y - H(l - x)\right]\left(1 + y'^2\right)^{\frac{3}{2}} = 0, \tag{5.2.6}$$

H being the transversal component of the reaction. The physical conditions also impose

$$y(0) = 0, \quad y'(0) = 0 \tag{5.2.7}$$

and

$$y(l) = 0. \tag{5.2.8}$$

4. *The cantilever bar with slight geometric imperfections.* Suppose that the bar is not perfectly built-in, making a small angle θ_0 with the ideal direction; let $\beta_0 = \tan\theta_0$. The mathematical model for this case is expressed by the B.-E. equation (5.2.2) and the Cauchy conditions

$$y(0) = 0, \quad y'(0) = \beta_0. \tag{5.2.9}$$

These four cases can be studied starting from the more general equation

$$\tilde{y}'' + \alpha^2 \tilde{y} \left[1 + (\tilde{y}' + c)^2\right]^{\frac{3}{2}} = 0, \tag{5.2.10}$$

c being a dimensionless parameter, conveniently chosen for each of the above cases.

The initial conditions will be general too

$$\tilde{y}(x_0) = a, \quad \tilde{y}'(x_0) = b, \tag{5.2.11}$$

given at an arbitrary point $x_0 \in [0, l]$. The parameters a and b will be also specified.

Our purpose is to get convenient LEM solutions for the general nonlinear Cauchy problem (5.2.10), (5.2.11). In order to simplify the computation, we shall replace the equation (5.2.10) with the polynomial ODS

$$\begin{aligned} y' &= z, \\ z' &= -\alpha^2 y w^3, \\ w' &= -\alpha^2 y (z + c) w^2, \end{aligned} \tag{5.2.12}$$

the new unknown functions z and w being defined by

$$z = y', \quad w = \sqrt{1 + (y' + c)^2}\,. \tag{5.2.13}$$

The Cauchy conditions (5.2.11) become

$$y(x_0) = a, \quad z(x_0) = b, \quad w(x_0) = \sqrt{1 + (b + c)^2}\,. \tag{5.2.14}$$

We shall apply LEM to this new Cauchy problem (5.2.12), (5.2.14).

The LEM mapping depends in this case on three parameters and is given by the exponential $v(x, \sigma, \xi, \zeta) = \mathrm{e}^{\sigma y + \xi z + \zeta w}$; the second LEM equivalent ODS reads

$$\begin{aligned} \frac{\mathrm{d}v_\gamma}{\mathrm{d}x} &= \gamma_1 v_{\gamma - \mathbf{e}_1 + \mathbf{e}_2} - \alpha^2 \gamma_2 v_{\gamma + \mathbf{e}_1 - \mathbf{e}_2 + 3\mathbf{e}_3} + \\ &\quad + \gamma_3 \left(\alpha^2 v_{\gamma + \mathbf{e}_1 + \mathbf{e}_2 + \mathbf{e}_3} + c v_{\gamma + \mathbf{e}_1 + \mathbf{e}_3} \right) \\ \mathbf{e}_j &= \left(\delta_i^j \right)_{i = \overline{1,3}}, \qquad |\gamma| \in \mathfrak{N}, \end{aligned} \tag{5.2.15}$$

where γ are indexes $\gamma = (\gamma_1, \gamma_2, \gamma_3)$ with three components, of length $|\gamma| = \gamma_1 + \gamma_2 + \gamma_3$, running through $\mathfrak{N}$. The associated by LEM Cauchy conditions are

$$v_\gamma(x_0) = a^{\gamma_1} b^{\gamma_2} \left[1 + (b + c)^2 \right]^{\frac{\gamma_3}{2}}, \quad |\gamma| \in \mathfrak{N}. \tag{5.2.16}$$

By theorem 3.3, the normal LEM representation of the first three components of the vector $\mathbf{V}(x) = \lfloor v_\gamma(x) \rfloor_{|\gamma| \in \mathfrak{N}}$ coincides with the solution of the polynomial Cauchy problem (5.2.12), (5.2.14).

1. In the case of *the cantilever bar*, we have $x_0 = 0$; introducing the change of function

$$\tilde{y}(x) = y(x) - f, \tag{5.2.17}$$

the equation (5.2.2) becomes

$$\tilde{y}'' + \alpha^2 \tilde{y}\left(1 + \tilde{y}'^2\right)^{\frac{3}{2}} = 0, \tag{5.2.18}$$

and the corresponding Cauchy conditions are

$$\tilde{y}(0) = -f, \quad \tilde{y}'(0) = 0. \tag{5.2.19}$$

Thus, in this case, the linear equivalent system (5.2.15) must be solved in the conditions (5.2.16), for $x_0 = 0, a = -f, b = 0, c = 0$. Passing over some details of computation, we give the fifth order approximation for the solution of the problem of bifurcation of the elastic equilibrium for the cantilever bar

$$y(x) \cong -f(\cos\alpha x - 1) - f^3\alpha^2\varphi(\alpha x) - f^5\alpha^4\psi(\alpha x); \tag{5.2.20}$$

the function $\psi(\alpha x)$ is analytic in αx and $\varphi(\alpha x)$ reads

$$\varphi(\alpha x) = \frac{1}{16}\left[\frac{3}{4}(\cos\alpha x - \cos 3\alpha x) - 3\alpha x \sin\alpha x\right]. \tag{5.2.21}$$

The condition of criticity for the cantilever bar is then obtained imposing that $y(l) = f$ to the above LEM solution, for a non-vanishing *y*; from the mathematical point of view, this means that the problem is reduced to the study of the critical points of $\phi(f) \equiv y(l, f) - f$ as a function depending on the Morse parameter *f*. This yields the following critical points for the cantilever bar acted upon by an axial force

$$\alpha l = (2k - 1)\frac{\pi}{2}, \quad k \in \mathfrak{N}. \tag{5.2.22}$$

Indeed, $\dfrac{\mathrm{d}\phi}{\mathrm{d}f}(0) = \cos\alpha l$, $\dfrac{\mathrm{d}^2\phi}{\mathrm{d}f^2}(0) = 0$, which involves (5.2.22).

2. For *the simply supported bar*, we consider the B.-E. equation (5.2.4) to which we add the Cauchy conditions

$$y(0) = 0, \quad y'(0) = \beta, \tag{5.2.23}$$

β being a real parameter. Thus, in this case, the general problem (5.2.10), (5.2.11) – hence its linear equivalent system – must be solved for $x_0 = 0$, $a = 0$, $b = \beta, c = 0$. We get the normal LEM solution

$$y(x,\beta) \cong \frac{\beta}{\gamma}\sin\gamma x + \frac{3\beta^3}{16\gamma\left(1+\beta^2\right)}\left[\sin 3\gamma x + 3\left(\sin\gamma x - \gamma x\cos\gamma x\right)\right] + \\ + \frac{\beta^5}{\gamma\left(1+\beta^2\right)^2} F(\gamma x, \beta), \tag{5.2.24}$$

where $\gamma = \alpha\left(1+\beta^2\right)^{\frac{3}{4}}$ and F is analytic with respect to β and γx. The Morse parameter is β, introduced in order to complete the Cauchy problem.

Formula (5.2.24) serves to deduce the bifurcation points for the simply supported bar. Indeed, applying the second two-point condition (5.2.8), we obtain $y(l,\beta) \equiv \chi(\beta) = 0$, which yields

$$\left.\frac{d\chi}{d\beta}\right|_{\beta=0} = \frac{\sin\alpha l}{\alpha}, \quad \left.\frac{d^2\chi(\beta)}{d\beta^2}\right|_{\beta=0} = 0. \tag{5.2.25}$$

Hence $\beta = 0$ is a critical point if $\sin\alpha l = 0$, which means

$$\alpha l = k\pi, \quad k \in \mathfrak{N}. \tag{5.2.26}$$

Remark. As mentioned above, the stability problem for the simply supported bar can be tackled by using energetic methods, e.g., as in [129], where Thompson and Hunt deduced the Euler-Lagrange equations for the study of the displacement of a straight bar. By using these equations, they obtained the coefficients of a series expansion of the solution y with respect to f. By Thompson's method, the development for the coefficient of f^3 is determined separately, for each criticity condition. By LEM, a global expansion for the coefficient of β^3 is straightforwardly obtained.

3. *The bar built-in at one end and simply supported at the other* is modeled, as previously shown, by the problem (5.2.6), (5.2.7), (5.2.8). We firstly use the change of function

$$\tilde{y}(x) = y(x) - \frac{H}{\alpha^2}(l - x), \tag{5.2.27}$$

that brings (5.2.6) into the form (5.2.10), with c given by

$$c = -\frac{H}{\alpha^2}. \tag{5.2.28}$$

The new unknown function $\tilde{y}(x)$ satisfies the conditions

$$\tilde{y}(0) = cl, \quad \tilde{y}'(0) = -c, \tag{5.2.29}$$

as well as

$$\tilde{y}(l) = 0. \tag{5.2.30}$$

As it can be easily seen, null conditions save a lot of computation; hence, we shall represent $\tilde{y}(x)$ around l, also adding the parametric Cauchy condition

$$\tilde{y}'(l) = \vartheta. \tag{5.2.31}$$

This means for $y(x)$

$$y'(l) = \vartheta - \frac{H}{\alpha^2}. \tag{5.2.32}$$

The general problem (5.2.10), (5.2.11) – therefore, its linear equivalent – must be solved in this case for $x_0 = l, a = 0, b = \vartheta, c = -\dfrac{H}{\alpha^2}$. After solving the finite linear systems, we find the following normal LEM solution for $y(x)$

$$\begin{aligned} y(x) &\cong -c(l - x) - \frac{\vartheta}{\gamma} \sin \gamma (l - x) + \\ &+ \frac{\vartheta^3}{\gamma\left(1 + (\vartheta + c)^2\right)} \Phi\left(\alpha(l - x), \vartheta, \sqrt[4]{1 + (\vartheta + c)^2}\right), \end{aligned} \tag{5.2.33}$$

in which $\gamma = \alpha\left(1 + (\vartheta + c)^2\right)^{\frac{3}{4}}$ and Φ is analytic with respect to its arguments.

By using this representation, we get $y(0)$ as a function of the parameters ϑ and c. As $y(0, \vartheta, c)$ must vanish by virtue of (5.2.7), the Morse criticity conditions are obtained by computing the Jacobian

$$J \equiv \left.\frac{D(y(0, \vartheta, c), y'(0, \vartheta, c))}{D(\vartheta, c)}\right|_{\substack{\vartheta=0 \\ c=0}} = 0; \tag{5.2.34}$$

we note that from the normal LEM solution (5.2.33) it follows that the second partial derivatives of $y(0, \vartheta, c)$ and $y'(0, \vartheta, c)$ with respect to ϑ and c are null for $\vartheta = 0,\ c = 0$. Computing J, we find

$$J = \begin{vmatrix} -\dfrac{\sin \alpha l}{\alpha l} & -1 \\ \cos \alpha l & 1 \end{vmatrix} = 0. \tag{5.2.35}$$

Thus, the critical points in this case are found by solving the functional equation

$$\tan \alpha l = \alpha l. \tag{5.2.36}$$

4. The mathematical model of *the cantilever bar with slight geometric imperfections* can also be brought to the form (5.2.10), (5.2.11) with the change of function

$$\tilde{y} = y - f. \tag{5.2.37}$$

In this case, $c = 0$. The Cauchy conditions (5.2.9) become

$$\tilde{y}(0) = -f, \quad \tilde{y}'(0) = \beta_0. \tag{5.2.38}$$

Thus, in this case, the nonlinear Cauchy problem (5.2.10), (5.2.11) must be solved for $x_0 = 0, a = -f,\ b = \beta_0, c = 0$. Applying the LEM techniques, we obtain the approximation of the parametric normal LEM solution

$$\begin{aligned} y(x) \cong {} & \frac{\beta_0}{\gamma_0} \sin \gamma_0 x + \\ & + \frac{9\beta_0^2}{16\gamma_0\left(1+\beta_0^2\right)}\left[\frac{1}{12}\left(\sin 3\gamma_0 x + 9 \sin \gamma_0 x\right) - \gamma_0 x \cos \gamma_0 x\right] + \\ & + f\left[1 - \cos \gamma_0 x - \frac{9\beta_0^2}{16\left(1+\beta_0^2\right)}\left(\gamma_0 x - \sin \gamma_0 x \cos \gamma_0 x\right) \sin \gamma_0 x\right], \end{aligned} \tag{5.2.39}$$

where $\gamma_0 = \alpha\left(1+\beta_0^2\right)^{\frac{3}{4}}$. Further, $y(l) = f$ implies

$$\frac{f}{f_0} = \frac{\tan \gamma_0 l}{\gamma_0 l} \frac{1 + \dfrac{9\beta_0^2}{16\left(1+\beta_0^2\right)}\left(1 - \dfrac{1}{3}\sin^2 \gamma_0 l - \gamma_0 l \cot \gamma_0 l\right)}{1 + \dfrac{9\beta_0^2}{16\left(1+\beta_0^2\right)}\left(\gamma_0 l - \sin \gamma_0 l \cos \gamma_0 l\right) \tan \gamma_0 l}. \tag{5.2.40}$$

In the above formula, $f_0 = \beta_0 l$ is the displacement due to the imperfect building-in. Observing that

$$\alpha l = \frac{\pi}{2}\sqrt{\frac{P}{P_{\mathrm{E}}}}, \tag{5.2.41}$$

in which Euler's loading P_{E} is

$$P_{\mathrm{E}} = \frac{\pi^2 EI}{4l^2}, \tag{5.2.42}$$

we claim that formula (5.2.40) gives a representation of the dimensionless ratio $\frac{f}{f_0}$ as a function of the ratio $\frac{P}{P_{\mathrm{E}}}$. Neglecting β_0^2 with respect to 1, we get the classical result obtained in the linear frame

$$\frac{f}{f_0} = \frac{\tan \alpha l}{\alpha l}, \tag{5.2.43}$$

leading to the critical value

$$P_{\mathrm{cr}} = P_{\mathrm{E}}, \tag{5.2.44}$$

corresponding to $\alpha l = \frac{\pi}{2}$.

Writing (5.2.40) in the form

$$\begin{aligned} &\frac{f_0}{f}\left\{1 + \frac{9\beta_0}{16\left(1+\beta_0^2\right)}\left[\beta_0\left(1 - \frac{1}{3}\sin^2\gamma_0 l\right) - \frac{f_0}{f}\gamma_0 l\left(\gamma_0 l - \sin\gamma_0 l\cos\gamma_0 l\right)\right]\right\} = \\ &= \frac{\gamma_0 l}{\tan\gamma_0 l}\left[1 + \frac{9\beta_0}{16\left(1+\beta_0^2\right)}\frac{f_0}{f}\right], \end{aligned} \tag{5.2.45}$$

we observe that the maximum deflection f, which by no means can exceed the bar length l, tends to infinity for $\gamma_0 l \to \frac{\pi}{2}$. It results that the critical load reads

$$P_{\mathrm{cr}} = \frac{P_{\mathrm{E}}}{\left(1+\beta_0^2\right)^{\frac{3}{2}}}, \tag{5.2.46}$$

or, approximately,

$$P_{\mathrm{cr}} \cong \frac{P_{\mathrm{E}}}{1+\frac{3}{2}\beta_0^2}. \tag{5.2.47}$$

COMMENT. This study of criticity emphasizes several significant facts.

We remark that in the first three cases 1, 2 and 3 the results coincide with those obtained in the linear case [68]; but the same criticity conditions are also obtained by standard linearization.

Yet, the case of a cantilever bar is usually treated starting from the simply supported bar and then using the symmetry of the structure [129]. On the contrary, the normal LEM representations (5.2.20), (5.2.17), (5.2.24) are independent from each other, such that the cases 1 and 2 can be studied separately, without restrictions concerning the bar symmetry or geometry. This advantage enabled the nonlinear study of criticity for the more complicated cases 3 and 4.

The cases 1, 2 and 3 correspond to a loss of stability by bifurcation.

In the case 4, things change. Although we neglected higher order terms in β_0 in formulae (5.2.39), (5.2.40) to the purpose of simplifying computation, it is not less true that by LEM a more complete image of the criticity could be obtained than that obtained by linearization: the postcritical bar aspect was obtained, as well as the loss of stability by divergence.

The case $\beta_0 = 0$ was actually separately treated as case 1. It should be mentioned that putting $\beta_0 = 0$ in (5.2.46), or even in (5.2.47), we obtain the value of Euler's critical loading corresponding to the case 1.

5.2.2. THE POSTCRITICAL BEHAVIOUR OF THE CANTILEVER BAR

In this postcritical study, we shall directly refer to the nonlinear ODS (5.2.1), that must be solved in null Cauchy conditions

$$y(0)=0, \quad \theta(0)=0, \tag{5.2.48}$$

We firstly apply the translation (5.2.17) and then the LEM mapping, depending in this case on two parameters

$$v(x,\sigma,\xi)=\mathrm{e}^{\sigma\tilde{y}(s)+\xi\theta(s)}; \tag{5.2.49}$$

this leads to the first linear equivalent of (5.2.1)

$$\frac{\partial v}{\partial x} = \sigma \sin \mathrm{D}_{\xi} v - \alpha^2 \xi \frac{\partial v}{\partial \sigma}, \qquad (5.2.50)$$

consistent on the Exp-type spaces, previously introduced in chapter 1.

By $\sin \mathrm{D}_{\xi}$, we mean the linear operator obtained by formally replacing the powers of θ from the McLaurin expansion of $\sin\theta$ with derivatives of the same order in σ and ξ.

Considering for v a series expansion in σ and ξ, we get the second LEM equivalent in the form of the linear first order ODS in s

$$\frac{\mathrm{d}v_{ij}}{\mathrm{d}s} = i\sum_{k=1}^{\infty} \frac{(-1)^{k+1}}{(2k-1)!} v_{i-1,j+2k-1} - j\alpha^2 v_{i+1,j-1}. \qquad (5.2.51)$$

Applying theorem 3.3, we get the following normal LEM representation

$$y(x) \cong -f(\cos\alpha x - 1) - f^3\alpha^2\Phi(\alpha x) - f^5\alpha^4\Psi(\alpha x), \qquad (5.2.52)$$

where $\Psi(\alpha x)$ is analytic in αx and $\Phi(\alpha x)$ reads

$$\Phi(\alpha x) = \frac{1}{16}\left[\frac{1}{4}(\cos 3\alpha x - \cos\alpha x) + \alpha x \sin\alpha x\right]. \qquad (5.2.53)$$

5.2.2.1. Postcritical formulae for the ratio f/l

The condition $y(l) = f$ implies the fact the bar length is conserved (it still remains l), if we neglect the shortening in the postcritical study.

Applying it to the LEM solution (5.2.52), we obtain

$$\cos\alpha l + (\alpha f)^2 \Phi(\alpha x) \cong 0, \qquad (5.2.54)$$

an approximate relation, depending on the parameters f and α.

This involves

$$\alpha f = \sqrt{-\frac{\cos\alpha l}{\Phi(\alpha x)}}; \qquad (5.2.55)$$

writing αf in the form $\alpha f = \frac{f}{l}\cdot(\alpha l)$, we get

$$\frac{f}{l} \cong \frac{4}{\alpha l \sqrt{\sin^2 \alpha l - \alpha l \tan \alpha l}} = \frac{4}{\alpha l} \sqrt{\frac{2 \cot \alpha l}{\sin 2\alpha l - 2\alpha l}} ; \qquad (5.2.56)$$

these relations hold true for $\frac{\pi}{2} < \alpha l < \pi$.

If $\alpha l = \frac{\pi}{2}$, i.e., if $\cot \frac{\pi}{2} = 0$, as $\alpha^2 = \frac{P}{EI}$, we obtain the first critical value of the axial force P in a nonlinear treatment – formula (5.2.42).

Let us note that, according to (5.2.41), formula (5.2.56) is in fact a direct LEM representation for the postcritical values of the ratio $\frac{f}{l}$ as a function of the supraunitary ratio $\frac{P}{P_{\text{cr}}}$.

Obviously, there is a natural connection between α and f that can be expressed in various ways, e.g. α allows the expansion

$$\alpha = \alpha_0 + \alpha_1 f + \alpha_2 \frac{f^2}{2!} + \ldots \equiv \sum_{j=0}^{\infty} \alpha_j \frac{f^j}{j!} . \qquad (5.2.57)$$

Introducing this in (5.2.54), a power series with respect to f is emphasized that must be identically null, hence all its coefficients must vanish. Cancelling the free term, therefore making $j = 0$, it is obtained

$$\cos \alpha_0 l = 0, \quad \alpha_0 l = (2k - 1)\frac{\pi}{2}, \quad k \in \mathfrak{N} ; \qquad (5.2.58)$$

this correspond to the criticity condition for the cantilever bar.

Further, determining the next two coefficients, we find $\alpha_1 = 0$, $\alpha_2 = \frac{\alpha_0^3}{8}$, which involves

$$\frac{\alpha}{\alpha_0} \cong 1 + 16\alpha_0^2 f^2 . \qquad (5.2.59)$$

Getting back to the criticity condition (5.2.58) for $k = 1$ and to the formula (5.2.41), also writing $\alpha_0 f = \frac{f}{l} \cdot (\alpha_0 l)$, we finally deduce

$$\frac{f}{l} \cong \frac{8}{\pi}\sqrt{\sqrt{\frac{P}{P_{cr}}} - 1}\,. \tag{5.2.60}$$

This is another approximating formula for the postcritical values of the ratio $\frac{f}{l}$, indirectly obtained by LEM.

The development (5.2.57) was previously used by Thompson and Hunt [129].

The connection between the dimensionless magnitudes αl and αf may be also expressed by the expansion

$$(\alpha f)^2 = \sum_{j=0}^{\infty} p_j \frac{(\alpha - \alpha_0)^j l^j}{j!}; \tag{5.2.61}$$

introducing it in formula (5.2.54), we again obtain a series in αl, whose coefficients must vanish. Stopping it at $j = 1$, we get the following approximating formula for αf

$$(\alpha f)^2 \cong 16(\alpha - \alpha_0)l; \tag{5.2.62}$$

taking now (5.2.54) into account, the very Schneider's formula [77] is obtained

$$\frac{f}{l} \cong \frac{8}{\pi}\sqrt{\frac{P_{\mathrm{cr}}}{P}}\sqrt{\sqrt{\frac{P}{P_{\mathrm{cr}}}} - 1}\,. \tag{5.2.63}$$

This formula was deduced by Schneider by a completely different method.

Anyway, it is seen that (5.2.63) is a second approximating formula for the postcritical behaviour of the cantilever bar, also deduced by LEM.

A well known in the literature postcritical formula is Grashof's [35]

$$\frac{f}{l} \cong \frac{8}{\pi}\sqrt{\frac{P}{P_{\mathrm{cr}}}}\sqrt{\sqrt{\frac{P}{P_{\mathrm{cr}}}} - 1}\,, \tag{5.2.64}$$

that was established starting from the integral form (5.1.37), written by using elliptic integrals

$$\frac{f}{l}=\frac{2k}{K(k)}, \quad k=\frac{\sin\theta l}{2}, \quad K(k)=\frac{\pi}{2}\sqrt{\frac{P}{P_{\mathrm{cr}}}}. \tag{5.2.65}$$

In the preceding formula, $K(k)$ is the complete elliptic integral of first kind, that was approximated by the first two terms of its power expansion

$$K(k)\cong\frac{\pi}{2}\left(1+\frac{k^2}{4}\right). \tag{5.2.66}$$

5.2.2.2. Other postcritical formulae

As $\left(\frac{\mathrm{d}x}{\mathrm{d}s}\right)^2+\left(\frac{\mathrm{d}\tilde{y}}{\mathrm{d}s}\right)^2=1$ and

$$\frac{\mathrm{d}}{\mathrm{d}s}\sqrt{1-\left(\frac{\mathrm{d}\tilde{y}}{\mathrm{d}s}\right)^2}=\frac{-\frac{\mathrm{d}\tilde{y}}{\mathrm{d}s}\frac{\mathrm{d}^2\tilde{y}}{\mathrm{d}s^2}}{\sqrt{1-\left(\frac{\mathrm{d}\tilde{y}}{\mathrm{d}s}\right)^2}}=\frac{-\frac{\mathrm{d}\tilde{y}}{\mathrm{d}s}\left(-\alpha^2\tilde{y}\right)\sqrt{1-\left(\frac{\mathrm{d}\tilde{y}}{\mathrm{d}s}\right)^2}}{\sqrt{1-\left(\frac{\mathrm{d}\tilde{y}}{\mathrm{d}s}\right)^2}}, \tag{5.2.67}$$

we have

$$\frac{\mathrm{d}x}{\mathrm{d}s}=1+\frac{\alpha^2}{2}\left(\tilde{y}^2-f^2\right), \tag{5.2.68}$$

where $\tilde{y}(0)=-f$.

From the normal LEM representation (5.2.52) it results

$$\tilde{y}(s)\cong -f\cos\alpha s-f^3\alpha^2\Phi(\alpha x), \tag{5.2.69}$$

with Φ given by (5.2.53). Thus, we obtain a first approximating formula for x

$$\frac{x(l)}{l}\cong 1-\frac{1}{4}(\alpha f)^2\left(1-\frac{\sin 2\alpha l}{\alpha l}\right). \tag{5.2.70}$$

Replacing here $\frac{f}{l}$ from (5.2.63), we deduce

$$\frac{x(l)}{l} \cong 1 + \frac{4}{\alpha l}\cot \alpha l\,. \tag{5.2.71}$$

Denote by $\delta = l - x(l)$ the displacement at the bar free end along its axis. The above formula becomes

$$\frac{\delta}{l} \cong -\frac{4}{\alpha l}\cot \alpha l\,. \tag{5.2.72}$$

A better approximation is

$$\begin{aligned}\frac{\delta}{l} \cong \frac{1}{4}(\alpha f)^2\left(1 - \frac{\sin 2\alpha l}{2\alpha l}\right) + \\ + \frac{1}{128}(\alpha f)^4\left[1 + 2\cos 2\alpha l - \frac{\sin 2\alpha l}{\alpha l}(2 + \cos 2\alpha l)\right];\end{aligned} \tag{5.2.73}$$

again taking into account (5.2.56), it results

$$\frac{\delta}{l} \cong -\frac{4}{\alpha l}\cot \alpha l\left[1 - \frac{(1 + \cos 2\alpha l)\cot \alpha l}{2\alpha l - \sin 2\alpha l}\right] + \frac{2\alpha l \sin 2\alpha l}{(2\alpha l - \sin 2\alpha l)^2}, \tag{5.2.74}$$

or

$$\frac{\delta}{l} \cong -\frac{8\cot \alpha l}{2\alpha l - \sin 2\alpha l}\left(1 + \frac{\sin 2\alpha l}{2\alpha l - \sin 2\alpha l} - \frac{3\cot \alpha l}{2\alpha l}\right). \tag{5.2.75}$$

It is noticed that formula (5.2.72) is a first approximation of (5.2.75). Numerical estimations show that the trigonometric expression put in brackets in the right member of (5.2.75) is much less than 1

$$\frac{\sin 2\alpha l}{2\alpha l - \sin 2\alpha l} - \frac{3\cot \alpha l}{2\alpha l} << 1, \tag{5.2.76}$$

such that we finally get an approximating formula, simpler than (5.2.75) and better than (5.2.72):

$$\frac{\delta}{l} \cong -\frac{8\cot \alpha l}{2\alpha l - \sin 2\alpha l}\,. \tag{5.2.77}$$

This last result can be compared with the exact solution

$$\frac{\delta}{l} \cong 2\left(1 - \frac{E(k)}{K(x)}\right), \tag{5.2.78}$$

where the ratio $\frac{\delta}{l}$ is expressed by the complete elliptic integrals of first and second kind respectively.

Using the first two terms of the expansions of both elliptic integrals,

$$E(k) \cong \frac{\pi}{2}\left(1 - \frac{k^2}{4}\right), \quad K(k) \cong \frac{\pi}{2}\left(1 + \frac{k^2}{4}\right), \tag{5.2.79}$$

a Grashof type formula is put into evidence:

$$\frac{\delta}{l} \cong 4\left(1 - \sqrt{\frac{P_{cr}}{P}}\right). \tag{5.2.80}$$

Further, integrating with respect to s the second equation (5.2.1), in which $y(s)$ was previously replaced by the normal LEM representation (5.2.52), we deduce, after some elementary computation

$$\frac{\theta(l)}{\alpha l} \cong \frac{f}{l} \operatorname{cosec} \alpha l \left(1 - \frac{2}{3} \frac{\sin 2\alpha l \sin^2 \alpha l}{2\alpha l - \sin 2\alpha l}\right); \tag{5.2.81}$$

neglecting the terms put in brackets, we get a simpler expression

$$\frac{\theta(l)}{\alpha l} \cong \frac{f}{l} \operatorname{cosec} \alpha l\,. \tag{5.2.82}$$

5.2.2.3. Numerical results

We shall compare the exact formula (EF) (5.2.65) of the ratio $\frac{f}{l}$ with the direct LEM approximating formula (5.2.56), denoted by LEM, as well as with the indirect LEM approximations (5.2.60), (5.2.63), denoted by LEM_1 and LEM_2 (Schneider's) accordingly. The last row of the table contains the values of $\frac{f}{l}$ computed by Grashof's formula (5.2.64), denoted by G.

The results of this comparison were realized for $1 < \frac{P}{P_{\mathrm{cr}}} < 1.3$ and they are presented in the table 5.18, being ordered following their precision. Thus, the mean square errors with respect to the exact solution are: 0.24% for LEM, 1.36% for LEM_1, 2.67% for LEM_2 and 4.22% for G.

Table 5.19 contains the values from the tables of $\frac{\delta}{l}$ expressed by elliptic integrals, denoted by FE, compared with those computed by the LEM formula (5.2.75), with its first approximation (5.2.71) denoted by LEM_1 and with the Grashof type formula (5.2.78), denoted by G. The corresponding mean square errors are: 0.30% for LEM, 1.93% for LEM_1 and 1.73% for G.

Table 5.18

Comparison among the values of f/l, computed by using three LEM variants, Grashof's formula and elliptic integrals

P/P_{cr} \ f/l	1.004	1.015	1.035	1.063	1.102	1.152	1.215	1.293
EF	0.110	0.220	0.324	0.422	0.514	0.594	0.662	0.720
LEM	0.110	0.220	0.324	0.422	0.516	0.601	0.676	0.741
LEM_1	0.116	0.220	0.329	0.435	0.541	0.642	0.738	0.829
LEM_2	0.114	0.220	0.335	0.448	0.563	0.689	0.814	0.942
G	0.114	0.221	0.341	0.462	0.596	0.740	0.898	1.072

Table 5.19

Comparison among the values of δ/l, computed by using two LEM variants, Grashof's formula and elliptic integrals

P/P_{cr} \ δ/l	1.004	1.015	1.035	1.063	1.102	1.152	1.215	1.293
FE	0.008	0.030	0.067	0.119	0.183	0.259	0.345	0.440
LEM	0.008	0.030	0.067	0.117	0.181	0.257	0.343	0.439
LEM_1	0.008	0.030	0.068	0.120	0.189	0.274	0.374	0.489
G	0.008	0.030	0.068	0.120	0.189	0.273	0.371	0.482

For $\theta(l)$ we set up table 5.20, where $\theta(l)$ was successively computed with formula (5.2.79), denoted by LEM, and with its first approximation, formula (5.2.80), denoted by LEM_1; the results were compared with the numerical values of elliptic integrals from tables, denoted by FE. The corresponding mean square errors are: 0.10% for LEM and 0.90% for LEM_1.

Table 5.20

Comparison among the values of $\theta(l)$, computed by using two LEM variants and numerical tables for elliptic integrals

P/P_{cr} \ $\theta(l)$	1.004	1.015	1.035	1.063	1.102	1.152	1.215	1.293
FE	10^0	20^0	30^0	40^0	50^0	60^0	70^0	80^0
LEM	$9^056'$	20^0	30^0	40^0	50^0	$60^010'$	$70^008'$	$80^027'$
LEM_1	$9^055'$	$19^057'$	$29^041'$	$39^012'$	$48^054'$	$58^027'$	$67^056'$	$77^037'$

In the last two tables, the comparison was also realized for $1 < \frac{P}{P_{cr}} < 1.3$.

5.2.2.4. Conclusions

From table 5.18 it ensues that LEM led to a simple formula that gives the values of $\frac{f}{l}$ with a satisfactory precision (the greatest error is under 2%); anyway, the LEM formula is better that Grashof's. Even the approximations LEM_1 and LEM_2 (this last one coincides with Schneider's formula) are more precise than Grashof's formula. Similar conclusions are valid also for $\frac{\delta}{l}$ (table 5.19) and for $\theta(l)$ (table 5.20).

The above values were computed only on the interval $1 < \frac{P}{P_{cr}} < 1.3$, corresponding to $\alpha l \in \left[\frac{\pi}{2}, 1.14\frac{\pi}{2}\right]$, or, numerically, to $\alpha l \in [1.57, 1.79]$. We made this choice not only because this interval is of interest, but also because the results obtained here are valid for $|\theta| < \frac{\pi}{2}$. For $\theta > \frac{\pi}{2}$, one can use in the system (5.2.1) a series expansion of $\sin\theta$, truncated up to 7th order; this leads to a polynomial system that can be also treated by LEM. The results obtained on this way will be consistent on the larger interval $\theta \in [0, \pi)$.

One can say that the method based on LEM that was proposed in this section for the study of the critical and postcritical study of the cantilever bar had several practical consequences; thus, some direct approximating postcritical formulae for $f/l, \delta/l$ and $\theta(l)$ were obtained, as well as critical values for loads, in various physical hypotheses.

It should be also mentioned that this method based on LEM does not depend on a certain mechanical interpretation. Using the same pattern, one can obtain similar results for various cases of load and support.

5.3. WOBBLE SOLITONS FOR HEAVY ELASTICA

The buckling of a bar acted upon by its own weight can be described in the frame of elastica only if the bar is sufficiently thin. Euler was the first to study the small deflections and the stability of a heavy elastica (column) [27]; Greenhill obtained the minimum unstable height for a column of given density and rigidity [36]. There are also recent studies of large deflections of heavy structures [10][170].

In [20], Veturia Chiroiu and Ligia Munteanu show, by using LEM, that a heavy cantilever bar, homogeneous and of constant cross section, allows wobble solitons. To prove this, they firstly establish the following mathematical model

$$\frac{d^2\theta}{d\sigma^2} = K^3(1-\sigma)\sin\theta, \quad \sigma = \frac{s}{L}, \tag{5.3.1}$$

where s is the arc length and $\theta(s)$ – the local angle of inclination.

The constant K reads

$$K = \left(\frac{\rho L^3}{EI}\right)^{\frac{1}{3}}, \tag{5.3.2}$$

where EI is the flexural rigidity, L – the bar length and ρ – its density.

We see that K represents the relative importance of density and length to flexural rigidity [4]. So, small $K << 1$ means a relatively rigid, and large $K >> 1$ means a relatively compliant material [345]. The boundary value conditions are

$$\theta(0) = \alpha, \quad \frac{d\theta}{d\sigma}(1) = 0. \tag{5.3.3}$$

Equation (5.3.1) is a second order nonlinear ODE with variable coefficients; it can be transformed in a nonlinear first order ODS with constant coefficients, having three unknown functions.

To this system one can apply LEM, more precisely, corollary 1.3. To compute the exponential matrix, the authors use Ramakrishna and Costa's technique, specially deduced for a class of 4×4 matrices; this method does not require spectral information, because it is based rather upon the matrix equation associated to the characteristic polynomial. This method can be extended to arbitrary 4×4 matrices, as well as to matrices with arbitrary dimensions.

Using Ramakrishna-Costa's method, the following LEM representation of the solution of the problem (5.3.1), (5.3.3) was established [20]

$$\begin{aligned}
\theta(s) = {} & \alpha + f^1(s)\frac{\sin\alpha}{6} + f^2(s)\frac{\sin\alpha\cos\alpha}{180} \\
& + f^3(s)\frac{\sin^3\alpha}{72} - f^4(s)\frac{\sin\alpha\cos^2\alpha}{60} \\
& + f^5(s)\frac{\sin^2\alpha\cos\alpha}{60} \\
& + 4\,\mathrm{Im}\left\{\ln\left[\mathrm{f}^6(s)\cosh\left(\frac{2}{3}\alpha K^{\frac{3}{2}}\right) + i f^7(s)\cosh\left(\frac{4}{3}\alpha K^{\frac{3}{2}}\right)\right]\right\} + \ldots
\end{aligned} \tag{5.3.4}$$

where

$$\begin{aligned}
& f^1(s) = 3kK^3 s(s-L)^2, \quad f^2(s) = kK^6\left(15s^4 - 15s^3 + 9s\right)(s-L)^2, \\
& f^3(s) = kK^9\left(\frac{s^7}{2} - \frac{9s^6}{10} + \frac{s^5}{4} - \frac{3s^4}{4} + \frac{3s}{4}\right)(s-L)^2, \\
& f^4(s) = kK^9\left(\frac{s^7}{6} - \frac{s^6}{3} + \frac{s^5}{4} - \frac{s^4}{4} + \frac{s^3}{2} - \frac{3s}{4}\right)(s-L)^2,
\end{aligned} \tag{5.3.5}$$

and

$$f^5(s) = (k-1)K^3\sqrt{3}\left(\frac{s^7}{504} + \frac{s^6}{180} + \frac{s^4}{12} + \frac{s^3}{6} + s\right)(s-L)^2,$$
$$f^6(s) = -3(k-1)s\,(s-L)^2\,\mathrm{e}^{-s}. \tag{5.3.6}$$

It can be seen that the LEM solution contains a function of the form $F = 4\,\mathrm{Im}(\ln(F + \mathrm{i}G))$, similar to the wobble function, called by Kälbermann wobble soliton.

By inverse scattering, Kälbermann [58] obtained another solution of the sine-Gordon equation, oscillatory and apparently stable, called by him *wobble soliton*, composed of a kink and a breather with its centre at rest; its behaviour is mainly due to the so called *wobble function*, present in the solution. The sine-Gordon equation, discovered at the end of 19th century, models many physical and mechanical phenomena (e.g., bending of thin bars, crystal dislocations, Bloch wall dynamics in ferromagnetics and ferroelectrics, fluxon propagation in long superconducting junctions, self-induced transparency in nonlinear optics, elementary particles, etc.)

The presence of the wobble function in the LEM solution certifies in fact the existence of the wobble soliton in the case of large cantilever bars; the well known results of Greenhill [36] and Wang [170] are particular cases of those obtained in [20].

The LEM solution (5.3.4) is fitted for a qualitative study of the phenomenon. It is represented, for arbitrary values of *K*, by wobble solitons, varying with *s* and α. The wobble soliton describes a flipping process of the system. This means that the cantilever is flipped to the other side since the torque changes sign. In the flipping process energy is dissipated through oscillations [170]. Wang has studied the asymptotic solutions for large K. His solutions, expressed in terms of polynomials and exponentials with respect to *s*, are contained in the wobble expression (5.3.4).

For $K < 0.5$, the solution is stable and θ differs only very little from α. The solution is stable and unique for $K \le 1.98653$; if this critical number is exceeded, then small variations of *K* will produce large variations in the shape of the solution.

In the neighbourhood of $\alpha = \pi$, three solutions occur for $1.98653 < K \le 3.82557$ and five – for $3.82557 < K \le 5.29566$; this result is concordant with Wang's too [170]. For $K >> 1$ there exist two exact solutions, given for $\theta = \alpha = 0$ and $\theta = \alpha = \pi$. The last one was also analysed by using the LEM solution [20].

5.4. THE NONLINEAR ELASTIC FRAME

The buckling of a two-bar frame may be considered a fundamental problem of the statics of constructions. For this reason, many papers in the literature, both theoretical and experimental, were dedicated to this subject [75]. The most relevant results are Koiter's [43], who set up a linear model, confirmed experimentally by Roorda [75].

Yet, because of the lack of an adequate mathematical tool, many of the models set up for this problem were significantly simplified, at the cost of accuracy.

In this section, a nonlinear model proposed by P.P.Teodorescu is presented, for a frame composed of two bars: one horizontal and one vertical; a qualitative study of this model was possible due to LEM.

5.4.1. STATEMENT OF THE PROBLEM

The nonlinear model was set up by P.P.Teodorescu in [108, 110].

A frame composed of two bars was considered, of constant cross section and of flexural rigidity EI in the frame plane (figure 5.4). Both bars have supposedly the same length l, though this restriction, introduced only to the purpose of simplifying computation, is not essential and can be easily removed. The frame is acted upon by a vertical force P, that can be responsible for the loss of stability of this structure.

The physical magnitudes corresponding to the vertical bar will be indexed by 1; the index 2 corresponds to those concerning the horizontal bar. Therefore, y_j represents the bar displacement, θ_j – the rotation of the bar cross section, each of them depending on the arc s_j, $j=1,2$.

The equations of the static equilibrium read

$$\begin{aligned} x_1' &= \cos\theta_1, \\ y_1' &= \sin\theta_1, \\ \theta_1' &= \widetilde{X}x_1 - \left(\alpha^2 - \omega^2\right)y_1, \end{aligned} \qquad (5.4.1)$$

$$\begin{aligned} x_2' &= \cos\theta_2, \\ y_2' &= \sin\theta_2, \\ \theta_2' &= -\omega^2 x_2 + \widetilde{X}y_2, \end{aligned} \qquad (5.4.2)$$

where

$$\alpha^2 = \frac{P}{EI}, \quad \tilde{X} = \frac{X}{EI}, \quad \omega^2 = \tilde{X} + \frac{f}{l}\alpha^2, \tag{5.4.3}$$

X being the statically indetermined reaction and f – the maximum deflection of the vertical bar (figure 5.4).

To complete the model, P.P.Teodorescu added the natural boundary value conditions

$$\begin{aligned} &x_j(0) = 0, \quad y_j(0) = 0, \quad j = 1,2, \\ &y_1(l) = f, \quad l - x_1(l) = y_2(l), \quad l - x_2(l) = f, \end{aligned} \tag{5.4.4}$$

as well as

$$\theta_1(l) = \theta_2(l) = -\theta_0, \quad \theta_0 > 0. \tag{5.4.5}$$

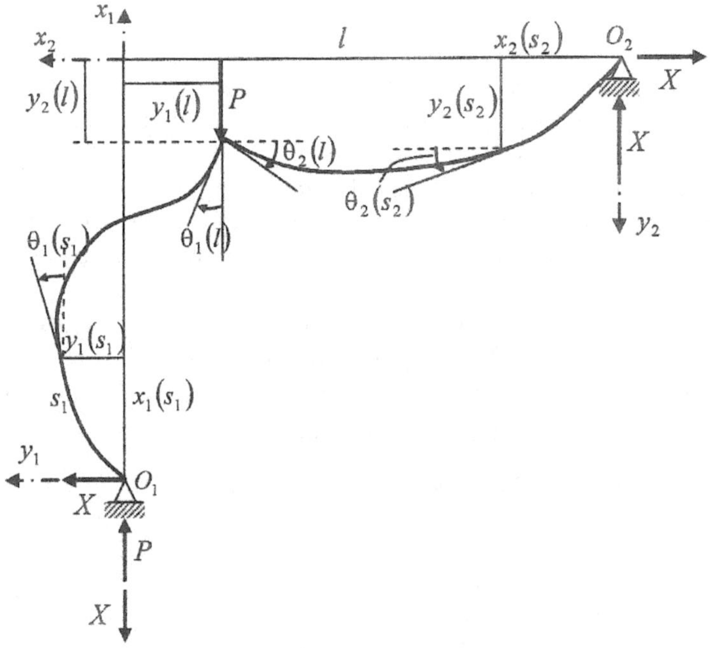

Figure 5.4. The nonlinear elastic frame

5.4.2. A MATHEMATICAL CORE AND ITS LEM SOLUTION

The above mentioned model is composed of two nonlinear ODSs, each of them depending on the stability parameters. To study the critic and post-critic behaviour of the frame, some qualitative representations of the magnitudes x_j, y_j, θ_j, $j=1,2$ are required; this kind of representations is offered by LEM.

We firstly observe that θ_1', θ_2' satisfy the same nonlinear ODE of polynomial type, with constant coefficients [116], [117]

$$\theta^{(4)}\theta' - \theta'''\theta'' = -\theta''\theta'^2 . \tag{5.4.6}$$

This equation is purely abstract and does not depend on any physical magnitude; this is why it was called *intrinsic*. Integrated once, the intrinsic equation leads to

$$\theta''' = k\theta' - \frac{1}{2}\theta'^3 , \tag{5.4.7}$$

where k is a constant that will be determined for θ_1 and θ_2 accordingly.

Taking into account the boundary value conditions (5.4.4), we deduce two Cauchy problems: one for θ_1

$$(\mathsf{P}_1):$$
$$\begin{gathered}\theta_1''' = -c_1\theta_1' - \frac{1}{2}\theta_1'^3, \quad c_1 = \widetilde{X}\sin\theta_1(0) + \left(\alpha^2 - \omega^2\right)\cos\theta_1(0), \\ \theta_1'(0) = 0, \quad \theta_1''(0) = \beta_1, \; \beta_1 = \widetilde{X}\cos\theta_1(0) - \left(\alpha^2 - \omega^2\right)\sin\theta_1(0),\end{gathered} \tag{5.4.8}$$

and another one for θ_2

$$(\mathsf{P}_2):$$
$$\begin{gathered}\theta_2''' = c_2\theta_2' - \frac{1}{2}\theta_2'^3, \qquad c_2 = \omega^2\sin\theta_2(0) + \widetilde{X}\cos\theta_2(0), \\ \theta_2'(0) = 0, \quad \theta_2''(0) = \beta_2, \quad \beta_2 = -\omega^2\cos\theta_2(0) + \widetilde{X}\sin\theta_2(0).\end{gathered} \tag{5.4.9}$$

Thus, the problem of the nonlinear elastic frame was reduced to two other nonlinear problems of polynomial type, (P_1) and (P_2), apparently independent one from the other. We shall apply LEM to both of them.

We shall return with details to the intrinsic equation in chapter 8, where the deduction of its LEM solution will be presented in detail. In this section, we give directly the LEM solution for both the above problems, written by using the same formula

$$\theta_j'(s_j) \cong \frac{\beta_j}{\sqrt{c_j}}\varphi_j\left(\sqrt{c_j}\,s_j\right) + \frac{\beta_j^3}{64c_j\sqrt{c_j}}\psi_j\left(\sqrt{c_j}\,s_j\right) \quad j=1,2 \tag{5.4.10}$$

where φ_j, ψ_j read

$$\begin{aligned}
&\varphi_1(\lambda) = \sin\lambda, \quad \varphi_1(\lambda) = \sinh\lambda, \\
&\psi_1(\lambda) = 3\sin\lambda - \sin 3\lambda + 12\left(\lambda\cos\lambda - \sin\lambda\right), \\
&\psi_2(\lambda) = -3\sinh\lambda + \sinh 3\lambda + 12\left(-\lambda\cosh\lambda + \sinh\lambda\right).
\end{aligned} \tag{5.4.11}$$

From (5.4.10), also considering the initial model of the frame, we easily find approximations in closed form for x_j, y_j, θ_j, $j = 1,2$.

5.4.3. CRITICITY CONDITIONS FOR THE ELASTIC FRAME

Imposing to the LEM formulae for x_j, y_j the conditions given at the joining point, we get four relations depending on two stability parameters, f and X, and on two constants of integration, θ_{10}, θ_{20}. To these relations, one can apply Morse's classical theory. Computing Morse's determinant, we get the following criticity condition for the nonlinear elastic frame [108, 110]

$$\tan \alpha l = \frac{\alpha l}{1 + \frac{1}{4}\alpha^2 l^2 + \frac{5}{144}\alpha^2 l^2}, \tag{5.4.12}$$

which yields the first critical load

$$\alpha_0 l \cong 1.11475\pi. \tag{5.4.13}$$

We shall compare this value with other results, previously obtained by other authors [43].

a) *Classical linear buckling analysis*

In the classical linear theory of elastic stability it is assumed that all the buckling deflections and the associated reactions are infinitesimal; the equations of the neutral equilibrium will be linear. Adding the corresponding boundary value conditions, one gets an eigenvalue problem that can be solved by standard methods. The resulting characteristic equation for the eigenvalues reads

$$\tan \alpha l = \frac{\alpha l}{1 + \frac{1}{3}\alpha^2 l^2}; \tag{5.4.14}$$

the first critical load is, in this case,

$$\alpha_0 l \cong 1.1861\pi . \tag{5.4.15}$$

We see that there is a small difference between the above two first critical loads, obviously due to the difference between the two corresponding formulae (5.4.12) and (5.4.14).

b) *Koiter's analysis*

Koiter's model for the elastic frame was obtained in a variational frame, minimizing an energetic functional of integral type. While the obtained ODEs are linear, the criticity condition is again (5.4.14), as in the linear case. The reason for this is that Koiter did not consider s_j as independent variables; the functions y_j were considered as depending on x_j. Should the linear part of the model (5.4.1), (5.4.2) coincide with the linear part of Koiter's model, the criticity conditions would coincide too. Certainly, this coincidence occurs in many other cases, e.g. in the buckling of a cantilever bar. In the buckling of the two-bar frame, the linear parts are different and so are the criticity conditions.

Yet, as noticed above, the values of the first critical loads deduced by the two models are very close to each other, therefore they are both in agreement with the experiments; only in a more refined analysis could this difference be significant.

In conclusion, the analysis by LEM of P.P.Teodorescu's nonlinear model for the elastic two-bar frame put into evidence not only criticity conditions, but also representation formulae for the displacements that can easily serve to a post-critical study.

Chapter 6

THE NONLINEAR PENDULUM

The purpose of this chapter is to study a class of problems connected with the nonlinear pendulum by using the linear equivalence method. Thus, the pendulum is considered in various hypostases: free, damped and acted upon by a periodic forcing, obtaining for each case LEM formulae for the displacement. Let us mention a new original method, based on LEM, of getting the pendulum period as a series with respect to the initial data; this is a specific method that, unlike the classical one, can be applied to other models. We exemplify this by getting such a period expansion for the Lotka-Volterra prey-predator model.

Using the averaging method, as well as the LEM basis presented in chapter 3, one can obtain new LEM solutions for the free and forced pendulum, that put better into evidence the periods of the solutions.

In section 6.7, a parallel between the Fermi, Pasta and Ulam's model of motion in a granular medium and the pendulum was established. This model was studied by Ştefania Donescu, who found solitons for this model as solutions of a nonlinear ODE of the same form as the B.-E. equation; thus, the pendulum may be viewed as a possible solitonic archetype. Section 6.8 presents the interesting results obtained by Ştefania Donescu, who applied LEM to the nonlinear coupled pendulum [23-25].

Recently, Veturia Chiroiu and Ligia Munteanu [56] successfully applied LEM to the double pendulum model; these results form the object of section 6.9. The last section contains the most recent results obtained by applying LEM to two neo-Hookean models: Bhattacharyya's simple pendulum and Doru Stănescu's model for neo-Hookean suspensions.

6.1. MATHEMATICAL MODELS FOR THE NONLINEAR PENDULUM

This section is devoted to the study of the nonlinear pendulum in three hypostases: free, damped and acted upon by a periodic forcing.

We shall firstly present the corresponding models and then we shall apply LEM.

Consider a rod of length *l*, fixed up at one end, carrying a bob of mass *m*, free to oscillate in a vertical plane. Neglecting frictions, dissipations and strains in the rod, the equation of motion for the pendulum reads [89]

$$ml\ddot{y} + mg \sin y = a \cos \theta t , \qquad (6.1.1)$$

if the pendulum is acted upon by a perturbation force $a \cos \theta t$; this equation was obtained by applying Newton's law in the above mentioned physical hypotheses.

The unknown function y is the deviation angle with respect to the vertical equilibrium position.

The equation (6.1.1) may be also written in the form

$$\ddot{y} + \omega^2 \sin y = A \cos \theta t, \quad \omega^2 = \frac{g}{l}, A = \frac{a}{ml}; \qquad (6.1.2)$$

it is known that this equation – of Duffing type – allows, if $\theta = 0$, an unique periodic solution, of period 2π, for arbitrary A, provided $\omega < 1$. If $A = 0$, we get the equation of the free pendulum.

The damped pendulum motion is governed by the nonlinear second order ODE [89]

$$\ddot{y} + \lambda \dot{y} + \omega^2 \sin y = 0 , \qquad (6.1.3)$$

where $\lambda = k' / 2m > 0, k'$ being the damping coefficient.

To get LEM solutions for these three models, we associate some arbitrary Cauchy conditions

$$y(0) = \alpha, \quad \dot{y}(0) = \beta . \qquad (6.1.4)$$

6.2. THE FREE PENDULUM FROM A LEM PERSPECTIVE

The oscillations of the nonlinear rigid pendulum are described, as mentioned above, by the equation (6.1.2), in which we take $A = 0$. We consider the associated first order equivalent ODS

$$\begin{aligned} \dot{y} &= z, \\ \dot{z} &= -\omega^2 \sin y, \end{aligned} \tag{6.2.1}$$

adopting for the sinus the classical MacLaurin series; this system is analytic with respect to the unknown functions y and z. The initial conditions (6.1.4) become

$$y(0) = \alpha, \quad z(0) = \beta . \tag{6.2.2}$$

Let us apply the linear equivalence method to this system. The LEM mapping depends on two parameters in this case

$$v(t, \sigma, \xi) = e^{\sigma y + \xi z} . \tag{6.2.3}$$

Introducing this in (6.2.2), we get the first LEM equivalent

$$\frac{\partial v}{\partial t} = \sigma \frac{\partial v}{\partial \xi} - \xi \omega^2 \sin(D_\sigma) v , \tag{6.2.4}$$

where we denoted by $\sin(D_\sigma)$ the formal linear operator

$$\sin(D_\sigma) \equiv \sum_{j=1}^{\infty} \frac{(-1)^{j-1}}{(2j-1)!} \frac{\partial^{2j-1}}{\partial \sigma^{2j-1}} , \tag{6.2.5}$$

consistent on the Exp-type space $\mathcal{E}_2^0(I)$; as this system has constant coefficients, I may be taken arbitrarily; e.g., one can take $I = [0, L]$.

According to the theorems of chapter 1, we get the second LEM equivalent, taking for v the expansion

$$v(t, \sigma, \xi) = 1 + \sum_{j+k=1}^{\infty} v_{jk}(t) \frac{\sigma^j}{j!} \frac{\xi^k}{k!} . \tag{6.2.6}$$

Let us note that the nonlinear operator is odd; it then follows that the LEM system can be "contracted" to

$$\frac{d\mathbf{V}}{dt} = \mathbf{A}\mathbf{V}(t), \tag{6.2.7}$$

V being the vector of the unknown functions, written by finite segments of odd indices

$$\mathbf{V}(t)=\left[\mathbf{V}_{2j-1}(t)\right]_{j\in\mathfrak{N}},\quad \mathbf{V}_{2j-1}(t)=\left[v_{2j-1-k,k}(t)\right]_{k=\overline{0,2j-1}},\tag{6.2.8}$$

and the LEM matrix **A**, which is constant, will contain only cells of odd indices

$$\mathbf{A}=\begin{bmatrix}\mathbf{A}_{11} & \mathbf{A}_{13} & \mathbf{A}_{15} & \ldots\\ \mathbf{0} & \mathbf{A}_{33} & \mathbf{A}_{35} & \ldots\\ \mathbf{0} & \mathbf{0} & \mathbf{A}_{55} & \ldots\\ \ldots & \ldots & \ldots & \ldots\end{bmatrix}.\tag{6.2.9}$$

The first diagonal cell $\mathbf{A}_{11}$ – the Jacobian matrix (the linear part of the operator), is, in this case

$$\mathbf{A}_{11}=\begin{bmatrix}0 & 1\\ -\omega^2 & 0\end{bmatrix}.\tag{6.2.10}$$

The other diagonal cells $\mathbf{A}_{2j-1,2j-1}$ have $2j$ rows and $2j$ columns and they depend only on the coefficients of $\mathbf{A}_{11}$

$$\mathbf{A}_{2j-1,2j-1}=\begin{bmatrix}0 & 2j-1 & 0 & 0 & \ldots & 0 & 0\\ -\omega^2 & 0 & 2j-2 & 0 & \ldots & 0 & 0\\ 0 & -2\omega^2 & 0 & 2j-3 & \ldots & 0 & 0\\ \ldots & \ldots & \ldots & \ldots & \ldots & \ldots & \ldots\\ 0 & 0 & 0 & 0 & \ldots & 0 & 1\\ 0 & 0 & 0 & 0 & \ldots & -(2j-1)\omega^2 & 0\end{bmatrix}.\tag{6.2.11}$$

The rectangular cells $\mathbf{A}_{2j-1,2(j+s)-1}$ have $2j$ rows, $2(j+s)$ columns and can be easily generated. Indeed,

$$\mathbf{A}_{2j-1,2(j+s)-1}=\frac{(-1)^{s+1}\omega^2}{(2s+1)!}\begin{bmatrix}0 & 0 & \ldots & 0 & 0 & \ldots & 0\\ 1 & 0 & \ldots & 0 & 0 & \ldots & 0\\ 0 & 2 & \ldots & 0 & 0 & \ldots & 0\\ \ldots & \ldots & \ldots & \ldots & \ldots & \ldots & 0\\ 0 & 0 & \ldots & (2j-1) & 0 & \ldots & 0\end{bmatrix}.\tag{6.2.12}$$

According to theorem 1.8, the eigenvalues of any diagonal cell $\mathbf{A}_{2j-1,2j-1}$ are

$$\pm\left[2\left(j-s\right)-1\right]\mathrm{i}\omega, \quad s=\overline{0, j-1}. \tag{6.2.13}$$

By virtue of theorem 3.3, the normal LEM representation for the free pendulum will contain only trigonometric functions, occasionally multiplied by polynomials in t. This also implies that this representation can be obtained by solving finite LEM ODSs, by identifying the coefficients. More precisely, one can prove

Theorem 6.1 [119-121, 126, 146]. *The LEM solution of the nonlinear system* (6.2.1) *with the arbitrary Cauchy conditions* (6.2.2) *reads*

$$y(t)=\sum_{m=1}^{\infty}\sum_{j=0}^{2m-1}\alpha^{2m-1-j}\beta^{j}u_{2m-1-j,j}(\omega t), \tag{6.2.14}$$

where $u_{10}(\tau)=\cos\tau$, $u_{01}(\tau)=\omega\sin\tau$ *and* $u_{2m-1-j,j}(\tau)$ *are given by*

$$\begin{aligned} u_{2m-1-j,j}(\tau)= &\sum_{k=1}^{m} A_{0,2k-1}^{j}\operatorname{trig}_{j}(2k-1)\tau+\tau\sum_{k=1}^{m-1} A_{1,2k-1}^{j}\operatorname{trig}_{j+1}(2k-1)\tau+ \\ &+\tau^{2}\sum_{k=1}^{m-2} A_{2,2k-1}^{j}\operatorname{trig}_{j+2}(2k-1)\tau+\ldots+\tau^{m-1}A_{j,1}^{j}\operatorname{trig}_{j+m-1}(\tau). \end{aligned} \tag{6.2.15}$$

The trig *functions are defined by*

$$\operatorname{trig}_{k}(\tau)=\begin{cases}\cos\tau & \text{for even } k,\\ \sin\tau & \text{for odd } k.\end{cases} \tag{6.2.16}$$

The coefficients $A_{i,j}^{k}$ are obtained step by step, by identification, from the linear ODS

$$\frac{\mathrm{d}\mathbf{U}_{2m-1}}{\mathrm{d}t}=\mathbf{A}_{2m-1,2m-1}^{\mathrm{T}}\mathbf{U}_{2m-1}+\omega^{2}\sum_{j=2}^{m}\frac{(-1)^{j}}{(2j-1)!}\mathbf{B}_{2m-2j+1}, \tag{6.2.17}$$

starting from $m=2$ *on; the vectors* **U** *and* **B** *have* $2m$ *components*

$$\begin{aligned} &\mathbf{U}_{2m-1}=\left(u_{2m-1,0}\quad u_{2m-2,1}\quad\ldots\quad u_{0,2m-1}\right)^{\mathrm{T}},\\ &\mathbf{B}_{2m-2j+1}=\left(u_{2m-2j,1}\quad 2u_{2m-2j-1,2}\ldots(2m-2j+1)u_{0,2m-2j+1}\quad 0\quad 0\ldots,0\right)^{\mathrm{T}}. \end{aligned} \tag{6.2.18}$$

For a better understanding, let us consider the particular case $\beta = 0$; we intend to get the 5th order approximation of the corresponding normal LEM solution.

According to theorem 3.3, we will obtain for $y(t)$ a series in odd powers of α.

To get the coefficient of α in this series, we must solve the linear ODS

$$\frac{d\mathbf{U}_1}{dt} = \mathbf{A}_{11}^{\mathrm{T}}\mathbf{U}_1, \quad \mathbf{U}_1 = \left[u_{10} \;\; u_{01}\right]^{\mathrm{T}}, \tag{6.2.19}$$

with the independent of α condition

$$\mathbf{U}_1(0) = \left[1 \;\; 0\right]^{\mathrm{T}}. \tag{6.2.20}$$

It results

$$\mathbf{U}_1(t) = \left[\cos\omega t \quad \frac{1}{\omega}\sin\omega t\right]^{\mathrm{T}}. \tag{6.2.21}$$

The coefficient of α in the LEM series of $y(t)$ is therefore

$$u_{10}(t) = \cos\omega t\,. \tag{6.2.22}$$

To obtain the coefficient of α in the LEM series of $z(t) = \dot{y}(t)$, one must also solve the system (6.2.19), but with the different conditions

$$\mathbf{U}_1(0) = \left[0 \;\; 1\right]^{\mathrm{T}}; \tag{6.2.23}$$

its solution is

$$\mathbf{U}_1(t) = \left[-\omega\sin\omega t,\; \cos\omega t\right]^{\mathrm{T}}. \tag{6.2.24}$$

Thus, the coefficient of α in the LEM series of $\dot{y}(t)$ is $-\omega\sin\omega t$.

Further, to get the coefficient $u_{30}(t)$ of α^3 from the LEM series of $y(t)$, one must solve the system

$$\frac{d\mathbf{U}_3}{dt} = \mathbf{A}_{33}^{\mathrm{T}}\mathbf{U}_3 + \mathbf{A}_{13}^{\mathrm{T}}\mathbf{U}_1, \qquad \mathbf{U}_3 = \left[u_{30}, u_{21}u_{12}, u_{03}\right]^{\mathrm{T}}, \tag{6.2.25}$$

in null Cauchy conditions

$$\mathbf{U}_3(0) = \left[0 \;\; 0 \;\; 0 \;\; 0\right]^{\mathrm{T}}, \tag{6.2.26}$$

with $\mathbf{U}_1$ given by (6.2.21); the eigenvalues of $\mathbf{A}_{33}$ are, by formula (6.2.13), $\pm i\omega, \pm 3i\omega$. We deduce

$$u_{30}(t) = \frac{1}{3 \cdot 2^6}\left(\cos \omega t - \cos 3\omega t + 12\omega t \sin \omega t\right). \tag{6.2.27}$$

The coefficient of α^3 from the LEM series of $\dot{y}(t)$ is also determined as first component of the solution of the linear ODS (6.2.25), that satisfies the null Cauchy conditions (6.2.26), but this time $\mathbf{U}_1$ will be given by (6.2.24). It follows

$$u_{30}(t) = \frac{\omega}{3 \cdot 2^6}\left(3\sin 3\omega t + 11\sin \omega t + 12\omega t \cos \omega t\right). \tag{6.2.28}$$

Finally, the coefficient of α^5 for $y(t)$ is $u_{50}(t)$ – the first component of the solution of the linear ODS

$$\begin{gathered}\frac{d\mathbf{U}_5}{dt} = \mathbf{A}_{55}^{\mathrm{T}}\mathbf{U}_5 + \mathbf{A}_{35}^{\mathrm{T}}\mathbf{U}_3 + \mathbf{A}_{15}^{\mathrm{T}}\mathbf{U}_1, \\ \mathbf{U}_5 = \left[u_{50}, u_{41}, u_{32}, u_{23}, u_{14}, u_{05}\right]^{\mathrm{T}},\end{gathered} \tag{6.2.29}$$

with $\mathbf{U}_1$ given by (6.2.21) and $\mathbf{U}_3$ determined from the corresponding to $y(t)$ linear Cauchy problem; the vector $\mathbf{U}_5$ also satisfies null Cauchy conditions

$$\mathbf{U}_5(0) = \begin{bmatrix}0 & 0 & 0 & 0 & 0 & 0\end{bmatrix}^{\mathrm{T}}. \tag{6.2.30}$$

We find

$$\begin{aligned}u_{50}(t) = {}& \frac{3\cos 5\omega t - 20\cos 3\omega t + 17\cos \omega t}{3 \cdot 5 \cdot 2^{12}} - \\ & - \frac{\omega t \sin 3\omega t}{2^{10}} - \frac{\omega^2 t^2 \sin \omega t}{2^9}.\end{aligned} \tag{6.2.31}$$

We see that the coefficient of α^5 from the LEM series of $\dot{y}(t)$ can be easily obtained by differentiating $u_{50}(t)$.

Let us mention that the coefficients of the powers of α from the above LEM series were obtained by using the LEM basis, presented in chapter 3 for ODEs that are odd with respect to the unknown functions.

Remark. The above computation also puts into evidence that, once computed, the coefficients of α^j do not change, no matter how far we get into the LEM series of the solution.

The 5th approximation of the normal LEM solution for the free pendulum is therefore

$$y(t) \cong \alpha u_{10}(t) + \alpha^3 u_{30}(t) + \alpha^5 u_{50}(t), \tag{6.2.32}$$

where $u_{10}(t)$ is given by (6.2.22), $u_{30}(t)$ – by (6.2.27), and $u_{50}(t)$ – by (6.2.31).

6.3. THE LEM METHOD FOR THE PENDULUM PERIOD

As we saw, the oscillations of the nonlinear free rigid pendulum are described by the equation

$$\ddot{y} + \omega^2 y = 0, \tag{6.3.1}$$

also called the equation of the mathematical pendulum; this equation was obtained by applying Newton's law of motion, under convenient physical hypotheses. The independent variable is the time, and the unknown function $y(t)$ represents the angle between the rod with its rest position. The constant ω^2 expresses the ratio

$$\omega^2 = \frac{g}{l}, \tag{6.3.2}$$

where g is the gravity acceleration and l – the rod length.

Multiplying the pendulum equation by $\dot{y}(t) \neq 0$, it can be integrated once, to give

$$\dot{y}^2(t) = 2\omega^2(\cos y - \cos\alpha), \tag{6.3.3}$$

α being the angle – between 0 and π – corresponding to the limit position of the pendulum, where the velocity is null.

6.3.1. CLASSICAL TREATMENT OF THE PROBLEM

With these preparations, we can now present the problem of the pendulum period in classical treatment. The presentation is inspired from [89], in which this problem is thoroughly studied.

From the first integral (6.3.3), it is noticed that the velocity depends only on the position $y(t)$ and it is periodic with respect to it; the equation (6.3.3) can be also considered as a first order ODE with separable variables. We deduce by integration

$$t = t_0 + \frac{1}{\omega\sqrt{2}} \int_{y_0}^{y} \frac{\mathrm{d}y}{\sqrt{\cos y - \cos\alpha}}, \quad \cos y > \cos\alpha, \tag{6.3.4}$$

where y_0 corresponds to the position at the arbitrary moment t_0 (that may differ from the initial moment). From (6.3.4) it is seen that the time interval $t - t_0$ depends only on the positions corresponding to the moments t, t_0; it then follows that the oscillatory motion is periodic, of period T. Changing the motion along the arc, the sense of the velocity also changes, its modulus remaining the same; consequently, the velocity is the same between two symmetric with respect to the rest position points. We thus deduce that the period T is expressed in the following integral form

$$T = \frac{2\sqrt{2}}{\omega} \int_{0}^{\alpha} \frac{\mathrm{d}y}{\sqrt{\cos y - \cos\alpha}}. \tag{6.3.5}$$

The obvious relation between T and α may be emphasized in various ways.

As

$$\cos y - \cos\alpha = 2\left(\sin^2\frac{y}{2} - \sin^2\frac{\alpha}{2}\right), \tag{6.3.6}$$

by using the notations

$$\sin\frac{y}{2} = k\sin\varphi, \quad k = \sin\frac{\alpha}{2}, \tag{6.3.7}$$

we can write

$$t = t_0 + \frac{1}{\omega} \int_{\varphi_0}^{\varphi} \frac{\mathrm{d}\psi}{\sqrt{1 - k^2\sin^2\psi}}, \tag{6.3.8}$$

with φ_0 given by

$$\sin\frac{y_0}{2} = k\sin\varphi_0. \tag{6.3.9}$$

Denoting by

$$\sin\varphi = z\,, \tag{6.3.10}$$

one can also write

$$t = t_0 + \frac{1}{\omega}\int_{z_0}^{z}\frac{\mathrm{d}\zeta}{\sqrt{(1-\zeta^2)(1-k^2\zeta^2)}}, \tag{6.3.11}$$

where z_0 satisfies

$$\sin\varphi_0 = z_0\,. \tag{6.3.12}$$

The two expressions (6.3.8) and (6.3.11) lead, by their form, to the elliptic integral of first kind

$$F(\varphi,k) = \int_0^{\varphi}\frac{\mathrm{d}\psi}{\sqrt{1-k^2\sin^2\psi}} = \int_0^{\sin\psi}\frac{\mathrm{d}z}{\sqrt{(1-z^2)(1-k^2z^2)}}, \tag{6.3.13}$$

φ being the amplitude and k – the modulus of the integral.

Thus, we obtain for t

$$t = t_0 + \frac{1}{\omega}\left[F(\varphi,k) - F(\varphi_0,k)\right], \tag{6.3.14}$$

or, putting $u = \omega t$,

$$u - u_0 = F(\varphi,k) - F(\varphi_0,k), \quad u_0 = \omega t_0\,. \tag{6.3.15}$$

Without restricting the generality, one can take $t_0 = 0$.

Admitting that $\varphi_0 = 0$, it results that $z_0 = 0$, $u_0 = 0$, as well as $F(\varphi_0,k) = 0$, hence

$$u = F(\varphi,k). \tag{6.3.16}$$

Following Abel, it is easier to express the angle φ as a function of u in the form

$$\sin\varphi = \operatorname{sn} u\,, \tag{6.3.17}$$

where sn is the symbol of the elliptic sine (the amplitude sinus), defined by Jacobi. One can also use the amplitude cosine, denoted by cn

$$\cos\varphi = \operatorname{cn} u\,. \tag{6.3.18}$$

Now, getting back to the period, formula (6.3.5) becomes, with all the above remarks

$$T = \frac{4}{\omega} K(k) = \frac{4}{\omega} \int_0^{\frac{\pi}{2}} \frac{\mathrm{d}\varphi}{\sqrt{1 - k^2 \sin^2 \varphi}} \\ = \frac{4}{\omega} \int_0^1 \frac{\mathrm{d}z}{\sqrt{\left(1 - z^2\right)\left(1 - k^2 z^2\right)}}. \tag{6.3.19}$$

In this formula, $K(k) = F\left(\frac{\pi}{2}, k\right)$ is the complete elliptic integral of first kind. As $k^2 < 1$, one can use Newton's binomial series

$$\left(1 - k^2 \sin^2 \varphi\right)^{-\frac{1}{2}} = 1 + \sum_{n=1}^{\infty} \frac{(2n)!}{2^{2n} (n!)^2} k^{2n} \sin^{2n} \varphi ; \tag{6.3.20}$$

this series is absolutely and uniformly convergent on $\left[0, \frac{\pi}{2}\right]$, therefore it can be integrated term by term with respect to φ. For the integrals in the right member one can use Wallis' formula [89]

$$\int_0^{\frac{\pi}{2}} \sin^{2n} \varphi \, \mathrm{d}\varphi = \frac{(2n)!}{2^{2n} (n!)^2} \frac{\pi}{2}, \tag{6.3.21}$$

getting for T the expansion

$$T = \frac{2\pi}{\omega} \left\{ 1 + \sum_{n=1}^{\infty} \frac{[(2n)!]^2}{2^{4n} (n!)^4} \sin^{2n} \frac{\alpha}{2} \right\}. \tag{6.3.22}$$

One can also expand $\sin^{2n} \frac{\alpha}{2}$ in an absolutely convergent series with respect to α, such that we finally obtain for the period T the following expansion

$$T = \frac{2\pi}{\omega} \left(1 + \frac{\alpha^2}{16} + \frac{11}{12} \frac{\alpha^4}{16^2} + \ldots \right). \tag{6.3.23}$$

Let us note that the ratio between the first two terms of this series is $\frac{\alpha^2}{16} < 1$; the ratio between the next two terms is

$$\frac{11}{12}\frac{\alpha^2}{16} < \frac{\alpha^2}{16} < 1. \tag{6.3.24}$$

It is seen that the series satisfies the requirement of D'Alembert's criterion, being even rapidly convergent; practically, we can take just the first two terms in this development

$$T \cong \frac{2\pi}{\omega}\left(1 + \frac{\alpha^2}{16}\right), \tag{6.3.25}$$

in order to obtain the period with a satisfactory precision. Indeed, if $\alpha = 0.4$ (corresponding to an angle of $22^0 55' 06''$), then the correction brought by the second term in the series is under 1%. The astronomical clocks, for example, have pendulums with amplitudes of $1^0 30'$, corresponding to a correction of about 0.005%.

For small oscillations, one can consider the linear part of the equation (6.3.1)

$$\ddot{y} + \omega^2 y = 0; \tag{6.3.26}$$

indeed, in this case $\sin y \cong y$. The general solution of this linear ODE with constant coefficients can be written in the form

$$y(t) = \alpha \cos(\omega t - \varphi), \tag{6.3.27}$$

the angle φ being determined from the initial conditions. The period is, in this case, given by Galilei's formula

$$T_0 = \frac{2\pi}{\omega}, \quad \omega = \sqrt{\frac{g}{l}}. \tag{6.3.28}$$

The series (6.3.23) may thus be written as

$$T = T_0\left[1 + \frac{\alpha^2}{16} + \frac{11}{12}\left(\frac{\alpha^2}{16}\right)^2 + \ldots\right], \tag{6.3.29}$$

putting into evidence the period T_0 of the solution corresponding to the linear part of the pendulum equation.

We shall see in what follows that this fact is significant for the relation linear-nonlinear .

6.3.2. THE TREATMENT BY LEM OF THE PENDULUM PERIOD

Let us now apply LEM to pendulum problem. In the previous section we saw that the LEM solution of the Cauchy problem

$$\ddot{y}+\omega^2 y=0,\quad y(0)=0,\quad \dot{y}(0)=\alpha \tag{6.3.30}$$

allows the normal LEM representation

$$y(t)\cong \alpha u_{10}(t)+\alpha^3 u_{30}(t)+\alpha^5 u_{50}(t), \tag{6.3.31}$$

where

$$\begin{aligned}
u_{10} &= \cos\omega t,\\
u_{30} &= \frac{1}{3\cdot 2^6}\left(-\cos 3\omega t+\cos\omega t+12\omega t\sin\omega t\right),\\
u_{50} &= \frac{1}{15\cdot 2^{12}}\left(3\cos 5\omega t-20\cos 3\omega t+17\cos\omega t\right)-\\
&\quad -\frac{1}{2^{10}}\left(\omega t\sin 3\omega t+2\omega^2 t^2\sin\omega t\right).
\end{aligned} \tag{6.3.32}$$

As previously mentioned, without restricting the generality, one can choose as initial moment $t=0$ that moment when the pendulum amplitude becomes maximum, also a moment when the velocity is null; this fully justifies the initial conditions from (6.3.30).

Imposing the periodicity condition to the LEM representation and its derivative, we get

$$\begin{aligned}
&\alpha(\cos\omega T-1)+\alpha^3 u_{30}(\omega T)+\alpha^5 u_{50}(\omega T)=0,\\
&-\alpha\sin\omega T+\alpha^3\frac{\mathrm{d}u_{30}}{\mathrm{d}t}(\omega T)+\alpha^5\frac{\mathrm{d}u_{50}}{\mathrm{d}t}(\omega T)=0.
\end{aligned} \tag{6.3.33}$$

Let us express the natural relation between the period T and α in the form

$$T(\alpha)=\sum_{j=0}^{\infty} T_j\alpha^j; \tag{6.3.34}$$

replacing this in (6.3.33), it results, by identifying the coefficients of the same powers of α, the series development of the free pendulum period

$$T(\alpha) \cong \frac{2\pi}{\omega}\left(1+\frac{\alpha^2}{16}+\frac{11}{3\cdot 2^{10}}\alpha^4\right), \tag{6.3.35}$$

where, as seen before, $T_0 = \frac{2\pi}{\omega}$ is the period corresponding to the solution of the linear part of the pendulum equation.

Remark. We see that the formula obtained by LEM for the pendulum period coincides with the expansion obtained in the classical frame; this should be expected. But the classical method is devoted to the pendulum problem and it cannot be applied to other models allowing periodic solutions. On the contrary, the method deduced by LEM is much more general; we shall see in the next subsection how we can apply it to other nonlinear models.

Formula (6.3.35) shows that the motion of the considered nonlinear oscillator depends on both the period of the linear part and on the initial data. Moreover, we see that the influence of α upon T is less for small α and increases with α. For a numerical comparison, from [129] it follows that the periods computed by using the Runge-Kutta method are in concordance with those deduced by formula (6.3.35); yet, for values of α around the instable critical point $\frac{\pi}{\omega}$, the formula cannot be anymore applied.

6.3.3. LEM PERIODS FOR THE LOTKA-VOLTERRA SYSTEM

We shall prove the generality of the above based on LEM method of getting periods in the case of a nonlinear model, well known in the literature: the Lotka-Volterra equations. They represent an ecosystem model, the so-called prey-predator model and allow periodic solutions. Their form is [129]

$$\begin{aligned} \dot{X} &= AX - XY, \\ \dot{Y} &= -BY + XY; \end{aligned} \tag{6.3.36}$$

where X is the prey population (say, rabbits) and Y is the predator population (say, foxes), A and B being known positive constants, significant for the model.

The system (6.3.36) describes the dynamics of these populations. The Lotka-Volterra equations were used for a long time in biological modelling, such as biological clocks or neurological nets.

A qualitative fact, common to both the pendulum and Lotka-Volterra systems, is that through any point of the associated phase space a closed curve passes, solution of the corresponding ODS. This means that both models allow periodic solutions.

To get periods for the solution of the Lotka-Volterra system, we firstly translate the system (6.3.36) from the stationary point

$$x_s = \frac{1}{B}, \quad y_s = \frac{1}{A}, \tag{6.3.37}$$

to the origin, by using the transformation

$$x = X + x_s, \quad y = Y + y_s; \tag{6.3.38}$$

we thus get a new first order homogeneous polynomial ODS with constant coefficients

$$\begin{aligned} \dot{x} &= -By - xy, \\ \dot{y} &= Ax + xy. \end{aligned} \tag{6.3.39}$$

We shall apply LEM to this system. After getting the first LEM equivalent – the linear PDE – we obtain the linear equivalent system in matrix form

$$\frac{\mathrm{d}\mathbf{V}}{\mathrm{d}t} = \mathbf{AV}, \tag{6.3.40}$$

the constant LEM matrix **A** being

$$\mathbf{A} = \begin{bmatrix} \mathbf{A}_{11} & \mathbf{A}_{12} & \mathbf{0} & \mathbf{0} & \dots \\ \mathbf{0} & \mathbf{A}_{22} & \mathbf{A}_{23} & \mathbf{0} & \dots \\ \mathbf{0} & \mathbf{0} & \mathbf{A}_{33} & \mathbf{A}_{34} & \dots \\ \dots & \dots & \dots & \dots & \dots \end{bmatrix}. \tag{6.3.41}$$

The diagonal cells $\mathbf{A}_{kk}$ are square tridiagonal matrices, generated by the coefficients of the linear part of the system. More precisely,

$$\mathbf{A}_{11} = \begin{bmatrix} 0 & -B \\ A & 0 \end{bmatrix}, \quad \mathbf{A}_{22} = \begin{bmatrix} 0 & -2B & 0 \\ A & 0 & -B \\ 0 & 2A & 0 \end{bmatrix}, \dots \tag{6.3.42}$$

and the super-diagonal cells $\mathbf{A}_{k,k+1}$ are defined by using the coefficients of the quadrates of the unknown functions, i.e.,

$$\mathbf{A}_{12} = \begin{bmatrix} 0 & -1 & 0 \\ 0 & 1 & 0 \end{bmatrix}, \quad \mathbf{A}_{23} = \begin{bmatrix} 0 & -2 & 0 & 0 \\ 0 & 1 & -1 & 0 \\ 0 & 0 & 2 & 0 \end{bmatrix}, \ldots \tag{6.3.43}$$

Denote by

$$\omega^2 = AB. \tag{6.3.44}$$

We see that the eigenvalues of $\mathbf{A}_{11}$, the Jacobian of the Lotka-Volterra system, are $\pm \mathrm{i}\omega$. By virtue of theorem 1.8, it follows that the eigenvalues of $\mathbf{A}_{22}$ are $\pm 2\mathrm{i}\omega$ and 0, those of $\mathbf{A}_{33}$ are $\pm \mathrm{i}\omega, \pm 3\mathrm{i}\omega$, etc.

Imposing now the initial conditions

$$x(0) = \alpha, \quad y(0) = 0, \tag{6.3.45}$$

similar to the pendulum and having a similar physical significance, we can apply theorem 3.3 to get the normal LEM solution corresponding to the considered model.

According to this theorem, the coefficient of α from the LEM solution for x is the first component of the vector $\mathbf{U}_1^1$, that satisfies the finite linear ODS with constant coefficients

$$\frac{\mathrm{d}\mathbf{U}_1}{\mathrm{d}t} = \mathbf{A}_{11}^{\mathrm{T}} \mathbf{U}_1, \tag{6.3.46}$$

and the Cauchy conditions

$$\mathbf{U}_1(0) = [1 \quad 0]^{\mathrm{T}}, \tag{6.3.47}$$

therefore

$$\mathbf{U}_1^1(t) \equiv \begin{bmatrix} u_{10}^1 \\ u_{01}^1 \end{bmatrix} = \begin{bmatrix} \cos \omega t \\ -\dfrac{\omega}{A} \sin \omega t \end{bmatrix}. \tag{6.3.48}$$

The coefficient of α^2 from the development of x is the first component of $\mathbf{U}_2^1$ – solution of the finite linear ODS

$$\frac{\mathrm{d}\mathbf{U}_2}{\mathrm{d}t} = \mathbf{A}_{22}^{\mathrm{T}} \mathbf{U}_2 + \mathbf{A}_{12}^{\mathrm{T}} \mathbf{U}_1^1, \tag{6.3.49}$$

in null Cauchy conditions

$$\mathbf{U}_2(0)=[0\quad 0\quad 0]^{\mathrm{T}}, \tag{6.3.50}$$

whence

$$u_{20}^1(t)=\frac{A^2}{3\omega^2}(\cos 2\omega t-\cos\omega t)-\frac{1}{3\omega}\sin\omega t+\frac{1}{6\omega}\sin 2\omega t. \tag{6.3.51}$$

Analogously,

$$\begin{aligned}u_{30}^1(t)=&\frac{37A^2-23\omega^2}{288\omega^4}\cos\omega t+\frac{\omega^2-2A^2}{9\omega^4}\cos 2\omega t\\&+\frac{3A^2-\omega^2}{32\omega^4}\cos 3\omega t+\frac{A^3}{9\omega^4}\cos 2\omega t\\&+\frac{A}{\omega^3}\left(\frac{7}{24}\sin\omega t-\frac{1}{3}\sin 2\omega t+\frac{1}{8}\sin 3\omega t\right)\\&+\frac{A^2+\omega^2}{24\omega^4}\left(\omega t\sin\omega t+\frac{1}{12}\cos\omega t\right)+\frac{A^2+\omega^2}{24\omega^4}\omega t\sin\omega t.\end{aligned} \tag{6.3.52}$$

Finally, we obtain for x the normal LEM approximation

$$x(t)\cong\alpha u_{10}^1(t)+\alpha^2u_{20}^1(t)+\alpha^3u_{30}^1(t), \tag{6.3.53}$$

and, accordingly, the similar LEM representation for y

$$y(t)\cong\alpha u_{10}^2(t)+\alpha^2u_{20}^2(t). \tag{6.3.54}$$

In (6.3.54), $u_{10}^2(t)$ is the first component of $\mathbf{U}_1^2$, the solution of the finite linear ODS (6.3.49), that also satisfies the Cauchy condition

$$\mathbf{U}_1(0)=[0\quad 1]^{\mathrm{T}}. \tag{6.3.55}$$

This involves

$$\mathbf{U}_1^2(t)\equiv\begin{bmatrix}u_{10}^1\\u_{01}^1\end{bmatrix}=\begin{bmatrix}\frac{A}{\omega}\sin\omega t\\\cos\omega t\end{bmatrix}. \tag{6.3.56}$$

Similarly, the coefficient of α^2 from the LEM solution for y is the first component of $\mathbf{U}_2^2$ – that satisfies the system (6.3.49) and null Cauchy conditions, therefore

$$u_{20}^2(t)=\frac{A^2}{3\omega^2}\left(\cos\omega t-\cos 2\omega t\right)+\frac{A^2}{6\omega^3}\left(\sin 2\omega t-2\sin\omega t\right). \qquad (6.3.57)$$

We now impose the periodicity condition to the above representations

$$x(0)=x(T)=\alpha, \quad y(0)=y(T)=0. \qquad (6.3.58)$$

Considering for T a development in the form $T(\alpha)=\sum_{j=0}^{\infty}T_j\alpha^j$, as in the case of the pendulum, we firstly get $T_0=\frac{2\pi}{\omega}$ and then the following approximating formula for the period of the solution of the Lotka-Volterra system [109, 126]

$$T\cong\frac{2\pi}{\omega}\left(1+\frac{A^2+\omega^2}{24\omega^4}\alpha^2\right). \qquad (6.3.59)$$

6.4. THE NONLINEAR FORCED PENDULUM

The mathematical model consists of equation (6.1.2), in which, for the sake of simplicity, we shall consider a new variable θt, re-noted by t; we thus have

$$\ddot{y}+\omega^2\sin y=A\cos t. \qquad (6.4.1)$$

As previously mentioned, this equation allows a unique periodic solution, of period 2π, for $\omega<1$ and arbitrary A.

The ODE (6.4.1) may be easily written in the form of a nonlinear homogeneous ODS with constant coefficients

$$\begin{aligned}\dot{y}&=z,\\ \dot{z}&=-\omega^2\sin y+u,\\ \dot{u}&=w,\\ \dot{w}&=-u,\end{aligned} \qquad (6.4.2)$$

allowing the origin as critical point.

The LEM mapping depends in this case of four parameters, therefore the components of the unknown vector from the associated

LEM linear ODS will have four indices. The corresponding to (6.4.2) LEM system is

$$\dot{v}_{ijkl} = iv_{i-1,j+1,kl} + j\left[-\omega^2 \sum_{m=1}^{\infty} \frac{(-1)^{m+1}}{(2m-1)!} v_{i+2m-1,j-1,kl} + v_{i,j-1,k+1,l}\right] + \quad (6.4.3)$$
$$+ kv_{ij,k-1,l} - lv_{ij,k+1,l-1}.$$

As in the case of the free pendulum, the nonlinear ODS is odd, therefore the LEM matrix will not contain cells with even indices.

To get the period of the solution, we consider the initial data

$$y(0) = \alpha, \quad z(0) = 0, \quad u(0) = A, \quad w(0) = 0, \quad (6.4.4)$$

with α indeterminate for the moment. From theorem 3.3, we deduce that $y(t)$ – the first component of the solution of the Cauchy problem (6.4.2), (6.4.4) – allows the normal LEM representation

$$y(t) = \sum_{i=1}^{\infty} \sum_{s=0}^{2i-1} \alpha^{2i-1-s} A^s \, u^1_{2i-1-s,0,s,0}(\omega t), \quad (6.4.5)$$

where $u^1_{2i-1-s,0,s,0}$ are components of the vectors $\mathbf{U}^k_{2i-1}(t)$ satisfying the ODSs with constant coefficients

$$\frac{\mathrm{d}\mathbf{U}^1_{2m-1}}{\mathrm{d}t} = \mathbf{A}^{\mathrm{T}}_{1,2m-1}\mathbf{U}^1_1 + \mathbf{A}^{\mathrm{T}}_{3,2m-1}\mathbf{U}^1_2 + \ldots \mathbf{A}^{\mathrm{T}}_{2m-1,2m-1}\mathbf{U}^1_{2m-1}, \quad m = \overline{1,i}, \quad (6.4.6)$$

as well as the independent of A and α Cauchy conditions

$$\mathbf{U}^1_1(0) = \left[\delta^1_k\right]_{k=\overline{1,4}}, \quad \mathbf{U}^1_{2m-1}(0) = \mathbf{0}, \quad m = \overline{2,i}. \quad (6.4.7)$$

In the above formulae, the matrices $\mathbf{A}^{\mathrm{T}}_{2k-1,2m-1}$ are the transposed of the corresponding LEM cells and they are straightforwardly determined from (6.4.3).

According to theorem 1.8, it follows that the eigenvalues of the diagonal cells $\mathbf{A}^{\mathrm{T}}_{2m-1,2m-1}$ from the transposed LEM matrix are

$$\rho_{ijkl} = i\rho_1 + j\rho_2 + k\rho_3 + l\rho_4, \quad i + j + k + l = 2m - 1, \quad (6.4.8)$$

where $\rho_1 = \mathrm{i}\omega, \rho_2 = -\mathrm{i}\omega, \rho_3 = \mathrm{i}, \rho_4 = -\mathrm{i}$ are the eigenvalues of $\mathbf{A}_{11}$, the matrix of the linear part of the ODS (6.4.2).

As these values are purely imaginary, the LEM solution will contain only sines and cosines, occasionally multiplied by powers of t. The

ODSs (6.4.6) may be solved step by step, starting from $m=1$; taking into account higher order effects does not modify the previously computed terms.

Finally, applying the Laplace transformation, we get the third order approximation of the LEM solution

$$y(t) \cong \alpha \cos \omega t + \frac{A}{\omega\left(1-\omega^2\right)}(\cos \omega t - \cos t) + \sum_{s=0}^{3} \alpha^{3-s} A^s \varphi_s(t) \tag{6.4.9}$$

where $\varphi_0(t)$ is precisely the first component $u_{30}^1(t)$ of $\mathbf{U}_3^1(t)$.

The functions $\varphi_s(t)$, $s=\overline{1,3}$, are of the form

$$\begin{aligned}\varphi_s(t) = {} & A_{s1} \operatorname{trig}_s(3\omega t) + \left(A_{s2} + t^2 A_{s3}\right)\operatorname{trig}_s(\omega t) + A_{s4} \operatorname{trig}_s((2\omega+1)t) + \\ & + A_{s5} \operatorname{trig}_s((\omega+2)t) + A_{s6} \operatorname{trig}_s((2\omega-1)t) + \\ & + A_{s7} \operatorname{trig}_s((\omega-2)t) + A_{s8} \operatorname{trig}_s(t) + \left(A_{s9} + t^2 A_{s,10}\right)\operatorname{trig}_s(t) + \\ & + A_{s,11}\, t \operatorname{trig}_{s+1}(\omega t) + A_{s,12}\, t \operatorname{trig}_{s+1}(t),\end{aligned} \tag{6.4.10}$$

the coefficients A_{sk} are computed by residues, as well as φ_0. The functions trig are given by (6.2.16).

The periodicity condition implies

$$\alpha \cong \alpha \cos 2\pi\omega + \frac{A}{\omega\left(1-\omega^2\right)}(\cos 2\pi\omega - 1) + \sum_{s=0}^{3} \alpha^{3-s} A^s \varphi_s(2\pi), \tag{6.4.11}$$

whence

$$\begin{aligned}& A^3 \varphi_3(2\pi) + \frac{A}{\omega\left(1-\omega^2\right)}(\cos 2\pi\omega - 1) + \alpha\left(\cos 2\pi\omega - 1 + A^2 \varphi_2(2\pi)\right) \\ & + \alpha^2 A \varphi_1(2\pi) + \alpha^3 \varphi_0(2\pi) \cong 0.\end{aligned} \tag{6.4.12}$$

The solution of this functional equations is unique for $\omega < 1$ and A arbitrary.

To get an approximation of α, we shall use LEM again, more precisely, corollary 4.4. Applying formula (4.4.49), we get for α the approximation

$$\alpha_3 = -\frac{f_0\left(f_1^2 - f_0 f_1\right)}{f_1^3 - 2 f_0 f_1 f_2 + f_3 f_0^2}, \tag{6.4.13}$$

where

$$f_0 = A^3\varphi_3(2\pi) + \frac{A}{\omega(1-\omega^2)}(\cos 2\pi\omega - 1), \qquad f_2 = A\varphi_1(2\pi),$$
$$f_1 = \cos 2\pi\omega - 1 + A^2\varphi_2(2\pi), \qquad f_3 = \varphi_0(2\pi). \tag{6.4.14}$$

6.5. THE NONLINEAR DAMPED PENDULUM

In this section, we give normal LEM representations for the damped pendulum. This allows the study of third order effects depending on the dimensionless damping factor $\chi = \lambda / \omega$.

Consequently, we shall consider the initial nonlinear problem (6.1.3), (6.1.4) in the following cases: *a)* $\chi < 1$, *b)* $\chi > 1$ and *c)* $\chi = 1$. For this last case, there are presented solutions allowing the study of 5th order effects.

The ODE (6.1.3) is odd, therefore the corresponding LEM matrix will contain only cells with odd indices.

With these preparations, we can apply theorem 3.3; it results that the normal LEM representation for the solution of the damped pendulum problem is of the form

$$y(t) = \sum_{m=2}^{\infty} \sum_{j=0}^{2m-1} \alpha^{2m-1-j} \beta^j u^1_{2m-1-j,j}(t), \tag{6.5.1}$$

where $u^1_{2m-1-j,j}$ are components of the vectors $\mathbf{U}^1_{2m-1}$, satisfying finite linear ODSs, corresponding to the ODS (3.3.13), and Cauchy conditions corresponding to (3.3.14).

To study third order effects, it is sufficient to take $m = 2$ in the LEM series (6.5.1). According to theorem 1.8, the eigenvalues of the transposed diagonal cells $\mathbf{A}^{\mathrm{T}}_{2m-1,2m-1}$ depend on the eigenvalues of the linear part of the ODE (6.1.3), given by

$$\mathbf{A}^{\mathrm{T}}_{11} \equiv \begin{bmatrix} 0 & -\omega^2 \\ 1 & -2\lambda \end{bmatrix}. \tag{6.5.2}$$

The eigenvalues of $\mathbf{A}^{\mathrm{T}}_{11}$ are complex-conjugate if $\chi < 1$, they are real and distinct for $\chi > 1$ and they coincide, if $\chi = 1$. These are precisely the above mentioned cases *a)*, *b)* and *c)*.

a) Subcritical damping. For $m = 2$ we get

$$y(t) \cong \alpha e^{-\lambda t}\left(\frac{\chi}{\sqrt{1-\chi^2}}\sin\mu t + \cos\mu t\right) + \beta e^{-\lambda t}\frac{\sin\mu t}{\mu} + \tag{6.5.3}$$
$$+\alpha^3 u_{30}^1(t) + \alpha^2\beta u_{21}^1(t) + \alpha\beta^2 u_{12}^1(t) + \beta^3 u_{03}^1(t),$$

where $\mu = \omega\sqrt{1-\chi^2}$, and the coefficients of the Cauchy data powers are

$$u_{30}^1(t) = \frac{e^{-\lambda t}}{3\delta}\left[-\chi\left(11+12\chi^2\right)\sqrt{1-\chi^2}\left(11+12\chi^2\right)\cos\mu t\right] +$$
$$+\frac{e^{-\lambda t}}{3\delta}\left(6+25\chi^2-12\chi^4\right)\sin\mu t + \varphi_1(\mu t),$$
$$u_{21}^1(t) = \frac{e^{-\lambda t}}{\omega\delta}\left[-\sqrt{1-\chi^2}\left(2+12\chi^2\right)\cos\mu t + 9\chi\sin\mu t\right] + \varphi_2(\mu t), \tag{6.5.4}$$
$$u_{12}^1(t) = \frac{e^{-\lambda t}}{\omega^2\delta}\left[-9\chi\sqrt{1-\chi^2}\cos\mu t + \left(2+3\chi^2\right)\sin\mu t\right] + \varphi_3(\mu t),$$
$$u_{03}^1(t) = \frac{e^{-\lambda t}}{\omega^3\delta}\left(-2\sqrt{1-\chi^2}\cos\mu t + \chi\sin\mu\right) + \varphi_4(\mu t).$$

In the above formulae, $\delta = 16\omega^2\left(4-3\chi^2\right)\sqrt{1-\chi^2}$, and

$$\varphi_j(\xi) = e^{-3\lambda t}\left(A_j\cos 3\xi + B_j\sin 3\xi + C_j\cos 3\xi + D_j\sin\xi\right). \tag{6.5.5}$$

The coefficients A_j, B_j, C_j, D_j are easily computed, taking into account the Cauchy conditions satisfied by $u_{ij}^1(t)$. For a better insight into the problem, we wrote them in tables 6.1 and 6.2, where the following notations were used

$$\delta_1 = 2^5\omega^7\chi\left(\chi^2-1\right), \qquad \delta_2 = 2^5\omega^7\left(\chi^2-1\right)\left(4-3\chi^2\right). \tag{6.5.6}$$

From the LEM formulae (6.5.3), (6.5.4) and (6.5.5) it is seen that for $\chi < 1$ (subcritical damping), the pendulum motion is qualitatively analogous with that corresponding to the linear case, i.e., it is pseudo-periodic and it even has the pseudo-period $\frac{2\pi}{\mu}$.

Table 6.1

The analytic expressions of the coefficients A_j, B_j, C_j, D_j for $j = 1,2$

j	1
$\delta_1 A_j$	$\frac{\omega^2}{3}\left(24\chi^4 - 20\chi^2 + 2\right)$
$\delta_1 B_j$	$\frac{\omega^2}{3}\frac{\left(24\chi^4 - 32\chi^2 + 9\right)}{\sqrt{1-\chi^2}}$
$\delta_2 C_j$	$-2\omega^3\chi$
$\delta_2 D_j$	$-\omega^3\frac{2\chi^2 - 1}{\sqrt{1-\chi^2}}$

Table 6.2.

The analytic expressions of the coefficients A_j, B_j, C_j, D_j for $j = 3,4$

j	3	4
$\delta_1 A_j$	$2\omega\left(3\chi^2 - 1\right)$	χ
$\delta_1 B_j$	$\omega\chi\frac{\left(6\chi^2 - 5\right)}{\sqrt{1-\chi^2}}$	$-\frac{3\chi^2 - 2}{3\sqrt{1-\chi^2}}$
$\delta_2 C_j$	$-4\omega\chi$	-1
$\delta_2 D_j$	$-\omega\frac{4\chi^2 - 1}{\sqrt{1-\chi^2}}$	$-\frac{\chi}{\sqrt{1-\chi^2}}$

For greater values of λ, the terms containing $e^{-(2j-1)\lambda t}$ may be neglected, at least for $j \geq 2$. Also, from the form of the LEM solutions, we see that all the coefficients, irrespective of the order of approximation, depend directly on the damping factor χ.

b) Supercritical damping. In this case, we obtain similar formulae, that can be also directly deduced from (6.5.3), (6.5.4), (6.5.5), replacing accordingly the trigonometric with the hyperbolic functions, and

$\sqrt{1-\chi^2}$ with $\sqrt{\chi^2-1}$; same rules hold true for φ_j. Moreover, the coefficients of φ_j are given in the same tables, with the only difference that $\sqrt{1-\chi^2}$ must be replaced by $\sqrt{\chi^2-1}$.

c) Critical damping. In view of a significant interpretation of the solution, as well as to the purpose of obtaining a simplified version of the formulae, we shall take $\alpha=0$. The 5th approximation ($m=3$) of the normal LEM solution is

$$y(t)\cong\frac{\beta}{\lambda}\mathrm{e}^{-\lambda t}\lambda t\left[1+\left(\frac{\beta}{4\lambda}\right)^2+\frac{34}{4^3}\left(\frac{\beta}{4\lambda}\right)^4+\ldots\right]-\\ -2\frac{\beta}{\lambda}\mathrm{e}^{-\lambda t}\left[\left(\frac{\beta}{4\lambda}\right)^2+\frac{201}{4^3}\left(\frac{\beta}{4\lambda}\right)^4+\ldots\right]. \tag{6.5.7}$$

In (6.5.7), the terms containing $\mathrm{e}^{-(2j-1)\lambda t}$ were neglected; this can be done for sufficiently great values of λ and for $j\geq 2$.

We conclude that the LEM series from (6.5.7) converges for $\left|\frac{\beta}{4\lambda}\right|<1$. Some numerical estimations show that, apparently, the damped pendulum never attains – at least in theory – the unstable equilibrium position; in other words, the values of $y(t)$ are far from π. Anyway, this happens for subunitary values of $\left|\frac{\beta}{4\lambda}\right|$.

6.6. PERIODIC LEM SOLUTIONS FOR THE NONLINEAR PENDULUM

Let us get back to the nonlinear rigid pendulum, to search for periodic solutions such that the period be better emphasized. Therefore, we consider again equation (6.1.2). We though take into consideration the conclusions of section 6.3 concerning the period; this is why we associate to (6.1.2) the Cauchy conditions

$$y(0)=\alpha,\quad \dot{y}(0)=0. \tag{6.6.1}$$

6.6.1. THE AVERAGED SYSTEM

To apply LEM, we firstly set up the equivalent to (6.1.2) first order ODS

$$\dot{y} = z, \quad \dot{z} = -\omega^2 \sin y + A\cos\theta t, \tag{6.6.2}$$

where θ is the frequency of the perturbation and $2\pi/\theta$ is its period. We expect a periodic solution, of period T close to that of the perturbation, hence we can try solutions of the form

$$y = u\cos\theta t - v\sin\theta t, \quad z = -u\sin\theta t - v\cos\theta t. \tag{6.6.3}$$

Introducing this in (6.6.2), we get a new nonlinear first order ODS, having u and v as unknown functions; we apply to the system the averaging method, in order to avoid secular terms and also the dependence on t of the coefficients. We get

$$\dot{u} = \left(\theta - \frac{1}{2}\right)v + \frac{\omega^2}{T}\int_0^T \sin y \sin\theta t\,\mathrm{d}t, \quad \dot{v} = -\left(\theta - \frac{1}{2}\right)u + \frac{\omega^2}{T}\int_0^T \sin y\cos\theta t\,\mathrm{d}t - \frac{A}{2}. \tag{6.6.4}$$

Let us consider for $\sin y = \sin(u\cos\theta t - v\sin\theta t)$ the expansion

$$\sin y = \sum_{k=1}^{\infty} \frac{(u\cos\theta t - v\sin\theta t)^{2k-1}}{(2k-1)!}(-1)^{k-1}. \tag{6.6.5}$$

We must compute

$$\frac{1}{T}\int_0^T \sin y\sin\theta t\,\mathrm{d}t = \frac{1}{T}\sum_{k=1}^{\infty}\int_0^T (u\cos\theta t - v\sin\theta t)^{2k-1}\sin\theta t\,\mathrm{d}t. \tag{6.6.6}$$

According to the tables [3], the coefficient of v^{2k-1} in the above expansion is

$$\frac{1}{T}\int_0^T \sin^{2k}\theta t\,\mathrm{d}t = \frac{1}{2\pi}\int_0^{2\pi}\sin^{2k}x\,\mathrm{d}x = \frac{(2k-1)!!}{2^k k!}. \tag{6.6.7}$$

Analogously,

$$\begin{aligned}\Im_{2k,2j} &\equiv \frac{1}{T}\int_0^T \cos^{2k-2j}\theta t \sin^{2j}\theta t \,\mathrm{d}t \\ &= \frac{(2k-2j-1)!!}{2k(2k-2)\cdots(2j+2)}\cdot\frac{1}{2\pi}\int_0^{2\pi}\sin^{2j}x\,\mathrm{d}x,\end{aligned} \tag{6.6.8}$$

which leads to the formula

$$\begin{aligned}\Im_{2k,2j} &= \frac{(2k-2j-1)!!}{2^{k-j}k(k-1)\cdots(j+1)}\cdot\frac{(2j-1)!!}{2^j j!} = \\ &= \frac{(2k-2j-1)!!(2j-1)!!}{2^k k!}.\end{aligned} \tag{6.6.9}$$

It results that, after averaging, the coefficient of v^{2k-1} from the expansion (6.6.5) is

$$\begin{aligned}\frac{1}{(2k-1)!}\cdot\frac{(2k-1)!!}{2^k k!} &= \frac{(2k-1)!!}{2^{k-1}\cdot 2^k k!(k-1)!(2k-1)!!} = \\ &= \frac{C_{k-1}^{j-1}}{2^{2k-1}k!(k-1)!}.\end{aligned} \tag{6.6.10}$$

We finally get

$$\begin{aligned}\frac{1}{T}\int_0^T \sin y \sin\theta t\,\mathrm{d}t &= \frac{v}{2}\sum_{k=1}^{\infty}\frac{(-1)^k}{2^{2k-1}(k-1)!k!}\sum_{j=1}^{k}C_{k-1}^{j-1}u^{2k-2j}\,v^{2j-2} \\ &= \frac{v}{2}\sum_{k=1}^{\infty}\frac{(-1)^k}{(k-1)!k!}\left(\frac{\sqrt{u^2+v^2}}{2}\right)^{2k-2},\end{aligned} \tag{6.6.11}$$

or else

$$\frac{1}{T}\int_0^T \sin y\sin\theta t\,\mathrm{d}t = -\frac{v}{\sqrt{u^2+v^2}}J_1\left(\sqrt{u^2+v^2}\right), \tag{6.6.12}$$

$J_1(x)$ being the Bessel function of first kind and index 1.

Consequently, the averaged system is of the form

$$\dot{u} = \left(\theta - \frac{1}{2}\right)v - \omega^2 \frac{v}{\sqrt{u^2 + v^2}} J_1\left(\sqrt{u^2 + v^2}\right),$$
$$\dot{v} = -\left(\theta - \frac{1}{2}\right)u + \omega^2 \frac{u}{\sqrt{u^2 + v^2}} J_1\left(\sqrt{u^2 + v^2}\right) - \frac{A}{2}; \tag{6.6.13}$$

the Cauchy data corresponding to (6.6.1) read

$$u(0) = \alpha, \quad v(0) = 0. \tag{6.6.14}$$

Remark. The real, accordingly imaginary part, of the complex function $we^{-\mathrm{i}\theta t}$ coincide precisely with the right members of the transformation (6.6.3). Hence, the averaged system (6.6.13) can be written in the more compact form

$$\dot{w} = -\mathrm{i}w\left[\left(\theta - \frac{1}{2}\right) - \frac{\omega^2}{|w|} J_1(|w|)\right] - \mathrm{i}\frac{A}{2}. \tag{6.6.15}$$

6.6.2. PERIODIC LEM SOLUTIONS FOR THE FORCED PENDULUM

In this case, the amplitude of the forcing does not vanish ($A \neq 0$), and the averaged system is non-homogeneous. In order to avoid cumbrous computation and also to put into evidence the significance of the LEM solutions, we shall translate the system such that it became homogeneous.

Consider therefore the functional equations

$$v\left[\left(\theta - \frac{1}{2}\right) - \frac{\omega^2}{\sqrt{u^2 + v^2}} J_1\left(\sqrt{u^2 + v^2}\right)\right] = 0,$$
$$-\left(\theta - \frac{1}{2}\right)u + \omega^2 \frac{u}{\sqrt{u^2 + v^2}} J_1\left(\sqrt{u^2 + v^2}\right) - \frac{A}{2} = 0, \tag{6.6.16}$$

by means of which one gets in fact the critical points of the system.

The square bracket from the first equation cannot be null, because, otherwise, the second equation would imply $A = 0$. So, $v = 0$ and the second equation (6.6.16) yields

$$-\left(\theta-\frac{1}{2}\right)u+\omega^2 J_1(u)+\frac{A}{2}=0. \tag{6.6.17}$$

The solutions of this functional equation depend on the amplitude of the forcing and they can be obtained numerically, or even geometrically, by intersecting the rightline

$$Y=\frac{1}{\omega^2}\left[\left(\theta-\frac{1}{2}\right)u-\frac{A}{2}\right] \tag{6.6.18}$$

with the graph of the function $Y=J_1(u)$ in a plane (u,Y). One can easily observe that the zeros u_0 of (6.6.17) form at most a finite set.

Let us perform the translation

$$U=u-u_0, \qquad V=v, \tag{6.6.19}$$

and then truncate the obtained ODS up to order 3. The truncated nonlinear ODS in the new unknown functions U, V reads

$$\begin{aligned}\dot U&=\varphi_1 V-2b_1UV-A_1U^2V-B_1V^3,\\ \dot V&=-\varphi_3U+a_1U^2+b_1V^2+A_1U^3+B_1UV^2,\end{aligned} \tag{6.6.20}$$

where the following notations were used

$$\begin{aligned}&a=\frac{u_0J_1'(u_0)-J_1(u_0)}{u_0^3}, \quad \varphi_1=\theta-\frac{1}{2}-\omega^2\frac{J_1(u_0)}{u_0},\\ &\varphi_3=\varphi_1-\omega^2 a u_0^2,\end{aligned} \tag{6.6.21}$$

$$A_1=\frac{\omega^2}{2}\left(a-\frac{J_1(u_0)}{u_0}\right), \quad B_1=\frac{\omega^2 a}{2}, \quad a_1=u_0A_1, \quad b_1=u_0B_1. \tag{6.6.22}$$

The Cauchy conditions associated to (6.6.20) are

$$U(0)=\beta, \quad V(0)=0, \qquad \beta=\alpha-u_0. \tag{6.6.23}$$

Let us also put

$$\Omega^2=\varphi_1\varphi_3. \tag{6.6.24}$$

The truncated of third order of the corresponding LEM matrix is

$$\mathbf{A} \equiv \begin{bmatrix} \mathbf{A}_{11} & \mathbf{A}_{12} & \mathbf{A}_{13} \\ \mathbf{0} & \mathbf{A}_{22} & \mathbf{A}_{23} \\ \mathbf{0} & \mathbf{0} & \mathbf{A}_{33} \end{bmatrix}, \tag{6.6.25}$$

where

$$\mathbf{A}_{11} = \begin{bmatrix} 0 & \varphi_1 \\ -\varphi_3 & 0 \end{bmatrix}, \ \mathbf{A}_{12} = \begin{bmatrix} 0 & -2b_1 & 0 \\ a_1 & 0 & b_1 \end{bmatrix}$$

$$, \qquad \mathbf{A}_{22} \equiv \begin{bmatrix} 0 & 2\varphi_1 & 0 \\ -\varphi_1 & 0 & \varphi_1 \\ 0 & -2\varphi_3 & 0 \end{bmatrix}, \tag{6.6.26}$$

and

$$\mathbf{A}_{23} = \begin{bmatrix} 0 & -4b_1 & 0 & 0 \\ a_1 & 0 & -b_1 & 0 \\ 0 & 2a_1 & 0 & 2b_1 \end{bmatrix}, \ \mathbf{A}_{33} = \begin{bmatrix} 0 & 3\varphi_1 & 0 & 0 \\ -\varphi_3 & 0 & 2\varphi_1 & 0 \\ 0 & -2\varphi_3 & 0 & \varphi_1 \\ 0 & 0 & -3\varphi_3 & 0 \end{bmatrix}, \tag{6.6.27}$$

$$\mathbf{A}_{13} = \begin{bmatrix} 0 & -A_1 & 0 & -B_1 \\ A_1 & 0 & B_1 & 0 \end{bmatrix}.$$

The normal LEM solution of the Cauchy problem (6.6.20), (6.6.23) reads, according to theorem 3.3,

$$\begin{aligned} U(t) &\cong \beta u_{10}^{(1)} + \beta^2 u_{20}^{(1)} + \beta^3 u_{30}^{(1)}, \\ V(t) &\cong \beta u_{10}^{(2)} + \beta^2 u_{20}^{(2)} + \beta^3 u_{30}^{(2)}, \end{aligned} \tag{6.6.28}$$

$u_{k0}^{(j)}, j = 1,2$ being the first two components of the vectors $\mathbf{U}_k(t)$ satisfying the finite linear ODS with constant coefficients

$$\begin{aligned} \dot{\mathbf{U}}_1^j &= \mathbf{A}_{11}^{\mathrm{T}} \mathbf{U}_1^j, \\ \dot{\mathbf{U}}_2^j &= \mathbf{A}_{22}^{\mathrm{T}} \mathbf{U}_2^j + \mathbf{A}_{12}^{\mathrm{T}} \mathbf{U}_1^j, \\ \dot{\mathbf{U}}_3 &= \mathbf{A}_{33}^{\mathrm{T}} \mathbf{U}_3^j + \mathbf{A}_{23}^{\mathrm{T}} \mathbf{U}_2^j + \mathbf{A}_{13}^{\mathrm{T}} \mathbf{U}_1^j, \end{aligned} \tag{6.6.29}$$

and the initial condition

$$\mathbf{U}_1^1(0)=\begin{bmatrix}1 & 0\end{bmatrix}^{\mathrm{T}}, \qquad \mathbf{U}_1^2(0)=\begin{bmatrix}0 & 1\end{bmatrix}^{\mathrm{T}} \quad \text{accordingly}, \tag{6.6.30}$$

as well as the null Cauchy conditions

$$\mathbf{U}_2^j(0)=\mathbf{0}, \quad \mathbf{U}_3^j(0)=\mathbf{0}, \quad j=1,2. \tag{6.6.31}$$

The eigenvalues of $\mathbf{A}_{11}^{\mathrm{T}}$ are $\pm \mathrm{i}\Omega$. By theorem 1.8, the eigenvalues of $\mathbf{A}_{22}^{\mathrm{T}}$ are $0, \pm 2\mathrm{i}\Omega$, those of $\mathbf{A}_{33}^{\mathrm{T}}$ being $\pm \mathrm{i}\Omega, \pm 3\mathrm{i}\Omega$. Using the Laplace transformation, with the method proposed in section 3.4, we can solve by blocks the finite linear ODS (6.6.29). We get

$$\begin{aligned}
&u_{10}^{(1)}(t)=\cos\Omega t,\\
&6\Omega^2 u_{20}^{(1)}(t)=3(b_1\varphi_3+a_1\varphi_1)+2a_1\varphi_1\cos\Omega t+(3b_1\varphi_3+a_1\varphi_1)\cos 2\Omega t,\\
&288\Omega^4 u_{30}^{(1)}(t)=-96\left(b_1\Omega^2+a_1\varphi_1^2\right)+\left[8\left(69b_1^2\varphi_3^2+24a_1b_1\Omega^2+47a_1^2\varphi_1^2\right)+\right.\\
&\qquad\left.+9\left(-3B_1\varphi_3^3-B_1\varphi_3\Omega^2+A_1\varphi_1\Omega^2+3A_1\varphi_3\Omega^2\right)\right]\cos\Omega t+\\
&\qquad+32\left(3a_1b_1\Omega^2+a_1^2\varphi_1^2\right)\cos 2\Omega t+\left[96\left(9b_1^2\varphi_3^2+10a_1b_1\Omega^2+a_1^2\varphi_1^2\right)+\right.\\
&\qquad\left.+144\left(3B_1\varphi_3^3+B_1\varphi_3\Omega^2-A_1\varphi_1\Omega^2-3A_1\varphi_3\Omega^2\right)\right]\cos 3\Omega t+\\
&\qquad+\left[48\!:\!\left(\cdot\cdot 9b_1^2\varphi_3^2+6a_1b_1\Omega^2+5a_1^2\varphi_1^2\right)\cdot\cdot+\right.\\
&\qquad\left.+18\left(\cdot\cdot 3B_1\varphi_3^3+B_1\varphi_3\Omega^2+3A_1\varphi_1\Omega^2+A_1\varphi_3\Omega^2\right)\right]\Omega t\sin\Omega t,
\end{aligned} \tag{6.6.32}$$

and also

$$\begin{aligned}
&\Omega u_{10}^{(2)}(t)=\varphi_3\sin\Omega t, \qquad 3\Omega u_{20}^{(2)}(t)=a_1(\sin\Omega t+\sin 2\Omega t),\\
&288\,\Omega^5 u_{30}^{(2)}(t)=\left[2\varphi_3\left(-81b_1^2\varphi_3^2-18a_1b_1\Omega^2+31a_1^2\varphi_1^2\right)+\right.\\
&\qquad\left.+144\left(3B_1\varphi_3^4+3B_1\varphi_3^2\Omega^2+8A_1\Omega^4\right)\right]\sin\Omega t-\\
&\qquad-64a_1^2\varphi_1\Omega^2\sin 2\Omega t+\\
&\qquad+\left[-18\left(b_1^2\varphi_3^3+a_1^2\varphi_1\Omega^2\right)+9\left(A_1\Omega^2-B_1\varphi_3^2\right)\left(\varphi_3^2+3\Omega^2\right)\right]\sin 3\Omega t+\\
&\qquad+\left[24\varphi_3\left(9b_1^2\varphi_3^2+6a_1b_1\Omega^2+5a_1^2\varphi_1^2\right)+\right.\\
&\qquad\left.+144\left(3B_1\varphi_3^4+B_1\varphi_3^2\Omega^2+3A_1\Omega^4+A_1\varphi_3^2\Omega^2\right)\right]\Omega t\sin\Omega t.
\end{aligned} \tag{6.6.33}$$

These expressions must now be replaced in (6.6.28), to obtain U and V. An elementary computation leads to the corresponding to u and v formulae. The final step is to introduce the LEM solutions found for u and v in (6.6.3), to get back to y and z; this represents the solution offered by LEM to the problem of the nonlinear forced pendulum.

6.6.3. PERIODIC LEM SOLUTIONS FOR THE FREE PENDULUM

In this case, $A=0$, therefore we consider equation (6.3.1), which is equivalent to the following first order nonlinear ODS

$$\dot{y}=z, \qquad \dot{z}=-\omega^2 \sin y. \tag{6.6.34}$$

to be solved under the Cauchy conditions deduced from (6.6.1)

$$y(0)=\alpha, \quad z(0)=0; \tag{6.6.35}$$

these conditions were chosen, as in the previous case, such that to better emphasize the period T of that motion by which the pendulum attains the point of null velocity at the angle α. The relation between T and α is expressed in form of the series

$$T(\alpha)=\frac{2\pi}{\omega}\left(1+\frac{\alpha^2}{16}+\frac{11}{3\cdot 2^{10}}\alpha^4+\ldots\right), \tag{6.6.36}$$

deduced either in a classical frame, or by LEM, as shown in section 6.3.

The series (6.6.36) is rapidly convergent, as $\alpha^2/16$ is subunitary; for convenient values of α, e.g., around 0.5, we can even approximate T only with the first two terms in the series.

Let be now

$$\varphi=\frac{2\pi}{T}. \tag{6.6.37}$$

We firstly apply the rotation

$$\begin{aligned} y &= u\cos\varphi t - v\sin\varphi t, \\ z &= -u\sin\varphi t - v\cos\varphi t; \end{aligned} \tag{6.6.38}$$

introducing these expressions in (6.6.34), we obtain a new nonlinear ODS, of unknown functions u and v. Applying, as above, the method of averaging, we find the first order nonlinear ODS with constant coefficients, analytic in u and v

$$\begin{aligned} \dot{u} &= \left(\varphi-\frac{1}{2}\right)v-\omega^2\frac{v}{\sqrt{u^2+v^2}}J_1\left(\sqrt{u^2+v^2}\right), \\ \dot{v} &= -\left(\varphi-\frac{1}{2}\right)u+\omega^2\frac{u}{\sqrt{u^2+v^2}}J_1\left(\sqrt{u^2+v^2}\right). \end{aligned} \tag{6.6.39}$$

Unlike the preceding case, this system is homogeneous and odd with respect to u and v. Hence, the associated LEM matrix will contain only cells with odd indices

$$\mathbf{A} = \begin{bmatrix} \mathbf{A}_{11} & \mathbf{A}_{13} & \mathbf{A}_{15} & \cdots \\ \mathbf{0} & \mathbf{A}_{33} & \mathbf{A}_{35} & \cdots \\ \mathbf{0} & \mathbf{0} & \mathbf{A}_{55} & \cdots \\ \cdots & \cdots & \cdots & \cdots \end{bmatrix}. \tag{6.6.40}$$

Let us use the notation

$$\varphi' = \varphi - \frac{\omega^2 + 1}{2}. \tag{6.6.41}$$

The square diagonal cells are

$$\mathbf{A}_{2j-1,2j-1} = \begin{bmatrix} 0 & (2j-1)\varphi' & 0 & \cdots & 0 & 0 \\ -\varphi' & 0 & (2j-2)\varphi' & \cdots & 0 & 0 \\ 0 & -2\varphi' & 0 & \cdots & 0 & 0 \\ \cdots & \cdots & \cdots & \cdots & \cdots & \cdots \\ 0 & 0 & 0 & \cdots & 0 & \varphi \\ 0 & 0 & 0 & \cdots & -(2j-1)\varphi' & 0 \end{bmatrix}. \tag{6.6.42}$$

The first cell

$$\mathbf{A}_{11} = \begin{bmatrix} 0 & \varphi' \\ -\varphi' & 0 \end{bmatrix} \tag{6.6.43}$$

has the eigenvalues

$$\lambda_{1,2} = \pm\, \mathrm{i}\varphi'. \tag{6.6.44}$$

Theorem 1.8 gives the eigenvalues of the matrices $\mathbf{A}_{2j-1,2j-1}$ in the form

$$\lambda_{1,2} = \pm\, \mathrm{i}\varphi', \quad \lambda_{3,4} = \pm\, 3\mathrm{i}\varphi', \ldots, \lambda_{2j-1,2j} = \pm\, (2j-1)\mathrm{i}\varphi'. \tag{6.6.45}$$

So, the corresponding normal LEM representation contains only sines and cosines, possibly multiplied by powers of t.

For five order effects, only the rectangular matrices

$$\mathbf{A}_{13}^{T}=\frac{\omega^2}{16}\begin{bmatrix}0 & -1\\ 1 & 0\\ 0 & -1\\ 1 & 0\end{bmatrix},\ \mathbf{A}_{15}^{T}=\frac{\omega^2}{384}\begin{bmatrix}0 & 1\\ -1 & 0\\ 0 & 2\\ -2 & 0\\ 0 & 2\\ -1 & 0\end{bmatrix},\ \mathbf{A}_{35}^{T}=\frac{\omega^2}{16}\begin{bmatrix}0 & -1 & 0 & 0\\ 3 & 0 & -2 & 0\\ 0 & 1 & 0 & -3\\ 3 & 0 & -1 & 0\\ 0 & 2 & 0 & -3\\ 0 & 0 & 1 & 0\end{bmatrix}. \tag{6.6.46}$$

are necessary.

To get the normal LEM solution, one must solve the finite linear ODS with constant coefficients

$$\begin{aligned}\frac{\mathrm{d}\mathbf{U}_1}{\mathrm{d}t}&=\mathbf{A}_{11}^{\mathrm{T}}\mathbf{U}_1,\\ \frac{\mathrm{d}\mathbf{U}_3}{\mathrm{d}t}&=\mathbf{A}_{13}^{\mathrm{T}}\mathbf{U}_1+\mathbf{A}_{33}^{\mathrm{T}}\mathbf{U}_3,\\ \frac{\mathrm{d}\mathbf{U}_5}{\mathrm{d}t}&=\mathbf{A}_{15}^{\mathrm{T}}\mathbf{U}_1+\mathbf{A}_{35}^{\mathrm{T}}\mathbf{U}_3+\mathbf{A}_{55}^{\mathrm{T}}\mathbf{U}_5.\end{aligned} \tag{6.6.47}$$

composed of three blocks, that can be solved separately, starting with the first one.

The Cauchy condition associated to the last blocks are

$$\mathbf{U}_3(0)=\mathbf{0},\quad \mathbf{U}_5(0)=\mathbf{0}, \tag{6.6.48}$$

for both u and v.

The first block must be solved with the condition

$$\mathbf{U}_1(0)=\begin{bmatrix}1 & 0\end{bmatrix}^{\mathrm{T}}, \tag{6.6.49}$$

to get u, and the corresponding to v conditions are

$$\mathbf{U}_1(0)=\begin{bmatrix}0 & 1\end{bmatrix}^{\mathrm{T}}. \tag{6.6.50}$$

We can use again the results of section 3.4, applying the Laplace transformation. We obtain the normal LEM solution of 5th order, for the averaged system, in the form

$$\begin{aligned}u(t)\cong{}&\alpha\cos\varphi' t-\alpha^3\frac{\omega^2}{16}t\sin\varphi' t+\\ &+\alpha^5\left[-\left(\frac{\omega^2}{16}\right)^2 t^2\cos\varphi' t+\frac{\omega^2}{384}t\sin\varphi' t\right],\end{aligned} \tag{6.6.51}$$

$$v(t) \cong -\alpha \sin \varphi' t - \alpha^3 \frac{\omega^2}{16} t \cos \varphi' t + \\ + \alpha^5 \left[\left(\frac{\omega^2}{16} \right)^2 t^2 \sin \varphi' t + \frac{\omega^2}{384} t \cos \varphi' t \right]. \tag{6.6.52}$$

Introducing this in (6.6.38), we deduce a new LEM solution for the free pendulum, that emphasizes better its period.

6.7. THE PENDULUM: A SOLITONIC ARHETYPE?

The solitary travelling wave discovered by Russel, together with Fermi, Pasta and Ulam's problem, mark a debut and, at the same time, constitutes a foundation for the modern theory of solitons, with multiple and effective applications in various domains.

In [24], Ştefania Donescu studied the three models proposed by Fermi, Pasta and Ulam for the motion in a granular medium and found solitons for two of them; she ascertained that the form of the soliton equation corresponding to the last model coincides with the B.-E. equation and established LEM representations of the solitons starting from those obtained for the B.-E. bar.

6.7.1. FERMI, PASTA AND ULAM'S MODELS

In the study of some differential equations renowned among scientists, such as sine-Gordon, Korteweg-de Vries, or Schrödinger's nonlinear equation, a class of solutions appeared, whose mathematical properties make them similar to particles; these are called solitons. The theory of solitons practically means to elaborate, to analyse and solve some nonlinear equations of motions, allowing solutions in the form a "localized" wave (i.e., big amplitude, small width, concentration without dispersion in a space point, as, for instance, the solitary travelling wave or the kink). The theory of solitons developed rapidly especially in the last half century, partly due to the computer "boom".

The Russel experiment was considered scientifically peculiar until 1955, when Fermi, Pasta and Ulam began to study in the Los Alamos laboratories the wave propagation in granular mediums, with applications to rocks and earths [28], extending then the obtained results to classes of materials with nonlinear dynamical behaviour (e.g.: composites armed with metals, cements, ceramics, sand, etc.).

Fermi, Pasta and Ulam used the following nonlinear dimensionless PDE, describing the motion of the granular medium in a one-dimensional model

$$\frac{\partial^2 y}{\partial t^2} = A^2\left(\frac{\partial y}{\partial t}, \frac{\partial y}{\partial x}\right)\frac{\partial^2 y}{\partial x^2},$$
$$y(x,0) = y_{(0)}(x), \quad \frac{\partial y}{\partial t}(x,0) = y_{(1)}(x); \tag{6.7.1}$$

$y(x,t)$ is the displacement, bounded for $x \to \infty, t \to \infty$, $A\left(\frac{\partial y}{\partial t}, \frac{\partial y}{\partial x}\right) > 0$ is the local propagation speed and $y_{(0)}(x)$, $y_{(1)}(x)$ are known functions.

In (6.7.1) we used the notations from [24].

The function A was empirically determined, leading to three types of dimensionless PDEs, for three types of models [25]. We consider here only one of these models, to which the following equation corresponds

$$\left(A^2 - 1\right)\frac{\partial e}{\partial x} - \eta y\left(1 + \delta e^2\right)^{3/2} = 0,$$
$$A = A(e, f), \quad e(x,0) = e_{(0)}(x); \tag{6.7.2}$$

here, $e_{(0)}(x)$ is known, η is one of the Lamé coefficients and δ is determined by the rotation degree of the granule.

Eliminating A between (6.7.1) and (6.7.2) and then searching for solutions of the form $y = y(kx + \omega t)$, where k is the wave number and ω – the circular frequency, Ştefania Donescu obtained the following nonlinear ODE

$$z'' + c^2 z\left(1 + z'^2\right)^{3/2} = 0, \quad z = \sqrt{k\delta}\, y, \tag{6.7.3}$$

where

$$z' = \frac{\mathrm{d}z}{\mathrm{d}u}, \quad z'' = \frac{\mathrm{d}^2 z}{\mathrm{d}u^2}, \quad u = kx + \omega t \tag{6.7.4}$$

and c is expressed as

$$c^2 = \frac{\eta\sqrt{\delta k}}{k^2 - \omega^2}. \tag{6.7.5}$$

This equation formally coincides with the B.-E. equation for the elastic bar displacement, as specified in chapter 5.

To (6.7.3), the following Cauchy conditions are associated

$$z(0) = 0, \quad z'(0) = z_0 . \tag{6.7.6}$$

By applying LEM, Ştefania Donescu proved

Theorem 6.2 [24]. *The solution of the Cauchy problem* (6.7.3), (6.7.6) *allows the following normal LEM representation*

$$y(u) = a_1 \sin du - a_2 u \cos du + a_3 \sin 3du + ... \tag{6.7.7}$$

where the dots represent higher order terms, the coefficients a_j *are given by*

$$a_1 = \frac{y_{(0)}}{d\sqrt{\delta k}} \left[1 + \frac{9 y_{(0)}^2}{16 \left(1 + y_{(0)}^2\right)} \right], \quad a_2 = \frac{9 y_{(0)}^3}{16\sqrt{\delta k} \left(1 + y_{(0)}^2\right)}, \tag{6.7.8}$$

$$a_3 = \frac{a_2}{3d},$$

and

$$d = c \left(1 + y_{(0)}^2\right)^{\frac{3}{4}} . \tag{6.7.9}$$

By using the LEM representation (6.7.7), in [24] there are put into evidence breather waves, appearing as a nonlinear interaction of two solitons represented by the function sech.

6.7.2. THE LINK WITH THE PENDULUM

Considering Kirkhhoff's analogy, we observe that the soliton equation (6.7.3) is equivalent with the nonlinear ODS with constant coefficients

$$\frac{\mathrm{d}\theta}{\mathrm{d}s} = -c^2 z, \quad \frac{\mathrm{d}z}{\mathrm{d}s} = \sin\theta, \quad \mathrm{d}s = \sqrt{1 + z'^2}\,\mathrm{d}u, \tag{6.7.10}$$

that obviously leads to the free pendulum equation

$$\frac{\mathrm{d}^2\theta}{\mathrm{d}s^2} + c^2 \sin\theta = 0 . \tag{6.7.11}$$

We associate to (6.7.10) the following Cauchy conditions, deduced from (6.7.6) by putting $z_0 = \tan\beta$

$$\theta(0)=\beta, \quad z(0)=0. \tag{6.7.12}$$

6.7.3 THE NORMAL LEM REPRESENTATION OF THE SOLUTION

We can straightforwardly apply the LEM formulae obtained for the nonlinear free pendulum.

We get [152, 153]

$$\begin{aligned}
\theta(s) &\cong \beta\cos cs + \frac{\beta^3}{3\cdot 2^6}\left(-\cos 3cs + \cos cs + 12cs\sin cs\right) + \\
&\quad + \frac{\beta^5}{15\cdot 2^{12}}\Theta(cs), \\
z(s) &\cong \frac{\beta}{c}\sin cs - \frac{\beta^3}{3\cdot 2^6 c}\left(3\sin 3cs + 11\sin cs + 12cs\cos cs\right) - \\
&\quad - \frac{\beta^5}{15\cdot 2^{12}c^2}Z(cs),
\end{aligned} \tag{6.7.13}$$

where

$$\begin{aligned}
\Theta(\xi) &= 3\cos 5\xi - 20\cos 3\xi + 17\cos\xi - 60\left(\xi\sin 3\xi + 2\xi^2\sin\xi\right), \\
Z(\xi) &= -15\sin 5\xi - 17\sin\xi - 60\left(3\xi\cos 3\xi + 4\xi\sin\xi + 2\xi^2\cos\xi\right).
\end{aligned} \tag{6.7.14}$$

This representation emphasizes effects up to order 5 inclusive.

The above considerations show that the pendulum may be considered as an abstract kernel of the solitonic phenomenon in Fermi, Pasta and Ulam's model. This fact restrains, on the one hand, the types of motion in the nonlinear frame and, on the other hand, points out the mathematical pendulum as a possible solitonic archetype.

Another benefit of this study is the practical possibility to get solitons, giving the physical constants beforehand computed values and then controlling the behaviour of the solution through their corresponding LEM representations.

6.8. THE COUPLED PENDULUM

This problem was studied in detail by Ştefania Donescu [21-25], who, in order to solve it, applied comparatively three methods: LEM, the cnoidal method and the analysis by wavelets. In the present section, there are presented the results obtained by Ştefania Donescu in connection with LEM.

To establish the mathematical model, a mechanical system was considered, composed of two physical pendulums connected to each other by an elastic arc; this pendulums, oscillating simultaneously, form the coupled pendulum (fig.6.1).

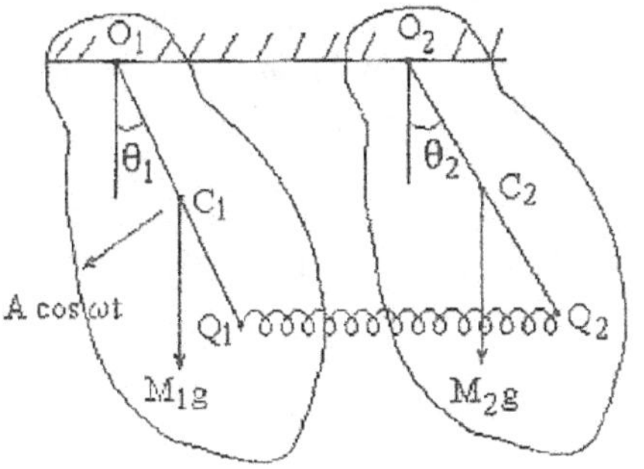

Figure 6.1. The coupled pendulum

The motion of this system was modelled according to energetic principles. The Euler-Lagrange theorem finally led to the dimensionless nonlinear ODS corresponding to the free coupled pendulum

$$\begin{aligned}&\ddot{\theta}_1 + w\sin\theta_1 + \alpha\gamma(\theta_1,\theta_2)[\cos\theta_1 + \xi\sin(\theta_2 - \theta_1)] = \delta\cos\tilde{\omega}\tau,\\&\ddot{\theta}_2 + \beta w\sin\theta_2 + \tilde{\alpha}\gamma(\theta_1,\theta_2)[\cos\theta_2 + \xi\sin(\theta_2 - \theta_1)] = 0,\end{aligned} \tag{6.8.1}$$

where θ_1, θ_2 are the unknown functions, representing accordingly the angles of the pendulum rods with the vertical equilibrium position and $\alpha, \tilde{\alpha}, \beta, w, \xi, \delta, \tilde{\omega}$ are known constants, determined by means of the physical constants of the system – masses, moments of inertia, lengths, etc. The function γ is given by

$$\begin{aligned}&\gamma(\theta_1,\theta_2) = \Phi^{-\frac{1}{2}}(\theta_1,\theta_2) - 1,\\&\Phi^{-\frac{1}{2}}(\theta_1,\theta_2) = 1 + 2\xi(\sin\theta_2 - \sin\theta_1) + 2\xi^2[1 - \cos(\theta_2 - \theta_1)].\end{aligned} \tag{6.8.2}$$

The system (6.8.1) was then reduced to a Bolotin type system

$$\frac{dz_j}{dt} = \sum_{k=1}^{6}\sum_{|\mu^{(k)}|=1}^{6} b_{\mu^{(k)}} z^{\mu^{(k)}}, \quad j = \overline{1,6}, \tag{6.8.3}$$

where $\mu^{(k)}$ are multiindices with k components and $b_{\mu^{(k)}}$ are known constants, perfectly determined. The associated initial conditions are

$$z_j(0) = z_{j0}, \qquad z_{j0} \in \Re,\ j = \overline{1,6}. \tag{6.8.4}$$

The Bolotin system (6.8.3) was treated by LEM, thus obtaining

Theorem 6.3 [21,23,24,25]. *The nonlinear polynomial ODS* (6.8.3) *with the initial conditions* (6.8.4), *modelling the motion of the free coupled pendulum provided* $\xi \le 0.3$, *allows a LEM solution, bounded for* $t \to \infty$, *in the form*

$$z_j(t) = \sum_{k,\eta} \left[B_{jk}(\eta) F_k(\mu t, \eta) + C_{jk}(\eta) \dot{F}_k(\mu t, \eta) \right], \tag{6.8.5}$$

where

$$F_k(\mu t, \eta) = C_k(\eta)(\mu t)^{k+1} \Phi_k(\mu t, \eta) \tag{6.8.6}$$

are Coulomb-type vibration functions and

$$\Phi_k(\mu t, \eta) = \sum_{m=k+1} (\mu t)^{m-k-1} A_m^{(k)}(\eta). \tag{6.8.7}$$

The constants $B_{jk}, B_{jk}, A_m^{(k)}$ *are obtained by recurrence and the sums from formulae* (6.8.5) *and* (6.8.7) *are computed according to the following correspondence rule*

$$\begin{aligned} &\eta = 0 \to k = \overline{0,9}, && \eta = 1 \to k = \overline{10,34}, \\ &\eta = 2 \to k = \overline{35,69}, && \eta = 3 \to k = \overline{70,159}. \\ &\eta = 4 \to k = \overline{160,289}. \end{aligned} \tag{6.8.8}$$

The LEM representation (6.8.5) is permanent and characterizes the long term behaviour of the solution of the coupled pendulum.

In [24], Ştefania Donescu established cnoidal representations for the coupled pendulum; she showed, on particular cases, that both LEM and cnoidal solutions lead to coinciding results.

6.9. THE LEM ANALYSIS FOR THE DOUBLE PENDULUM MOTION

The nonlinear pendulum represents, as previously noticed, a reference example of the classical mechanics. The double pendulum, despite his apparent simplicity, shows a complex and rich behaviour, very suitable for a qualitative study of chaotic motions in a nonlinear frame. Four centuries after Galilei, it became again an object of study, this time as a chaotic system.

Veturia Chiroiu, Ligia Munteanu and Traian Badea had the idea to apply LEM to the nonlinear model of the double pendulum [56]. Staring from the first linear equivalent, the authors established LEM solutions in the form of a superposition of Coulomb type vibrations; using these solutions, they could emphasize chaotic motions.

The obtained LEM solutions are of a remarkable accuracy and they capture the influence of all the coefficients of the nonlinear operator modelling the phenomenon.

Along with the qualitative study based on LEM solutions, which is itself a first, the authors initiated by this paper a new method of getting LEM solutions starting from the first linear equivalent, also valid for other nonlinear models, similar to that of the double pendulum.

In this section, we shall present the results obtained by Veturia Chiroiu, Ligia Munteanu and Traian Badea in [56], using the notations and the ideas of this paper.

6.9.1. THE EQUATIONS OF MOTION

A double pendulum is considered, subjected to a periodic non-conservative load. The pendulum is composed of two straight rods $\overline{O_1O_2}$ and $\overline{O_3O_4}$, of lengths l_1, l_2, of masses M_1, M_2 and mass centres C_1, C_2 respectively; the rods are articulated in O_3 and suspended in O_1, freely oscillating without friction in the vertical plane xO_1y (fig.6.2).

The forces acting upon the rods are, firstly, the weight bars; the corresponding generalised forces are

$$G_1 = -M_1 g l_1 \sin\theta_1 - M_1 g l \sin\theta_1, \qquad G_2 = -M_2 g l_2 \sin\theta_2. \qquad (6.9.1)$$

The authors consider three cases of non-conservative loads:

A. a generalised force $A\cos\omega t$, in O_4, $[A] = \dfrac{Kg \cdot \mathrm{m}^2}{\mathrm{s}^2}$;

B. a generalised force $P\cos\omega t$ along $\overline{O_3O_4}$, $[P]=\dfrac{Kg\cdot m}{s^2}$;

C. a vertical force $P\cos\omega t$, in C_2.

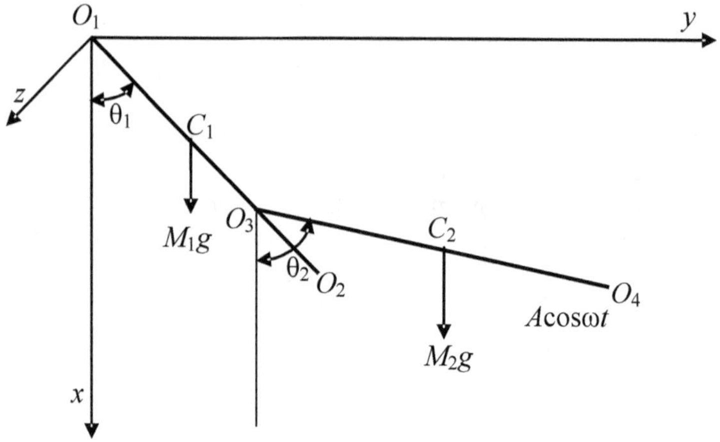

Figure 6.2. The double pendulum

The expressions of these generalised forces are given in table 6.3.

Table 6.3

The expressions of the generalised forces in the cases A, B, C.

	A	**B**	**C**
Q_1	**0**	$Pl\sin(\theta_2-\theta_1)\cos\omega t$	$-Pl\sin\theta_1\cos\omega t$
Q_2	$A\cos\omega t$	**0**	$-Pl_2\sin\theta_2\cos\omega t$

The variational model involves Lagrange's equations

$$\begin{aligned}
&\frac{\mathrm{d}}{\mathrm{d}t}\left(\frac{\partial T}{\partial\dot\theta_1}\right)-\frac{\partial T}{\partial\theta_1}=G_1+Q_1,\\
&\frac{\mathrm{d}}{\mathrm{d}t}\left(\frac{\partial T}{\partial\dot\theta_2}\right)-\frac{\partial T}{\partial\theta_2}=G_2+Q_2.
\end{aligned}\tag{6.9.2}$$

After computation, there are obtained the following dimensionless equations of motions

$$\begin{aligned}
&\ddot{\theta}_1 + \alpha\left[\ddot{\theta}_2 \cos(\theta_2 - \theta_1) - \dot{\theta}_2^2 \sin(\theta_2 - \theta_1)\right] + \beta \sin\theta_1 = q_1, \\
&\ddot{\theta}_2 + \gamma\left[\ddot{\theta}_1 \cos(\theta_2 - \theta_1) + \dot{\theta}_1^2 \sin(\theta_2 - \theta_1)\right] + \sin\theta_2 = q_2,
\end{aligned} \tag{6.9.3}$$

where the independent variable t is also dimensionless and the dimensionless coefficients α, β, γ depend on the rod lengths and on the massic quantities of the system; their full expressions can be found in [56]. The terms q_1, q_2 are given in table 6.4.

Table 6.4

The expressions of q_1, q_2 in the cases A, B, C

	A	B	C
q_1	**0**	$\nu \sin(\theta_2 - \theta_1)\cos\omega t$	$-\nu \sin\theta_1 \cos\omega t$
q_2	$\delta \cos\omega t$	**0**	$-\tilde{\mu} \sin\theta_2 \cos\omega t$

In this table,

$$\delta = \frac{A}{M_2 g l_2}, \quad \tilde{\mu} = \frac{P}{M_2 g}, \quad \nu = \frac{4}{3}\alpha\mu. \tag{6.9.4}$$

By using the change of functions

$$\begin{aligned}
&z_1 = \theta_1, \quad z_2 = \theta_2, \quad z_3 = \dot{\theta}_1, z_4 = \dot{\theta}_2, \\
&z_5 = \cos\omega t, \quad z_6 = -\omega \sin\omega t,
\end{aligned} \tag{6.9.5}$$

the implicit nonlinear ODS (6.9.3) can be written explicitly

$$\begin{aligned}
&\dot{z}_1 = z_3, \quad \dot{z}_2 = z_4, \\
&\dot{z}_3 = \frac{1}{h(\mathbf{z})}\left[-\beta \sin z_1 + \alpha \sin z_2 \cos(z_2 - z_1) + \alpha z_4^2 \sin(z_2 - z_1) + \right. \\
&\quad \left. + \alpha\gamma z_3^2 \sin(z_2 - z_1)\cos(z_2 - z_1) - A_3(\mathbf{z}) + \alpha A_4(\mathbf{z})\cos(z_2 - z_1)\right] \\
&\dot{z}_4 = \frac{1}{h(\mathbf{z})}\left[-\sin z_2 + \beta\gamma \sin z_1 \cos(z_2 - z_1) - \gamma z_3^2 \sin(z_2 - z_1) - \right. \\
&\quad \left. - \alpha\gamma z_4^2 \sin(z_2 - z_1)\cos(z_2 - z_1) - A_4(\mathbf{z}) + \gamma A_3(\mathbf{z})\cos(z_2 - z_1)\right] \\
&\dot{z}_5 = z_6, \quad \dot{z}_6 = -\omega^2 z_5.
\end{aligned} \tag{6.9.6}$$

Here, $\mathbf{z} = \left[z_j\right]^{\mathrm{T}}_{j=\overline{1,6}}$ and

$$h(\mathbf{z}) \equiv 1 - \alpha\gamma\cos^2(z_2 - z_1), \quad h(\mathbf{z}) \neq 0; \tag{6.9.7}$$

the expressions of $A_3(\mathbf{z})$, $A_4(\mathbf{z})$ are given in table 6.5.

The ODS (6.9.6) can also be written in the standard form, easier handled

$$\dot{z}_n = \sum_{p=1}^{6} A_{np} z_p + G_n(\mathbf{z}), \qquad G_n(\mathbf{z}) = \frac{g_n(\mathbf{z})}{h(\mathbf{z})}, \quad n = \overline{1,6}, \tag{6.9.8}$$

A_{np}, g_n being obtained by identification from the ODS (6.9.6).

Table 6.5

The expressions of $A_3(\mathbf{z}), A_4(\mathbf{z})$ in the cases A, B, C

	A	**B**	**C**
$A_3(\mathbf{z})$	**0**	$-\nu z_5 \sin(z_2 - z_1)$	$\nu z_5 \sin z_1$
$A_4(\mathbf{z})$	$-\delta z_5$	**0**	$\tilde{\mu} z_5 \sin z_2$

There are also considered the initial conditions of the form

$$\begin{aligned} &z_1(0) = z_{10}, \quad z_2(0) = z_{20}, \quad z_3(0) = z_{30}, \\ &z_4(0) = z_{40}, \quad z_5(0) = 1, \quad\;\; z_6(0) = 0\,. \end{aligned} \tag{6.9.9}$$

With these preparations, one applies LEM to the ODS (6.9.8).

6.9.2. THE LEM SOLUTION

The LEM mapping depends on six parameters $\boldsymbol{\sigma} = \left[\sigma_j\right]_{j=\overline{1,6}}$

$$v(t,\boldsymbol{\sigma}) = \mathrm{e}^{\langle \boldsymbol{\sigma}, \mathbf{z} \rangle}, \tag{6.9.10}$$

and the first LEM equivalent reads

$$\frac{\partial v}{\partial t} = \langle \boldsymbol{\sigma}, \mathbf{G}(D_{\boldsymbol{\sigma}}) \rangle\, v, \qquad \langle \boldsymbol{\sigma}, \mathbf{G}(D_{\boldsymbol{\sigma}}) \rangle \equiv \sum_{n=1}^{6} \sigma_n G_n(D_{\boldsymbol{\sigma}}). \tag{6.9.11}$$

The Cauchy conditions (6.9.9) become for v

$$v(0,\boldsymbol{\sigma})=\mathrm{e}^{\sum_{j=1}^{4}\sigma_j z_{j0}+\sigma_5} . \tag{6.9.12}$$

Veturia Chiroiu had the idea to search solutions for the first linear equivalent in the form

$$\begin{aligned} v(t,\boldsymbol{\sigma}) = {} & 1+\sum_{k=1}^{6}\sum_{i=1} A_k^i \frac{\sigma_k^i}{i!}+\sum_{\substack{k,l=1\\ k\neq l}}^{6}\sum_{i,j=1} A_k^i A_l^j \frac{\sigma_k^i}{i!}\frac{\sigma_l^j}{j!}+ \\ & +\sum_{\substack{k,l,m=1\\ k\neq l\neq m}}^{6}\sum_{i,j,r=1} A_k^i A_l^j A_m^r \frac{\sigma_k^i}{i!}\frac{\sigma_l^j}{j!}\frac{\sigma_m^r}{r!}+\ldots \end{aligned} \tag{6.9.13}$$

This expression is introduced in (6.9.11), after the series development of the trigonometric functions; the unknown coefficients $A_n(t),\, n=\overline{1,6}$ are then determined by identifying the same powers of σ and t.

It should be mentioned that, although the corresponding LEM equivalent system is not used here, the Jacobian matrix becomes relevant also in this procedure

$$\mathbf{A}=\begin{bmatrix} 0 & 0 & 1 & 0 & 0 & 0 \\ 0 & 0 & 0 & 1 & 0 & 0 \\ -\beta\zeta & \alpha\zeta & 0 & 0 & -\alpha\delta\zeta & 0 \\ \beta\gamma\zeta & -\zeta & 0 & 0 & \delta\zeta & 0 \\ 0 & 0 & 0 & 0 & 0 & 1 \\ 0 & 0 & 0 & 0 & -\omega^2 & 0 \end{bmatrix}, \quad \zeta=1+\alpha\gamma . \tag{6.9.14}$$

The associated characteristic equation is

$$\left[\lambda^4+(\beta+1)\zeta\lambda^2+\beta\zeta^2(1-\alpha\gamma)\right]\left(\lambda^2+\omega^2\right)=0, \tag{6.9.15}$$

therefore the eigenvalues of **A** are all purely imaginary

$$\pm \mathrm{i}p_1, \quad \pm \mathrm{i}p_2, \quad \pm \mathrm{i}\omega . \tag{6.9.16}$$

Denoting by

$$\mu = \sum_{j=1}^{6} \alpha_j \lambda_j, \quad \sum_{j=1}^{6} \alpha_j = \eta, \quad \eta = \overline{0,6}, \tag{6.9.17}$$

the authors of [56] eventually obtained the following LEM representation for the solution of the nonlinear double pendulum problem

$$z_n(t) \equiv A_n(t) = \sum_{\substack{k=\overline{0,k_{\max}} \\ \eta=0,6}} \left[B_{nk}(\eta) F_k(\mu t, \eta) + C_{nk}(\eta) \dot{F}_k(\mu t, \eta) \right]. \tag{6.9.18}$$

In these formulae, the functions $F_k(\mu t, \eta)$ have the form of *Coulomb vibrations* [3] and the constants $B_{nk}(\eta), C_{nk}(\eta)$ depend on the initial data and on the coefficients of the nonlinear ODS. The maximum value of the index k was chosen, via numerical experiment, such that the solution capture all the significant contributions of the coefficients. It resulted $k_{\max} = 33$; for $k_{\max} > 33$, no significant differences were found in the behaviour of the solution.

The first terms of the sum in the right member of (6.9.18) represent the solution of the linear part of the system (6.9.8), corresponding to small oscillations ($n = 1,2,\ k = 0,1,\ \eta = 0,\ \omega = \delta = \tilde{\mu} = 0$)

$$z_n = \sum_{j=0}^{1} B_{nj} \sin p_j t + C_{nj} \cos p_j t, \quad n = 1,2; \tag{6.9.19}$$

here

$$B_{1j} = \left(1 - \frac{p_1^2}{\gamma}\right) B_{2j}, \quad C_{1j} = \left(1 - \frac{p_2^2}{\gamma}\right) C_{2j}, \quad j = 0,1, \tag{6.9.20}$$

the constants $B_{2j}, C_{2j}, j = 0,1$, being determined from the first four initial conditions (6.9.9). For $\eta = 0$ and $k > 2$, the solution will contain terms of the form

$$\sin(i_1 p_1 + i_2 p_2)t, \quad \cos(i_1 p_1 + i_2 p_2)t, \quad i_1 + i_2 = 1, 2, 3, \ldots, \tag{6.9.21}$$

and for $\eta \neq 0$, the solution will also show other terms, of the form

$$F((i_1 p_1 + i_2 p_2)t, \eta), \quad i_1 + i_2 = 1, 2, 3, \ldots \tag{6.9.22}$$

6.9.3. NUMERICAL STUDIES AND CONCLUSIONS

The authors of [56] proved the efficiency of the LEM solution on concrete numerical cases. For the Cauchy data for the unknown functions $\theta_1, \theta_2, \dot{\theta}_1, \dot{\theta}_2$ it was chosen the interval of variation $[-1.5, 1.5]$. Numerical simulations showed that for $\omega = \delta = \tilde{\mu} = 0$ the solution is bounded and stable. For certain nonzero values of the parameters $\omega, \delta, \tilde{\mu}$, the solution is bounded, but there were detected certain domains of variation of these parameters for which the motion becomes chaotic.

For instance, for $\delta = 1.1$ in case A, $\tilde{\mu} = 1.2$ in case B and $\tilde{\mu} = 0.9$ in case C, the solutions were found stable, bounded and multiperiodic on the chosen time interval; their behaviour does not change for greater values of *t*.

For $\delta < 1.1$ in case A, $\tilde{\mu} < 1.2$ in case B and $\tilde{\mu} < 0.9$ in case C, the solutions are multiperiodic, but they are still stable and bounded for $t \to \infty$; the expression (6.9.18) allows to emphasize the jump from a periodic solution to one with a double, triple or even higher period.

For $\delta > 1.1$ in case A, $\tilde{\mu} > 1.2$ in case B and $\tilde{\mu} > 0.9$ in case C, the solutions are unstable and not bounded for $t \to \infty$; in [56], a typical transition from regular to non-regular motions, as time increases, was emphasized. The presence of the cascade doubling period suggests a route to chaos; the amplitude of the oscillations increases and the system looses its stability. Starting from these numerical studies, the authors put into evidence chaotic attractors, which, at first sight, seem to be axiom A attractors. The authors also set up the corresponding Poincaré sections.

The LEM solutions were then compared with those obtained by applying the standard Runge-Kutta method of order 4. The error function

$$e(t) = \sqrt{\left[\theta_1^{LEM}(t) - \theta_1^{R\text{-}K}(t)\right]^2 + \left[\theta_2^{LEM}(t) - \theta_2^{R\text{-}K}(t)\right]^2} \qquad (6.9.23)$$

takes values under 10^{-4}.

In conclusion, this original procedure of determining the LEM solution for the double pendulum, presented in [56], fully proved its effectiveness. The LEM solution, expressed in the form of a superposition of Coulomb type vibrations, has the capability to explain the deep causes for which the oscillations are loosing their stability, becoming chaotic. Moreover, this procedure of getting LEM solutions may be easily applied to other nonlinear dynamical systems, whose solutions show complex behaviour.

6.10. LEM SOLUTIONS FOR TWO NEO-HOOKEAN MODELS

In this section, the normal LEM representations for the solutions of two neo-Hookean models will be established: Doru Stănescu 's model for a quarter of an automobile and the neo-Hookean simple pendulum, built up by Bhattacharryya [7].

In [83-85], Doru Stănescu set up several new neo-Hookean models, having as a starting point Bhattacharyya's results. He proposes a generalization of Bhattacharryya's model for the neo-Hookean simple pendulum and sets up two completely new, original models: the neo-Hookean double pendulum and the model for neo-Hookean suspensions. For all these models, Doru Stănescu realized a detailed study of stability, including the study of the small oscillations around the stable equilibrium positions and a comparison with the linear case.

The neo-Hookean suspensions were treated by LEM in [156,160].

6.10.1. NEO-HOOKEAN MODELS AND THEIR EQUIVALENT ODSs

The model of the neo-Hookean solid extends Hooke's law to the class of large deformations, being applicable to the plastic or rubber-like materials, also called hyperelastic. More precisely, unlike linear elastic materials, the response to an applied stress of neo-Hookean ones is nonlinear .

The first neo-Hookean solid model was proposed by Ronald Rivlin in 1948 [72]; he assumed that the extra stresses due to deformation are proportional to Finger's tensor.

Rivlin may be considered an initiator of the neo-Hookean theories; in fact, the Mooney-Rivlin solid [53] is perhaps the most important generalization of the neo-Hookean solid model, in which the strain energy is a linear combination of two invariants of Finger's tensor.

In time, the neo-Hookean models grew more and more important, mainly due to their numerous and various applications.

Basically, for the neo-Hookean models treated here, one can consider as typical the following nonlinear ODS

$$\begin{aligned} \ddot{u} &= F(u, z_1, \dot{u}, \dot{z}_1) + \frac{1}{u} a_1(u, z_1, \dot{u}, \dot{z}_1) + \frac{1}{u^2} b_1(u, z_1, \dot{u}, \dot{z}_1), \\ \ddot{z}_1 &= G(u, z_1, \dot{u}, \dot{z}_1) + \frac{1}{u} a_2(u, z_1, \dot{u}, \dot{z}_1) + \frac{1}{u^2} b_2(u, z_1, \dot{u}, \dot{z}_1), \end{aligned} \tag{6.10.1}$$

where $F(u, z_1, \dot{u}, \dot{z}_1), G(u, z_1, \dot{u}, \dot{z}_1)$ and $a_j(u, z_1, \dot{u}, \dot{z}_1), b_j(u, z_1, \dot{u}, \dot{z}_1)$, $j = 1,2$, are analytic with respect to their variables, each of them running through $\Re$.

One can transform this system such that it contain no denominators. We associate the corresponding first order ODS

$$\begin{aligned}
&\dot{u} = w, \\
&\dot{w} = F(u, z_1, w, y) + \frac{1}{u} a_1(u, z_1, w, y) + \frac{1}{u^2} b_1(u, z_1, w, y), \\
&\dot{z}_1 = y, \\
&\dot{y} = G(u, z_1, w, y) + \frac{1}{u} a_2(u, z_1, w, y) + \frac{1}{u^2} b_2(u, z_1, w, y).
\end{aligned} \tag{6.10.2}$$

Now, let us introduce a new unknown function, tightly connected with u

$$v = \frac{1}{u}. \tag{6.10.3}$$

We obviously have

$$\dot{v} = -wv^2, \tag{6.10.4}$$

therefore the nonlinear ODS (6.10.2) takes the form of a first order polynomial ODS of degree 3, with five equations and five unknown functions

$$\begin{aligned}
&\dot{u} = w, \\
&\dot{w} = F(u, z_1, w, y) + a_1(u, z_1, w, y)v + b_1(u, z_1, w, y)v^2, \\
&\dot{z}_1 = y, \\
&\dot{y} = G(u, z_1, w, y) + a_2(u, z_1, w, y)v + b_2(u, z_1, w, y)v^2, \\
&\dot{v} = -wv^2.
\end{aligned} \tag{6.10.5}$$

Note that, by introducing some Cauchy data, one can see that the two systems (6.10.2) and (6.10.5) are equivalent. This is immediately seen from the first and last equations of (6.10.5).

6.10.2.. NEO-HOOKEAN SUSPENSIONS

6.10.2.1. A Model for Neo-Hookean Suspensions

Doru Stănescu created the model of a quarter of automobile [83] in the form of two solids of masses m_1, m_2, oscillating in a vertical plane (fig.6.3). The mass m_1 stands for the wheel and all the connected elements and m_2 represents the quarter of the automobile.

The mass m_1 is related to the road by the linear arc of elastic constant *k*; the two masses m_1, m_2 are connected to each other by the neo-Hookean element *ENH*, for which the elastic force is expressed as

$$F_e = k_1 z - \frac{k_2}{z^2}, \tag{6.10.6}$$

where k_1, k_2 are strictly positive constants and z is the elongation of the neo-Hookean element.

Doru Stănescu isolated the two masses and then wrote for each of them Newton's law, finally establishing the system of equations of motion

$$\begin{aligned} m_1 \ddot{z}_1 &= -kz_1 + k_1(z_2 - z_1) - \frac{k_2}{(z_2 - z_1)^2} + m_1 g, \\ m_1 \ddot{z}_2 &= -k_1(z_2 - z_1) + \frac{k_2}{(z_2 - z_1)^2} + m_2 g. \end{aligned} \tag{6.10.7}$$

This is a nonlinear ODS that can be brought to the form (6.10.1).

Let us note that, from physical reasons, the difference $(z_2 - z_1)$ does not vanish.

6.10.2.2. The Equivalent Polynomial System

By the previously described procedure, we can transform (6.10.7) in a polynomial system.

We firstly use the change of function $u = z_2 - z_1$. The system in z_1 and $u = z_2 - z_1$ reads

$$\ddot{u} = \frac{k}{m_1} z_1 - \frac{k_1}{m} u + \frac{k_2}{mu^2},$$
$$\ddot{z}_1 = -\frac{k}{m_1} z_1 + \frac{k_1}{m_1} u - \frac{k_2}{m_1 u^2} + g. \tag{6.10.8}$$

In these equations, we used the notation

$$m = \frac{m_1 m_2}{m_1 + m_2}. \tag{6.10.9}$$

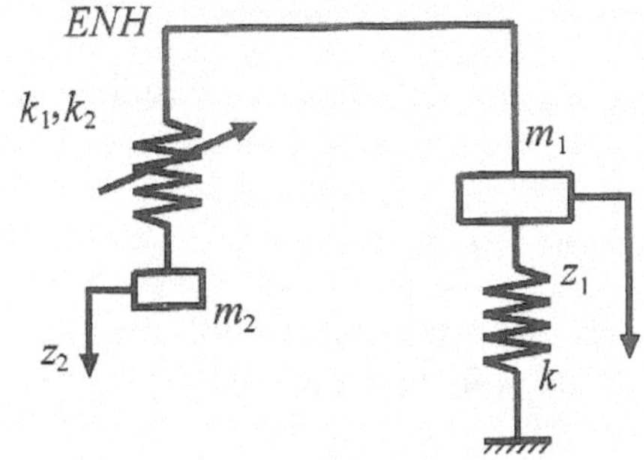

Figure 6.3. The model of a quarter of an automobile [83]

By considering the new function (6.10.3), the above nonlinear system may be finally written in the form of a polynomial first order ODS of degree 3, with five equations and five unknown functions

$$\dot{u} = w,$$
$$\dot{w} = -\frac{k_1}{m} u + \frac{k}{m_1} z_1 + \frac{k_2}{m} v^2,$$
$$\dot{z}_1 = y, \tag{6.10.10}$$
$$\dot{y} = \frac{k_1}{m_1} u - \frac{k}{m_1} z_1 - \frac{k_2}{m_1} v^2 + g,$$
$$\dot{v} = -wv^2.$$

In [83, 84], Doru Stănescu proved that the nonlinear ODS (6.10.7) has a unique critical point, given by

$$\left(z_0, 0, \xi, 0 \right), \tag{6.10.11}$$

where

$$z_0 = \frac{m_1 + m_2}{k} g, \tag{6.10.12}$$

and ξ is the unique real solution of the third degree algebraic equation

$$k_1(\xi - z_0)^3 - m_2 g(\xi - z_0)^2 - k_2 = 0. \tag{6.10.13}$$

This yields immediately the critical point for the polynomial system (6.10.10)

$$\left(u_0, 0, z_0, 0, \frac{1}{u_0} \right), \tag{6.10.14}$$

u_0 being the unique real (strictly positive) solution of the third degree equation

$$k_1 u^3 - m_2 g u^2 - k_2 = 0. \tag{6.10.15}$$

Let us perform a translation to the polynomial ODS (6.10.10), in order to move the critical point to the origin

$$U = u - u_0, W = w, Z = z_1 - \frac{m_1 + m_2}{k} g, Y = y, V = v - \frac{1}{u_0}. \tag{6.10.16}$$

With this change of functions, the ODS (6.10.10) becomes a homogeneous third degree differential operator of polynomial type

$$\begin{aligned}
\dot{U} &= W, \\
\dot{W} &= -\frac{k_1}{m} U + \frac{k}{m_1} Z + 2\frac{k_2}{m u_0} V + \frac{k_2}{m} V^2, \\
\dot{Z} &= Y, \\
\dot{Y} &= \frac{k_1}{m_1} U - \frac{k}{m_1} Z - 2\frac{k_2}{m_1 u_0} V - \frac{k_2}{m_1} V^2, \\
\dot{V} &= -\frac{1}{u_0^2} W - \frac{2}{u_0} WV - WV^2.
\end{aligned} \tag{6.10.17}$$

6.10.2.3. The normal LEM representations

Let us apply LEM to the system (6.10.17), more precisely, the theorem 3.3, in order to get the corresponding normal LEM solution. In this case, the LEM exponential mapping will depend on five parameters. The equivalent LEM system is written in the form

$$\begin{aligned}
\dot{v}_{ijsln} &= iv_{i-1,j+1,sln} \\
&+ j\left(\frac{k}{m_1}v_{i,j-1,s+1,ln} - \frac{k_1}{m}v_{i+1,j-1,sln} + 2\frac{k_2}{mu_0}v_{i,j-1,sl,n+1} + \right. \\
&\left. + \frac{k_2}{m}v_{i,j-1,sl,n+2}\right) + sv_{ij,k-1,l+1,n} \\
&+ l\left(-\frac{k}{m_1}v_{ij,s+1,l-1,n} + \frac{k_1}{m_1}v_{i+1,js,l-1,n} - \right. \\
&\left. - 2\frac{k_2}{m_1 u_0}v_{ijs,l-1,n+1} - \frac{k_2}{m_1}v_{ijs,l-1,n+2}\right) + \\
&+ n\left(-\frac{1}{u_0^2}v_{i,j+1,sl,n-1} - \frac{2}{u_0}v_{i,j+1,sln} - v_{i,j+1,sl,n+1}\right).
\end{aligned} \tag{6.10.18}$$

The LEM matrix is therefore tridiagonal

$$\mathbf{A} \equiv \begin{bmatrix} \mathbf{A}_{11} & \mathbf{A}_{12} & \mathbf{A}_{13} & \mathbf{0} & \mathbf{0} & \ldots \\ \mathbf{0} & \mathbf{A}_{22} & \mathbf{A}_{23} & \mathbf{A}_{24} & \mathbf{0} & \ldots \\ \mathbf{0} & \mathbf{0} & \mathbf{A}_{33} & \mathbf{A}_{34} & \mathbf{A}_{35} & \ldots \\ \ldots & \ldots & \ldots & \ldots & \ldots & \ldots \end{bmatrix}, \tag{6.10.19}$$

$\mathbf{A}_{11}$ is the linear part of the polynomial operator

$$\mathbf{A}_{11} = \begin{bmatrix} 0 & 1 & 0 & 0 & 0 \\ \frac{k_1}{m} & 0 & \frac{k}{m_1} & 0 & \frac{2k_2}{mu_0} \\ 0 & 0 & 0 & 1 & 0 \\ \frac{k_1}{m_1} & 0 & -k & 0 & -\frac{2k_2}{m_1 u_0} \\ 0 & -\frac{1}{u_0^2} & 0 & 0 & 0 \end{bmatrix}, \tag{6.10.20}$$

and the 5×15 matrix $\mathbf{A}_{12}$ is given by

$$\mathbf{A}_{12} = \begin{bmatrix} 0 & \dots & 0 & 0 & 0 & 0 & 0 \\ 0 & \dots & 0 & 0 & 0 & 0 & \frac{k_2}{m} \\ 0 & \dots & 0 & 0 & 0 & 0 & 0 \\ 0 & \dots & 0 & 0 & 0 & 0 & -\frac{k_2}{m_1} \\ 0 & \dots & 0 & -\frac{2}{u_0} & 0 & 0 & 0 \end{bmatrix}. \tag{6.10.21}$$

The matrices $\mathbf{A}_{kk}$ are also generated by the coefficients of the linear part of the operator and $\mathbf{A}_{13}$ contains only one non-zero term, -1, situated on the 5th row and 28th column.

Let us stop to second order effects; this is sufficient to bring in the contribution of the coefficients that really matter in the solution. According to theorem 3.3, one must solve the linear system with constant coefficients

$$\begin{aligned} &\frac{\mathrm{d}\mathbf{U}_1}{\mathrm{d}t} = \mathbf{A}_{11}^{\mathrm{T}}\mathbf{U}_1, \\ &\frac{\mathrm{d}\mathbf{U}_2}{\mathrm{d}t} = \mathbf{A}_{22}^{\mathrm{T}}\mathbf{U}_2 + \mathbf{A}_{12}^{\mathrm{T}}\mathbf{U}_1, \quad \mathbf{U}_s(t) = \left[u_\gamma(t)\right]_{|\gamma|=s}, \\ &\gamma = (i, j, s, l, n), \quad s = 1,2, \end{aligned} \tag{6.10.22}$$

with $\mathbf{U}_1$ satisfying the Cauchy conditions

$$\mathbf{U}_1(0) = \mathbf{e}_1^5 \equiv \begin{bmatrix} 1 & 0 & 0 & 0 & 0 \end{bmatrix}^{\mathrm{T}}, \quad \mathbf{U}_2(0) = \mathbf{0}, \; s = \overline{2,l}, \tag{6.10.23}$$

for the coefficients of the Cauchy data in U and

$$\mathbf{U}_1(0) = \mathbf{e}_3^5 \equiv \begin{bmatrix} 0 & 0 & 1 & 0 & 0 \end{bmatrix}^{\mathrm{T}}, \quad \mathbf{U}_2(0) = \mathbf{0}, \; s = \overline{2,l}, \tag{6.10.24}$$

for the coefficients of the Cauchy data in Z.

Let us take the following Cauchy conditions for the system (6.10.10)

$$\begin{aligned} &u(0) = u_0 + \alpha, \quad v(0) = 0, \quad z_1(0) = \frac{m_1 + m_2}{k} g + \beta, \\ &y(0) = 0, \quad v(0) = \frac{1}{u_0 + \alpha}, \end{aligned} \tag{6.10.25}$$

with α, β sufficiently small.

This results in the following initial data for the homogeneous system (6.10.17)

$$U(0)=\alpha,\quad W(0)=0,\quad Z(0)=\beta,$$
$$Y(0)=0,\quad V(0)=\eta=-\frac{\alpha}{u_0(u_0+\alpha)}. \tag{6.10.26}$$

The eigenvalues of $\mathbf{A}_{11}^{\mathrm{T}}$ – the transposed of the Jacobian matrix of (6.10.17) – are solutions of the polynomial equations

$$p\left[p^4+\left(\frac{a}{m}+\frac{k}{m_1}\right)p^2+\frac{ak}{m_1 m_2}\right]=0, \tag{6.10.27}$$

where m is given by (6.10.9) and a by

$$a=k_1+\frac{k_2}{u_0^3}, \tag{6.10.28}$$

using the same notation as in [83].

Therefore, one has

$$p_1=0,\quad p_{2,3}=\pm\,\mathrm{i}\omega,\quad p_{4,5}=\pm\,\mathrm{i}\Omega, \tag{6.10.29}$$

where

$$\omega^2=b-\sqrt{b^2-c},\qquad b=\frac{a}{m}+\frac{k}{m_1},\quad c=\frac{4ak}{m_1\,m_2}. \tag{6.10.30}$$
$$\Omega^2=b+\sqrt{b^2-c},$$

The last four eigenvalues obviously coincide with those obtained in [83].

To solve the second block in (6.10.22) we need the eigenvalues of $\mathbf{A}_{22}^{\mathrm{T}}$. This is already done by theorem 1.8. In this case, the 15×15 matrix $\mathbf{A}_{22}^{\mathrm{T}}$ has the following eigenvalues

$$p_{1,2,3}=0,\quad p_{4,5}=\pm\,\mathrm{i}\omega,\quad p_{6,7}=\pm\,\mathrm{i}\Omega,\quad p_{8,9}=\pm\,2\mathrm{i}\omega,$$
$$p_{10,11}=\pm\,2\mathrm{i}\Omega,\quad p_{12,13}=\pm\,\mathrm{i}(\Omega-\omega),\quad p_{14,15}=\pm\mathrm{i}\,(\Omega+\omega). \tag{6.10.31}$$

Knowing all these eigenvalues immediately shapes the solution of the auxiliary system (6.10.22). This system can be solved by blocks. The first block gives

$$\mathbf{U}_1(t) = c_1 \begin{bmatrix} 1 \\ 0 \\ 0 \\ 0 \\ \dfrac{1}{u_0^2} \end{bmatrix} + c_2 \begin{bmatrix} \cos \omega t \\ a_2^{\omega} \sin \omega t \\ a_3^{\omega} \cos \omega t \\ a_4^{\omega} \sin \omega t \\ a_5^{\omega} \cos \omega t \end{bmatrix} + c_3 \begin{bmatrix} \cos \Omega t \\ a_2^{\Omega} \sin \omega t \\ a_3^{\Omega} \cos \Omega t \\ a_4^{\Omega} \sin \omega t \\ a_5^{\Omega} \cos \Omega t \end{bmatrix}. \qquad (6.10.32)$$

In the above formulae,

$$a_2^{\mu} = -m_2 \mu \left(1 - \frac{m_1}{k} \mu^2 \right) f(\mu), \quad a_3^{\mu} = -m_2 \mu^2 f(\mu),$$

$$a_4^{\mu} = -m_2 \mu f(\mu), \quad a_5^{\mu} = \frac{2k_2}{k_1 u_0}, \quad f(\mu) \equiv \frac{1}{k_1 \left(1 - \dfrac{z_0}{g} \mu^2 \right)}. \qquad (6.10.33)$$

The coefficients $c_j, j = \overline{1,3}$ satisfy the system

$$\mathbf{M} \begin{bmatrix} c_1^U \\ c_2^U \\ c_3^U \end{bmatrix} = \mathbf{e}_1^5, \qquad \mathbf{M} = \begin{bmatrix} 1 & 1 & 1 \\ 0 & a_3^{\omega} & a_3^{\Omega} \\ \dfrac{1}{u_0^2} & a_5^{\omega} & a_5^{\Omega} \end{bmatrix}, \qquad (6.10.34)$$

for the representation of U and the system

$$\mathbf{M} \begin{bmatrix} c_1^Z \\ c_2^Z \\ c_3^Z \end{bmatrix} = \mathbf{e}_3^5, \qquad (6.10.35)$$

in the case of Z.

By theorem 3.3, the coefficients of α, β, η in the normal LEM representation of U and Z are, accordingly, the first, the third and the last component of $\mathbf{U}_1(t)$ given by (6.10.32).

Let us pass to the quadratics of the data. This means to solve the second block of the linear system (6.10.22). As we know the eigenvalues of the matrix $\mathbf{A}_{22}^{\mathrm{T}}$ and as this matrix has a special symmetry, we immediately see that we can search for vector solutions in trigonometric form, of components $A\cos\lambda t$ or $B\sin\lambda t$, λ taking every time one of the values

$$\begin{aligned}&\phi_1=\omega,\quad \phi_2=\Omega,\quad \phi_3=2\omega,\quad \phi_4=2\Omega,\\ &\phi_5=(\Omega-\omega),\quad \phi_6=(\Omega+\omega).\end{aligned}\tag{6.10.36}$$

The involved components are determined by identification. As ω and Ω are double roots, we shall also have terms of the form $t\cos\omega t, t\sin\omega t$ and, respectively, of the form $t\cos\Omega t, t\sin\Omega t$ in the LEM solution. The arbitrary constants multiplying the vector solutions are determined by using the null Cauchy data.

It should be mentioned that the LEM solution of Z will not contain constants or terms in the form of second degree polynomials in t, as the involved determinants obtained from $\det\left(\mathbf{A}_{22}^{\mathrm{T}}-\lambda\mathbf{E}\right)$ cancel at $\lambda=0$ together with their first and second derivative in λ. For the same reasons, the LEM solution of U will contain just constants and no polynomials in t.

We can now write the normal LEM solution of the system (6.10.17)

$$\begin{aligned}U(t)\cong{}&\alpha\left(c_1^U+c_2^U\cos\omega t+c_3^U\cos\Omega t\right)+\\&+\beta\left(c_2^U a_3^{\omega}\cos\omega t+c_3^U a_3^{\Omega}\cos\Omega t\right)+\\&+\eta\left(c_1^U\frac{1}{u_0^2}+c_2^U a_5^{\omega}\cos\omega t+c_3^U a_5^{\Omega}\cos\Omega t\right)+\\&+\alpha^2u_{20000}^U+\alpha\beta u_{11000}^U+\beta^2u_{00200}^U+\alpha\eta u_{10001}^U\\&+\beta\eta u_{00101}^U+\eta^2u_{00002}^U,\\ Z(t)\cong{}&\alpha\left(c_2^Z\cos\omega t+c_3^Z\cos\Omega t\right)+\\&+\beta\left(c_2^Z a_3^{\omega}\cos\omega t+c_3^Z a_3^{\Omega}\cos\Omega t\right)+\\&+\eta\left(c_2^Z a_5^{\omega}\cos\omega t+c_3^Z a_5^{\Omega}\cos\Omega t\right)+\\&+\alpha^2u_{20000}^Z+\alpha\beta u_{11000}^Z+\beta^2u_{00200}^Z+\\&+\alpha\eta u_{10001}^Z+\beta\eta u_{00101}^Z+\eta^2u_{00002}^Z,\end{aligned}\tag{6.10.37}$$

where u_{ijsln}^U, u_{ijsln}^Z are the components of the solution of the second block in (6.10.22), computed for U and Z accordingly. They can be shaped as follows

$$u_\gamma^U(t) = A_{0\gamma}^U + \sum_{j=1}^{6} A_{j\gamma}^U \cos\phi_j t, \qquad \gamma = (i, j, s, l, n), \tag{6.10.38}$$

for the coefficients of $\alpha^2, \alpha\beta, \beta^2, \alpha\eta$ in U and by

$$u_\gamma^Z(t) = \sum_{j=1}^{6} A_{j\gamma}^Z \cos\phi_j t, \qquad \gamma = (i, j, s, l, n) \tag{6.10.39}$$

for the corresponding coefficients in the LEM solution for Z. The remaining two coefficients have the form

$$u_\gamma^U(t) = \sum_{j=1}^{6} A_{j\gamma}^U \cos\phi_j t + A_\gamma^U t \sin\omega t + B_\gamma^U t \sin\Omega t, \qquad \gamma = (i, j, s, l, n), \tag{6.10.40}$$

in the LEM solution of U and similar form in that of Z.

Note that $c_1^Z = 0$.

The normal LEM solution thus obtained can now be used to the accurate study, both numerical and qualitative, of the neo-Hookean suspension model.

6.10.2.4. The generalized model of neo-Hookean suspensions

Doru Stănescu [83] generalized the above model considering, instead of the form (6.10.6) of the elastic force, the expression

$$F_e = k_1 z^j - \frac{k_2}{z^n}. \tag{6.10.41}$$

with j, n positive integers. This resulted in the following system of equation of motion

$$\begin{aligned} m_1\ddot{z}_1 &= -kz_1 + k_1(z_2 - z_1)^j - \frac{k_2}{(z_2 - z_1)^n} + m_1 g, \\ m_1\ddot{z}_2 &= -k_1(z_2 - z_1)^j + \frac{k_2}{(z_2 - z_1)^n} + m_2 g. \end{aligned} \tag{6.10.42}$$

Let us take $j = 1$. Performing the same steps as before, we are led to the following polynomial ODS

$$
\begin{aligned}
&\dot{u} = w,\\
&\dot{w} = -\frac{k_1}{m}u + \frac{k}{m_1}z_1 + \frac{k_2}{m}v^n,\\
&\dot{z}_1 = y,\\
&\dot{y} = \frac{k_1}{m_1}u - \frac{k}{m_1}z_1 - \frac{k_2}{m_1}v^n + g,\\
&\dot{v} = -wv^2.
\end{aligned}
\tag{6.10.43}
$$

This system allows the same critical point (6.10.14), but u_0 now satisfies the algebraic equation

$$k_1 u^{n+1} - m_2 g u^n - k_2 = 0\,, \tag{6.10.44}$$

instead of (6.10.15). As the coefficients of this equation are positive, one can suppose that (6.10.15) allows at least one positive solution; let it be u_0. Performing the translation (6.10.16) around the critical point, we are led, taking into account only second order effects, to the homogeneous polynomial ODS

$$
\begin{aligned}
&\dot{U} = W,\\
&\dot{W} = -\frac{k_1}{m}U + \frac{k}{m_1}Z + \frac{nk_2}{mu_0^{n-1}}V + \frac{K}{m}V^2,\\
&\dot{Z} = Y,\\
&\dot{Y} = \frac{k_1}{m_1}U - \frac{k}{m_1}Z - \frac{nk_2}{m_1 u_0^{n-1}}V - \frac{K}{m_1}V^2,\\
&\dot{V} = -\frac{1}{u_0^2}W - \frac{2}{u_0}WV,
\end{aligned}
\tag{6.10.45}
$$

where

$$K = \frac{n(n-1)k_2}{2u_0^{n-2}}\,. \tag{6.10.46}$$

The eigenvalues of the $\mathbf{A}_{11}^{\mathrm{T}}$ – the transposed of the Jacobian matrix of (6.10.45) – are the same as in formulae (6.10.29), (6.10.30), in which a is now given by

$$a = k_1 + \frac{nk_2}{u_0^{n+1}}. \tag{6.10.47}$$

Again, the 5×15 matrix $\mathbf{A}_{12}$ has only three non-vanishing entries; more precisely,

$$\mathbf{A}_{12} = \begin{bmatrix} 0 & \dots & 0 & 0 & 0 & 0 & 0 \\ 0 & \dots & 0 & 0 & 0 & 0 & \dfrac{K}{m} \\ 0 & \dots & 0 & 0 & 0 & 0 & 0 \\ 0 & \dots & 0 & 0 & 0 & 0 & -\dfrac{K}{m_1} \\ 0 & \dots & 0 & -\dfrac{2}{u_0} & 0 & 0 & 0 \end{bmatrix}. \tag{6.10.48}$$

Obviously, the Cauchy conditions are the same as (6.10.26).

Taking into account theorem 3.3, we can say that the normal LEM solution for U and Z, corresponding to the generalized model of suspensions, is given by formulae similar to (6.10.37); this basically means that its behaviour may be considered as being topologically equivalent with that corresponding to the model (6.10.7).

6.10.3. THE NEO-HOOKEAN SIMPLE PENDULUM

This model was established by Bhattacharyya, who also studied its stability. We shall briefly present this model and then we transform it, such that it could be easier treated by linear equivalence. The scope of this section is to get the normal LEM solution corresponding to the neo-Hookean simple pendulum.

6.10.3.1. Bhattacharyya's Neo-Hookean Pendulum

Bhattacharyya [7] considered a simple pendulum with neo-Hookean rod (fig. 6.3). Denoting by l_0 the length of the undeformed pendulum and by x the displacement, the relative and the equilibrium displacements are, accordingly

$$\lambda = \frac{l_0 + x_{st} + x}{l_0}, \quad \lambda_s = \frac{l_0 + x_{st}}{l_0}, \tag{6.10.49}$$

x_{st} being the static displacement; they are both positive.

The axial force is

$$T(\lambda) = A_0 G\left(\lambda - \frac{1}{\lambda^2}\right), \tag{6.10.50}$$

G being the shearing modulus of elasticity.

Denote by $Y(t)$ the displacement of the abutment; the corresponding equations of motion read

$$\begin{aligned} &\ddot{\lambda} - \lambda\dot{\theta}^2 + H\left(\lambda - \frac{1}{\lambda^2}\right) = \left(\omega_0^2 + \ddot{\bar{Y}}\right)\cos\theta, \\ &\lambda\ddot{\theta} + 2\dot{\lambda}\dot{\theta} + \left(\omega_0^2 + \ddot{\bar{Y}}\right)\sin\theta = 0, \end{aligned} \tag{6.10.51}$$

where

$$H = \frac{A_0 G}{m l_0}, \quad \omega_0^2 = \frac{g}{l_0}, \quad \bar{Y} = \frac{Y}{l_0}. \tag{6.10.52}$$

If

$$\bar{Y} = BH\lambda, \qquad B \in \Re, \tag{6.10.53}$$

then the equations of motion become

$$\begin{aligned} &\ddot{\lambda} - \lambda\dot{\theta}^2 + H\left[(1 - B\cos\theta)\lambda - \frac{1}{\lambda^2}\right] = \omega_0^2\cos\theta, \\ &\lambda\ddot{\theta} + 2\dot{\lambda}\dot{\theta} + \left(\omega_0^2 + BH\lambda\right)\sin\theta = 0. \end{aligned} \tag{6.10.54}$$

The critical points are given by $\theta = 0, \theta = \pi$. Let us focus on $\theta = 0$. For this case, Doru Stănescu proved in [83] that if $B < 1$, then the above system allows a unique equilibrium point. Consider now $B \in (0,1)$. This is the case to which we will apply LEM.

6.10.3.2. Normal LEM Solutions for the Neo-Hookean Pendulum

Let us divide the second equation of (6.10.54) by λ

$$\begin{aligned} &\ddot{\lambda} = \lambda\dot{\theta}^2 - H\left[(1 - B\cos\theta)\lambda - \frac{1}{\lambda^2}\right] + \omega_0^2\cos\theta, \\ &\ddot{\theta} = -2\dot{\lambda}\dot{\theta}\frac{1}{\lambda} - \left(\frac{\omega_0^2}{\lambda} + BH\right)\sin\theta; \end{aligned} \tag{6.10.55}$$

this system is, obviously, of the type (6.10.1). The associated first order ODS is

$$\begin{aligned}
&\dot{\lambda} = w, \\
&\dot{w} = \lambda y^2 - H\left[\left(1 - B\cos\theta\right)\lambda - \frac{1}{\lambda^2}\right] + \omega_0^2 \cos\theta, \\
&\dot{\theta} = y, \\
&\dot{y} = -2wy\frac{1}{\lambda} - \left(\frac{\omega_0^2}{\lambda} + BH\right)\sin\theta.
\end{aligned} \tag{6.10.56}$$

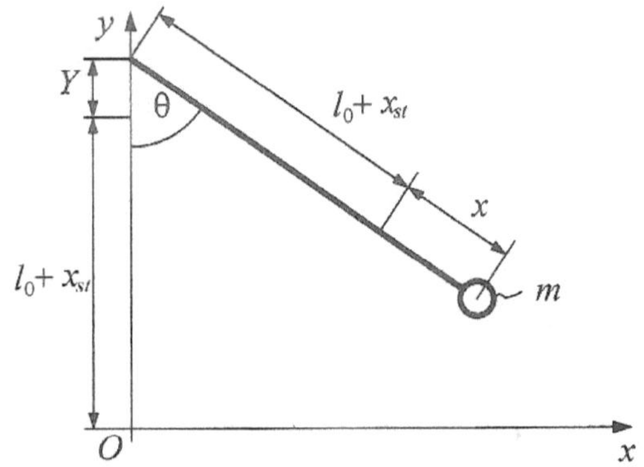

Figure 6.4. The neo-Hookean simple pendulum [85]

We introduce the new unknown function similar to (6.10.3)

$$v = \frac{1}{\lambda}, \qquad \dot{v} = -wv^2. \tag{6.10.57}$$

This yields

$$\begin{aligned}
&\dot{\lambda} = w, \\
&\dot{w} = \lambda y^2 - H\left[\left(1 - B\cos\theta\right)\lambda - v^2\right] + \omega_0^2 \cos\theta, \\
&\dot{\theta} = y, \\
&\dot{y} = -2vwy - \left(\omega_0^2 v + BH\right)\sin\theta, \\
&\dot{v} = -wv^2.
\end{aligned} \tag{6.10.58}$$

The unique critical point of this system reads

$$\left(\lambda_0, 0, 0, 0, \frac{1}{\lambda_0}\right), \tag{6.10.59}$$

where λ_0 is the unique real solution of the third degree algebraic equation

$$\lambda^3 - \frac{\omega_0^2}{H(1-B)}\lambda^2 - \frac{1}{(1-B)} = 0, \tag{6.10.60}$$

Perform now the following translation around the critical point

$$\Lambda = \lambda - \lambda_0, \quad W = w, \quad \Theta = \theta, \quad Y = y, \quad V = v - \frac{1}{\lambda_0}. \tag{6.10.61}$$

Introducing this in (6.10.58), we get

$$\begin{aligned}
&\dot{\Lambda} = W, \\
&\dot{W} = (\Lambda + \lambda_0)Y^2 - H\left[(1 - B\cos\Theta)(\Lambda + \lambda_0) - \left(V + \frac{1}{\lambda_0}\right)^2\right] + \\
&\quad + \omega_0^2 \cos\Theta, \\
&\dot{\Theta} = Y, \\
&\dot{Y} = -2WY\left(V + \frac{1}{\lambda_0}\right) \left[\omega_0^2\left(V + \frac{1}{\lambda_0}\right) + BH\right]\sin\Theta, \\
&\dot{V} = -W\left(V + \frac{1}{\lambda_0}\right)^2.
\end{aligned} \tag{6.10.62}$$

Let us take into account only effects up to second order, as for the above models. This means that we must take

$$\sin\Theta \approx \Theta, \quad \cos\Theta \approx 1 - \frac{\Theta^2}{2}. \tag{6.10.63}$$

After some elementary computation, we get a homogeneous polynomial ODS

$$\begin{aligned}
&\dot{\Lambda} = W, \\
&\dot{W} = -H(1-B)\Lambda - \frac{1}{2}\left[H(1-B)\lambda_0 + \omega_0^2\right]\Theta^2 + \lambda_0 Y^2, \\
&\dot{\Theta} = Y, \\
&\dot{Y} = -\frac{\omega_0^2 + BH\lambda_0}{\lambda_0}\Theta - \omega_0^2 V\Theta - \frac{2}{\lambda_0}WY, \\
&\dot{V} = -\frac{1}{\lambda_0^2}W - \frac{2}{\lambda_0}WV.
\end{aligned} \tag{6.10.64}$$

If

$$\begin{aligned}
&\lambda(0) = \lambda_0 + \alpha, \quad w(0) = 0, \quad \theta(0) = \beta, \\
&y(0) = 0, \quad v(0) = \frac{1}{\lambda_0 + \alpha},
\end{aligned} \tag{6.10.65}$$

with α, β sufficiently small, are some Cauchy conditions for the ODS (6.10.58), then the corresponding Cauchy conditions for (6.10.64) read

$$\begin{aligned}
&\Lambda(0) = \alpha, \quad W(0) = 0, \quad \Theta(0) = \beta, \\
&Y(0) = 0, \quad V(0) = \eta = -\frac{\alpha}{\lambda_0(\lambda_0 + \alpha)}.
\end{aligned} \tag{6.10.66}$$

In this case too, the LEM mapping will depend on five parameters. The second LEM equivalent corresponding to (6.10.64) may be written in the form

$$\begin{aligned}
&\dot{v}_{ijsln} = i v_{i-1,j+1,sln} + \\
&+ j\left[-H(1-B)v_{i+1,j-1,sln} - \Phi^2 v_{i,j-1,s+2,ln} + \lambda_0 v_{i,j-1,s,l+2,n}\right] + \\
&+ s v_{ij,s-1,l+1,n} + \\
&+ l\left(-\omega_1^2 v_{ij,s+1,l-1,n} - \omega_0^2 v_{ij,s+1,l-1,n+1} - \frac{2}{\lambda_0} v_{i,j+1,s,ln}\right) + \\
&+ n\left(-\frac{1}{\lambda_0^2} v_{i,j+1,sl,n-1} - \frac{2}{\lambda_0} v_{i,j+1,sln}\right),
\end{aligned} \tag{6.10.67}$$

where we used the notations

$$\omega^2 = \frac{\omega_0^2 + BH\lambda_0}{\lambda_0}, \quad \Phi^2 = \frac{1}{2}\left[\omega_0^2 + H(1-B)\lambda_0\right]. \tag{6.10.68}$$

These notations are fully justified, as the involved quantities are positive.

We truncate the LEM matrix up to order 2, to get

$$\mathbf{A} \equiv \begin{bmatrix} \mathbf{A}_{11} & \mathbf{A}_{12} \\ \mathbf{0} & \mathbf{A}_{22} \end{bmatrix}. \tag{6.10.69}$$

The first cell $\mathbf{A}_{11}$ is the linear part of the polynomial operator

$$\mathbf{A}_{11} = \begin{bmatrix} 0 & 1 & 0 & 0 & 0 \\ -H(1-B) & 0 & 0 & 0 & \dfrac{2H}{\lambda_0} \\ 0 & 0 & 0 & 1 & 0 \\ 0 & 0 & -\omega^2 & 0 & 0 \\ 0 & -\dfrac{1}{\lambda_0^2} & 0 & 0 & 0 \end{bmatrix}, \tag{6.10.70}$$

and the 5×15 sparse matrix $\mathbf{A}_{12}$ is given by

$$\mathbf{A}_{12} = \begin{bmatrix} 0 & \dots & 0 & 0 & 0 & 0 & 0 & 0 & 0 & 0 & 0 & 0 & 0 \\ 0 & \dots & 0 & -\Phi^2 & 0 & 0 & 0 & \lambda_0 & 0 & 0 & 0 & 0 & 0 \\ 0 & \dots & 0 & 0 & 0 & 0 & 0 & 0 & 0 & 0 & 0 & 0 & 0 \\ 0 & \dots & 0 & 0 & 0 & -\dfrac{2}{\lambda_0} & 0 & 0 & 0 & 0 & \omega_0^2 & 0 & 0 \\ 0 & \dots & 0 & 0 & 0 & 0 & 0 & 0 & 0 & -\dfrac{2}{\lambda_0} & 0 & 0 & 0 \end{bmatrix}. \tag{6.10.71}$$

The matrix $\mathbf{A}_{22}$ is also generated by the coefficients of the linear part of the operator.

According to Theorem 3.3, we must solve the same linear system (6.10.22) with the initial conditions (6.10.23) for the LEM representation of Λ and the initial problem (6.10.22), (6.10.24) to get Θ. Naturally, in this case the matrix $\mathbf{A}_{11}$ is given by (6.10.70) and $\mathbf{A}_{12}$ by (6.10.71).

To solve the finite linear system, we also need the eigenvalues of the diagonal matrices $\mathbf{A}_{11}$ and $\mathbf{A}_{22}$. The characteristic equation for $\mathbf{A}_{11}$ is

$$p\left(p^2+\Omega^2\right)\left(p^2+\omega^2\right)=0\,, \tag{6.10.72}$$

with

$$\Omega^2=H(1-B)+\frac{2H}{\lambda_0^3}\,. \tag{6.10.73}$$

The eigenvalues of $\mathbf{A}_{22}$ are exactly those from (6.10.31), only with ω expressed by (6.10.68) and Ω by (6.10.73).

We apply the Laplace transform, in order to solve the corresponding to (6.10.22) LEM system.

Let us note that the system for $\mathbf{U}_1(t)$ is decoupled. With this remark, the coefficients of the data in the LEM solution of Λ are given by the vector

$$\mathbf{U}_1^{\Lambda}(t)=\begin{bmatrix} \dfrac{1}{\Omega^2}\left[H(1-B)\cos\Omega t+\dfrac{2H}{\lambda_0^3}\right] \\ 0 \\ 0 \\ 0 \\ \dfrac{2H}{\lambda_0\Omega^2}(1-\cos\Omega t) \end{bmatrix}, \tag{6.10.74}$$

and those for Θ are components of the vector

$$\mathbf{U}_1^{\Theta}(t):=\begin{bmatrix} 0 \\ 0 \\ \cos\omega t \\ 0 \\ 0 \end{bmatrix}. \tag{6.10.75}$$

Let us apply now the Laplace transform to the second block in (6.10.22). We thus get the coefficients of the quadrates of the data in the LEM expansions.

Therefore, for Λ we obtain the Laplace transform of the coefficient of α^2 in the form

$$-\frac{6H^3(1-B)^2}{\lambda_0^3}\frac{f(p^2)}{\Delta_\Lambda(p)}, \tag{6.10.76}$$

where we used the notations

$$\begin{aligned} f(p) &= p^4 + 2p^2\left[\omega^2 + H(1-B)\right] + \left[\omega^2 - H(1-B)\right]^2, \\ \Delta_\Lambda(p) &= p\left(p^2 + \Omega^2\right)^2 \delta(p), \\ \delta(p) &= \left(p^2 + 4\Omega^2\right)\left[p^2 + (\Omega-\omega)^2\right]\left[p^2 + (\Omega+\omega)^2\right]. \end{aligned} \tag{6.10.77}$$

The coefficient of β^2 reads

$$\frac{\left[p^2 + H(1-B)\right]\left[p^2 + 4H(1-B)\right]}{\left(p^2 + 4\omega^2\right)}\left[2\omega^4\lambda_0 - \Phi^2\left(p^2 + 2\omega^2\right)\right]\frac{f(p^2)}{\Delta_\Lambda(p)}. \tag{6.10.78}$$

There are no other terms containing quadrates of the data in the LEM solution of Λ.

Due to the form of the above Laplace transform, we see that the normal LEM solution corresponding to Λ has the form

$$\begin{aligned} \Lambda(t) \cong{}& \frac{\alpha}{\Omega^2}\left[H(1-B)\cos\Omega t + \frac{2H}{\lambda_0^3}\right] + \frac{2H\eta}{\lambda_0\Omega^2}(1-\cos\Omega t) + \\ &+ \alpha^2\left[a_0^\Lambda + a_1^\Lambda\cos\Omega t + a_2^\Lambda t\sin\Omega t + a_3^\Lambda\cos 2\Omega t\right. \\ &\left.+ a_4^\Lambda\cos(\Omega-\omega)t + a_5^\Lambda\cos(\Omega+\omega)t\right] + \\ &+ \beta^2\left[b_0^\Lambda + b_1^\Lambda\cos\Omega t + b_2^\Lambda t\sin\Omega t + b_3^\Lambda\cos 2\Omega t + \right. \\ &\left.+ b_4^\Lambda\cos(\Omega-\omega)t + b_5^\Lambda\cos(\Omega+\omega)t + b_6^\Lambda\cos 2\omega t\right]. \end{aligned} \tag{6.10.79}$$

In the above formulae, the coefficients a_j^Λ, b_j^Λ are determined by residues.

Same way, we get the LEM coefficients of the quadrates of the data in Θ. The Laplace transform of the coefficients of $\alpha\beta, \beta\eta$ are, accordingly

$$\frac{H(1-B)}{\lambda_0^2}\left[p^2 - 3\omega^2 + H(1-B)\right]\frac{g(p)}{\Delta_\Theta(p)}, \tag{6.10.80}$$

and

$$-\,\omega_0^2\,\frac{f(p)g(p)}{\Delta_\Theta(p)}, \tag{6.10.81}$$

where we used the following notations

$$\begin{aligned} g(p) &= p\left[p^2 + H(1-B)\right]\left[p^2 + 4H(1-B)\right], \\ \Delta_\Theta(p) &= \left(p^2+\omega^2\right)^2\left(p^2+\Omega^2\right)\delta(p). \end{aligned} \tag{6.10.82}$$

As previously, there are no other terms containing quadrates of the data in the normal LEM solution of Θ. Again, the special form of the Laplace transforms (6.10.80), (6.10.81) yield the following LEM solution for Θ

$$\begin{aligned} \Theta(t) \cong\ & \beta\cos\omega t + \\ & + \alpha\beta\left[a_1^\Theta \cos\omega t + a_2^\Theta t\sin\omega t + a_3^\Theta \cos\Omega t + a_4^\Theta \cos 2\Omega t + \right. \\ & \left. + a_5^\Theta \cos(\Omega-\omega)t + a_6^\Theta \cos(\Omega+\omega)t\right] + \\ & + \beta\eta\left[b_1^\Theta \cos\omega t + b_2^\Theta t\sin\omega t + b_3^\Theta \cos\Omega t + b_4^\Theta \cos 2\Omega t + \right. \\ & \left. + b_5^\Theta \cos(\Omega-\omega)t + b_6^\Theta \cos(\Omega+\omega)t\right], \end{aligned} \tag{6.10.83}$$

the coefficients a_j^Θ, b_j^Θ being determined by residues.

6.10.4. CONCLUSIONS

The above treated neo-Hookean models present some similarities. Indeed, both of them contain a term of the form (6.10.6). If we stick to second order effects, then we see that the corresponding LEM solutions are analogous, from the mathematical point of view; we even emphasized this analogy by using the same notations for the eigenvalues of the linear part of both the corresponding systems (6.10.17) and (6.10.64). This analogy is, in fact, an effect of theorem 3.3. Indeed, by the normal LEM representation, the solutions of nonlinear ODSs with constant coefficients are viewed as elements of some Clifford algebra with respect to the data. From the same theorem, it follows that the coefficients strongly depend on the eigenvalues of the linear part of the differential operator. We see that, in both considered models, the Jacobian matrices have purely imaginary eigenvalues, due to the physical conditions imposed to the parameters in the case of the neo-Hookean suspensions and to the condition $0 < B < 1$ in the case of the pendulum. With these specifications, we see that the normal LEM solutions (6.10.37) on the one hand and (6.10.79), (6.10.83) on the other hand, suggest in this case that one can consider both motions structurally equivalent around the corresponding critical points. One can also extend this analogy to the generalized neo-Hookean model for suspensions (6.10.42), as previously pointed out. Naturally, for a better insight into the effective behavior of the LEM solutions, one can consider particular cases of motions to which one can easily apply the corresponding LEM formulae established here.

6.11. TWO OGDEN TYPE MODELS TREATED BY LEM

In this section, there are considered two models of deformation of some rubbery shells of Ogden type [62], established by Dana Petroşanu in [65-67]; from the mathematical point of view, the model results in solving a two-point boundary value problem for a second order nonlinear ODE. Basically, such problems cannot be solved analitically and, consequently, they are treated numerically, thus loosing the qualitative behaviour of the solution.

The aim of this section is to apply LEM to the above mentioned nonlinear models, emphasizing its benefits, by comparison to numerical methods; thus, we continuate the series of hyperelastic models treated by LEM in section 6.10.

We firstly present the hyperelastic models, as they were established in [64-66]. We then apply LEM to obtain approximate analytic expressions for the solution of the considered nonlinear two-point problems and also approximate parametric expressions for the value of its derivative at the left end of the considered interval. Finally, this analytical approximation of the solution is tested numerically and computed pointwisely.

6.11.1. THE MODELS

The behaviour of the isotropic elastic materials is basically described by nonlinear laws, most often extensions of Hooke's law; particularly, unlike the linear elastic, the rubbery materials show a nonlinear response to an applied stress. Ogden's is a hyperelastic material model used to describe the nonlinear stress-strain behaviour of hyperelastic materials. Ogden's model, like other hyperelastic models, assumes that the material behaviour can be described by means of a strain energy density function, from which the stress-strain relationships can be derived; these materials can generally be considered to be isotropic, incompressible and strain-rate independent.

Conveniently fitting the material parameters, Ogden's model accurately describes the rubber-like material behaviour. For particular values of material constants, this model will reduce to either the neo-Hookean or the Mooney-Rivlin model. Let us also mention that Ogden's models are isothermic and, generally, quasi-statical.

Dana Petroşanu establishes in [64,66] an Ogden-type model for the large deformations of a rubbery spherical and cylindrical thin shells.

a) *The hyperelastic sphere*. Consider a rubbery spherical thin shell, of internal radius A and external radius B. If **X** is a point of the undeformed sphere and **x** characterizes its strain state, then, admitting that the deformation is spherically simmetric, one has

$$\mathbf{x} = f(R)\cdot \mathbf{X}, \quad R \in [A, B]. \tag{6.11.1}$$

The function f is generally unknown. The equilibrium equation, involving the strain energy density W and $f(R)$, leads, for a particular W chosen by Ogden [61], to the following second order nonlinear ODE for f [64]

$$f'' = -\frac{f'}{R}\,\frac{8(2k+1)R^2 f'^2 + 2(17k+10)Rff' + 4\ (4k+1)f^2 - 12}{3(2k+1)R^2 f'^2 + 6(2k+1)Rff' + (4k+1)f^2 - 3}, \quad R \in [A, B]. \tag{6.11.2}$$

Denoting by a, b the radii of the deformed sphere, this equation will be solved in the conditions

$$f(A) = \frac{a}{A}, \quad f(B) = \frac{b}{B}. \tag{6.11.3}$$

This is a nonlinear two-point problem for which, basically, one cannot find analytic solutions. Dana Petroşanu proved that this problem allows a unique solution and tried to solve it numerically. The methods in use are: shooting, collocations and finite differences. They all lead to an algebraic equation that must be solved also numerically, e.g., by using Newton-type methods or dichotomy. In [65], the shooting method was prefered, combined with the dichotomic procedure.

b) *The hyperelastic cylinder*. Consider a rubbery thin shell in form of a circular cylinder, of height L and of basis radii A and B. To simplify computation, in [67] the height L was considered unmodified after deformation. This is possible if the height L is big enough, compared with the radii A and B, such that the deformed height be negligible with respect to the deformed radii. Dana Petroşanu [67] found the following equation for the unknown function $f = f(R)$

$$f''(R) = -\frac{f'}{R} \cdot \frac{7(2k+1)R^2 f'^2 + (29k+1)Rff' + 3(5k+4)f^2 - 3(k+2)}{3(2k+1)R^2 f'^2 + 6(2k+1)Rff' + (5k+4)f^2 - (k+2)}, \qquad (6.11.4)$$

where $k = a_1 / a_2$ is a dimensionless material constant. She also proved that the denominator of (6.11.4) does not vanish within the considered domain.

The boundary conditions were chosen as follows

$$f(A) = \frac{a}{A}, \quad f(B) = \frac{b}{B}. \qquad (6.11.5)$$

In [67] it was proved that this problem allows a unique solution.

6.11.2. LEM SOLUTIONS FOR THE HYPERELASTIC SPHERE

The model consists of the two-point problem (6.11.2), (6.11.3), allowing a unique solution, as shown in [65]. We can apply the LEM results for differential operators analytic with respect to the unknown functions and their derivatives. One approximates the analytic member in the right side of (6.11.2) by the corresponding McLaurin expansion with respect to f, f'. The third order aproximation reads

$$f'' = -\frac{4f'}{R} - k_1 ff'^2 - k_2 Rf'^3, \qquad R \in [A, B], \qquad (6.11.6)$$

where we used the notations

$$k_1 = \frac{2}{3}(7k+2), \qquad k_2 = \frac{4}{3}(2k+1). \qquad (6.11.7)$$

This is a second order polynomial ODE, with variable coefficients, odd with respect to the unknown function f and its derivative.

The LEM mapping will depend on two parameters. The second LEM equivalent is

$$v'_{ij} = iv_{i-1,j+1} + j\left(-\frac{4}{R}v_{ij} - k_1 v_{i+1,j+1} - k_2 Rv_{i,j+2}\right), \quad i,j \in \mathfrak{N}. \qquad (6.11.8)$$

Going as far as third order effects, we are led to the (finite) truncated ODS

$$\frac{\mathrm{d}\mathbf{V}^{(3)}}{\mathrm{d}R} = \mathbf{A}^{(3)}\mathbf{V}^{(3)},$$

$$\mathbf{V}^{(3)} = \begin{bmatrix} \mathbf{V}_1^{(3)} \\ \mathbf{V}_3^{(3)} \end{bmatrix}, \quad \mathbf{V}_1^{(3)} = \begin{bmatrix} v_{10}^{(3)} \\ v_{01}^{(3)} \end{bmatrix}, \quad \mathbf{V}_3^{(3)} = \begin{bmatrix} v_{30}^{(3)} \\ v_{21}^{(3)} \\ v_{12}^{(3)} \\ v_{03}^{(3)} \end{bmatrix}. \tag{6.11.9}$$

The truncated LEM matrix is

$$\mathbf{A}^{(3)} = \begin{bmatrix} \mathbf{A}_{11} & \mathbf{A}_{13} \\ \mathbf{0} & \mathbf{A}_{33} \end{bmatrix}, \tag{6.11.10}$$

where

$$\mathbf{A}_{11} = \begin{bmatrix} 0 & 1 \\ 0 & -\dfrac{4}{R} \end{bmatrix}, \qquad \mathbf{A}_{33} = \begin{bmatrix} 0 & 3 & 0 & 0 \\ -\dfrac{4}{R} & 0 & 2 & 0 \\ 0 & -\dfrac{8}{R} & 0 & 1 \\ 0 & 0 & -\dfrac{12}{R} & 0 \end{bmatrix}, \tag{6.11.11}$$

and

$$\mathbf{A}_{13} = \begin{bmatrix} 0 & 0 & 0 & 0 \\ 0 & 0 & -k_1 & -k_2 \end{bmatrix}. \tag{6.11.12}$$

We observe that, by virtue of the oddness of (6.11.7), the LEM matrix will be "contracted", containing only cels with odd indices.

Let us now apply theorems 1.6 and 2.2. We chose the following Cauchy conditions

$$f(A) = \frac{a}{A}, \quad f'(A) = \alpha, \tag{6.11.13}$$

α being a parameter that will be determined from the boundary conditions.

Let us denote by f_0 the solution of the linear part of the ODS (6.11.7), i.e.

$$f'' = -\frac{4f'}{R}, \qquad R \in [A, B], \tag{6.11.14}$$

subjected to the initial conditions (6.11.13). Obviously,

$$f_0 = \frac{\alpha A}{3}\left(1 - \frac{A^3}{R^3}\right) + \frac{a}{A}. \tag{6.11.15}$$

Its derivative f_0' with respect to R is

$$f_0' = \alpha \frac{A^4}{R^4}. \tag{6.11.16}$$

According to theorem 2.2,

$$\mathbf{V}_3^{(3)} = \begin{bmatrix} f_0^3 & f_0^2 f_0' & f_0 f_0'^2 & f_0'^3 \end{bmatrix}^{\mathrm{T}}. \tag{6.11.17}$$

The vector $\mathbf{V}_1^{(3)}$ is the solution of the non-homogeneous linear ODS

$$\frac{\mathrm{d}\mathbf{V}_1^{(3)}}{\mathrm{d}R} = \mathbf{A}_{11}^{(3)}\mathbf{V}_1^{(3)} + \mathbf{A}_{13}^{(3)}\mathbf{V}_3^{(3)}, \tag{6.11.18}$$

or, explicitely,

$$\begin{aligned} v_{10}' &= v_{01}, \\ v_{01}' &= -\frac{4}{R}v_{01} - k_1 f_0 f_0'^2 - k_2 f_0'^3. \end{aligned} \tag{6.11.19}$$

This results in the following linear non-homogeneous second order ODE

$$v_{10}'' + \frac{4}{R}v_{10}' = -k_1 R f_0 f_0'^2 - k_2 R f_0'^3, \tag{6.11.20}$$

to be solved under the initial conditions

$$v_{10}(A) = \frac{a}{A}, \quad v_{10}'(A) = \alpha. \tag{6.11.21}$$

Replacing f_0, f_0' by their expressions (6.11.15), (6.11.16) and integrating once (6.11.20), we get

$$v'_{10}(\alpha, R) = v_{01}(\alpha, R) = \frac{c_{01}}{R^4} + \frac{k_1}{3} a\alpha^2 \frac{A^7}{R^7}$$
$$+ \alpha^3 \frac{A^2}{18}\left[2k_1 \frac{A^7}{R^7} + (3k_2 - k_1)\frac{A^{10}}{R^{10}}\right]. \tag{6.11.22}$$

The integration constant c_{01} is determined from the Cauchy conditions (6.11.21); we deduce

$$c_{01} = \alpha A^4 - \frac{k_1}{3} a\alpha^2 A^4 - \alpha^3 \frac{A^6}{18}(3k_2 + k_1). \tag{6.11.23}$$

Integrating once more (6.11.22), it is obtained

$$v_{10}(\alpha, R) = c_{10} - \frac{c_{01}}{3R^3} - \frac{k_1}{3} a\alpha^2 \frac{A^7}{R^6} -$$
$$- \alpha^3 \frac{A^3}{162}\left[3k_1 \frac{A^6}{R^6} + (3k_2 - k_1)\frac{A^9}{R^9}\right]. \tag{6.11.24}$$

The LEM solution must satisfy the two-point conditions

$$v_{10}(A) = \frac{a}{A}, \quad v_{10}(B) = \frac{b}{B}. \tag{6.11.25}$$

Applying them to (6.11.24), we deduce

$$c_{10} - \frac{c_{01}}{3A^3} - \frac{k_1}{3} a\alpha^2 A - \alpha^3 \frac{A^3}{162}(3k_2 + 2k_1) = \frac{a}{A},$$
$$c_{10} - \frac{c_{01}}{3A^3}(1 - r^3) - \frac{k_1}{3} a\,\alpha^2 A r^6 -$$
$$- \alpha^3 \frac{A^3}{162}\left[3k_1 r^6 + (3k_2 - k_1) r^9\right] = \frac{b}{B}, \tag{6.11.26}$$

where

$$r = \frac{A}{B}. \tag{6.11.27}$$

Note that the above procedure leads to the unique solution of the problem, according to theorems 4.6, 4.7 and corollary 4.2 Subtracting the above two relations (6.11.26) and replacing c_{01} by its expression (6.11.23), we are led to the following algebraic equation for α

$$\alpha a_1 + \alpha^2 a_2 + \alpha^3 a_3 = a_0, \tag{6.11.28}$$

the coefficients a_j being given by

$$a_0 = \frac{\frac{b}{B} - \frac{a}{A}}{1 - r^3}, \quad a_1 = \frac{A}{3}, \quad a_2 = -\frac{k_1}{18} aA\left(1 - r^3\right),$$
$$a_3 = \frac{A^3}{162}\left[3k_1\left(1 + r^3\right) + \left(3k_2 - k_1\right)\left(1 + r^3 + r^6\right) - 9k_2 - 3k_1\right]. \tag{6.11.29}$$

To compute solution for this equation, we can apply corollary 4.3. Once α found, we replace it in the above formulae.

Table 6.6

The values of $f'(A)$ for the hyperelastic sphere

Nr.	A	B	a	b	α (LEM)	α (linear)
1.	0.1	0.200	0.20	0.250	-14.3985	-25.7142
2.	0.1	0.200	0.15	0.200	-11.8483	-17.1428
3.	0.1	0.200	0.15	0.250	-6.3628	-8.5714
4.	0.1	0.200	0.10	0.220	5.1817	3.4285
5.	0.1	0.150	0.15	0.200	-5.6606	-7.1052
6.	0.1	0.150	0.11	0.170	1.5329	1.4210
7.	0.2	0.225	0.19	0.215	0.2826	0.2779
8.	0.2	0.225	0.20	0.224	-0.2223	-0.2239
9.	0.2	0.250	0.30	0.350	-2.6356	-3.0737
10.	0.2	0.500	0.25	0.450	-4.2275	-5.6089
11.	0.2	0.500	0.25	0.550	-1.9848	-2.4038
12.	0.2	0.300	0.30	0.400	-2.8303	-3.5526
13.	0.2	0.300	0.22	0.300	-1.8888	-2.1315
14.	0.2	0.400	0.20	0.440	2.5901	1.7142
15.	1.0	1.500	1.10	1.700	0.15329	0.1421
16.	2.0	3.000	2.20	3.000	-0.1888	-0.2131
17.	2.0	4.000	2.00	4.400	0.2590	0.1714

Table 6.6 gives, for a range of a, b, A, B the corresponding values of α, compared with those for the linear case. It is seen that, for the chosen values of a, b, A, B, the absolute values of α are, most of them, greater in the linear case than in the nonlinear one. This means that the solution is increasing or decreasing smoother in the nonlinear than in the linear case. It is also noted that the values of α seem to depend, in a

certain way, mostly on the ratios $a/A, b/B$ and r (see positions 6 and 15, also 13 and 16).

6.11.3. LEM SOLUTIONS FOR THE HYPERELASTIC CYLINDER

The model consists of the two-point problem (6.11.4), (6.11.5). The existence and uniqueness of the solution is ensured in this case too [66,67]. We can apply the techniques from the previous section, considering the series expansion of the ratio in the right member with respect to f, f'. The third order approximation gives

$$f'' = -\frac{3f'}{R} - k_1 f f'^2 - k_2 R f'^3, \qquad R \in [A, B], \tag{6.11.30}$$

where we used the notations

$$k_1 = \frac{7k+17}{k+2}, \qquad k_2 = \frac{2(2k+1)}{k+2}. \tag{6.11.31}$$

This is a second order polynomial ODE, with variable coefficients, odd with respect to the unknown function f and its derivative, similar to that corresponding to the hyperelastic sphere.

The LEM mapping will depend on two parameters in this case too. The second LEM equivalent reads

$$v'_{ij} = i v_{i-1,j+1} + j\left(-\frac{3}{R} v_{ij} - k_1 v_{i+1,j+1} - k_2 R v_{i,j+2}\right), \qquad i,j \in \mathcal{N}. \tag{6.11.32}$$

Going as far as third order effects, we are led to the (finite) truncated ODS

$$\frac{\mathrm{d}\mathbf{V}^{(3)}}{\mathrm{d}R} = \mathbf{A}^{(3)}\mathbf{V}^{(3)},$$

$$\mathbf{V}^{(3)} = \begin{bmatrix} \mathbf{V}_1^{(3)} \\ \mathbf{V}_3^{(3)} \end{bmatrix}, \quad \mathbf{V}_1^{(3)} = \begin{bmatrix} v_{10}^{(3)} \\ v_{01}^{(3)} \end{bmatrix}, \quad \mathbf{V}_3^{(3)} = \begin{bmatrix} v_{30}^{(3)} \\ v_{21}^{(3)} \\ v_{12}^{(3)} \\ v_{03}^{(3)} \end{bmatrix}. \tag{6.11.33}$$

The truncated LEM matrix has the form (6.11.10), with

$$\mathbf{A}_{11} = \begin{bmatrix} 0 & 1 \\ 0 & -\frac{3}{R} \end{bmatrix}, \qquad \mathbf{A}_{33} = \begin{bmatrix} 0 & 3 & 0 & 0 \\ -\frac{3}{R} & 0 & 2 & 0 \\ 0 & -\frac{6}{R} & 0 & 1 \\ 0 & 0 & -\frac{9}{R} & 0 \end{bmatrix}, \tag{6.11.34}$$

$\mathbf{A}_{13}$ being given by (6.11.12), with the only difference that, in this case, the constants k_1, k_2 are expressed by (6.11.34).

Again, by virtue of the oddness of (6.11.30), the LEM matrix will be "contracted", containing only cells with odd indices.

Let us now apply theorems 1.6 and 2.2. We chose the following Cauchy conditions

$$f(A) = \frac{a}{A}, \quad f'(A) = \alpha, \tag{6.11.35}$$

α being a parameter that will be determined from the boundary conditions.

Denote now by f_0 the solution of the linear part of the ODS (6.11.7), i.e.

$$f'' = -\frac{3f'}{R}, \qquad R \in [A, B], \tag{6.11.36}$$

subjected to the initial conditions (6.11.35). We first get

$$f_0' = \alpha \frac{A^3}{R^3}; \tag{6.11.37}$$

obviously,

$$f_0 = \frac{\alpha A}{2}\left(1 - \frac{A^2}{R^2}\right) + \frac{a}{A}. \tag{6.11.38}$$

As prevoiously, according to theorem 2.2, the vector $\mathbf{V}_3^{(3)}$ is of the form (6.11.17). The vector $\mathbf{V}_1^{(3)}$ is the solution of the non-homogeneous linear ODS

$$\frac{\mathrm{d}\mathbf{V}_1^{(3)}}{\mathrm{d}R} = \mathbf{A}_{11}^{(3)}\mathbf{V}_1^{(3)} + \mathbf{A}_{13}^{(3)}\mathbf{V}_3^{(3)}, \tag{6.11.39}$$

or, explicitely,

$$\begin{aligned} v_{10}' &= v_{01}, \\ v_{01}' &= -\frac{3}{R}v_{01} - k_1 f_0 f_0'^2 - k_2 f_0'^3. \end{aligned} \tag{6.11.40}$$

This leads to the following linear non-homogeneous second order ODE

$$v_{10}'' + \frac{3}{R}v_{10}' = -k_1 R f_0 f_0'^2 - k_2 R f_0'^3, \tag{6.11.41}$$

to be solved under the initial conditions

$$v_{10}(A) = \frac{a}{A}, \quad v_{10}'(A) = \alpha. \tag{6.11.42}$$

Replacing f_0, f_0' by their expressions (6.11.37), (6.11.38) and integrating once (6.11.41), we get

$$\begin{aligned} v_{10}'(\alpha, R) = v_{01}(\alpha, R) = \frac{c_{01}}{R^3} + \frac{k_1}{2}a\alpha^2\frac{A^5}{R^5} + \\ + \alpha^3\frac{A^2}{8}\left[2k_1\frac{A^7}{R^7} + (2k_2 - k_1)\frac{A^7}{R^7}\right]. \end{aligned} \tag{6.11.43}$$

Applying the Cauchy conditions (6.11.42), we get the integration constant c_{01} in the form

$$c_{01} = \alpha A^3 - \frac{k_1}{2}a\alpha^2 A^3 - \alpha^3\frac{A^5}{8}(2k_2 + k_1). \tag{6.11.44}$$

The LEM solution is obtained integrating once more (6.11.43)

$$\begin{aligned} v_{10}(\alpha, R) = c_{10} - \frac{c_{01}}{2R^2} - \frac{k_1}{8}a\alpha^2\frac{A^5}{R^4} \\ - \alpha^3\frac{A^3}{48}\left[3k_1\frac{A^4}{R^4} + (2k_2 - k_1)\frac{A^6}{R^6}\right]; \end{aligned} \tag{6.11.45}$$

it must satisfy the two-point conditions

$$v_{10}(A)=\frac{a}{A}, \quad v_{10}(B)=\frac{b}{B}. \tag{6.11.46}$$

Applying them to (6.11.45), we get, as before, the algebraic equation for α

$$\alpha a_1 + \alpha^2 a_2 + \alpha^3 a_3 = a_0, \tag{6.11.47}$$

where $r = A/B$, the coefficients a_j being given by

$$a_0 = \frac{\frac{b}{B} - \frac{a}{A}}{1-r^2}, \quad a_1 = \frac{A}{2}, \quad a_2 = -\frac{k_1}{8} aA\left(1-r^2\right),$$
$$a_3 = \frac{A^3}{48}\left[3k_1\left(1+r^2\right) + \left(2k_2 - k_1\right)\left(1+r^2+r^4\right) - 6k_2 - 3k_1\right]. \tag{6.11.48}$$

It should be mentioned that, in order to justify the above procedure, we made use again of theorems 4.6, 4.7 and corollary 4.2.

Table 6.7

The values of $f'(A)$ for the hyperelastic cylinder

Nr.	A	B	a	b	α (LEM)	α (linear)
1.	0.1	0.200	0.20	0.250	-6.8377	-20.0000
2.	0.1	0.200	0.15	0.200	-5.9571	-13.3333
3.	0.1	0.200	0.15	0.250	-3.6994	-6.6666
4.	0.1	0.200	0.10	0.220	5.4733	2.6666
5.	0.1	0.150	0.15	0.200	-3.7337	-6.0000
6.	0.1	0.150	0.11	0.170	1.4850	1.2000
7.	0.2	0.225	0.19	0.215	0.2706	0.2647
8.	0.2	0.225	0.20	0.224	-0.2080	-0.2117
9.	0.2	0.250	0.30	0.350	-1.9533	-2.7777
10	0.2	0.500	0.25	0.450	-2.2286	-4.1666
11	0.2	0.500	0.25	0.550	-1.1979	-1.7857
12	0.2	0.300	0.30	0.400	-1.8668	-3.0000
13	0.2	0.300	0.22	0.300	-1.3592	-1.8000
14	0.2	0.400	0.20	0.440	2.7366	1.3333
15	1.0	1.500	1.10	1.700	1.4850	1.2000
16	2.0	3.000	2.20	3.000	-0.1359	-0.1800
17	2.0	4.000	2.00	4.400	0.2736	0.1333

To get solutions for (6.11.47), we can apply again corollary 4.3; after finding α, we replace it in the above formulae of the LEM solution.

Table 6.7 gives, for the same range of values for *a*, *b*, *A*, *B*, the corresponding values of α, in the case of the hyperelastic cylinder, compared with those for the linear case.

This table emphasizes almost the same conclusions as in the case of the hyperelastic sphere. Even the dependence of α on the ratios $a/A, b/B$ and r is put into evidence by positions 6 and 15, also by 13 and 16. Comparing both tables 6.6 and 6.7, we see that, for the same values of a, b, A, B, the corresponding absolute values of α seem to be, in general, greater for the sphere than for the cylinder.

6.11.4. CONCLUSIONS

In this section, we set up a method based on LEM to get solutions for some Ogden type hyperelastic models involving nonlinear two-point problems. In the case of such models, the greatest difficulty is to get the value of the derivative of the solution at one of the interval ends. This is a simple task in the linear case, but by no means easy in the nonlinear one, where only numerical methods can give a quantitative answer. It is seen that, unlike by pure numerical methods, by LEM we get analytic approximations for the solution and also analytic formulae for the derivative of the solution at one of the ends. More precisely, this method can be applied without difficulty at the right end *B*, getting analytic approximations for $f'(B)$ and corresponding analytic expressions for the LEM solution. It should be mentioned that the analytic approximations have the advantage of emphasizing the qualitative behaviour of the solution.

Let us also mention that the approximating formula

$$f'(A)=\frac{a_0}{a_3\alpha_1\alpha_2+a_2\alpha_2+a_1},\quad \alpha_1=\frac{a_0}{a_1},\alpha_2=\frac{a_0}{a_2\alpha_1+a_1},\tag{6.11.49}$$

where a_0, a_1, a_2, a_3 are given by (6.11.29) for the sphere and by (6.11.48) for the cylinder, is satisfactory for the considered range of values of a, b, A, B.

Finally, we observe that both the above hyperelastic models have a similar form; this induces a certain pattern in their solving. Together with an extended study of the LEM solutions for such models, this common pattern represents a perspective for future researches.

Chapter 7

OTHER APPLICATIONS OF LEM TO PHYSICS, CHEMISTRY AND ENGINEERING

The scope of the present chapter is to analyse by LEM other nonlinear dynamical systems, used as models in mechanics, physics, engineering, chemistry and biology, also by using the LEM representation, introduced in chapter 3 to the purpose of getting solutions of nonlinear ODS with polynomial coefficients. In the first section, LEM is applied to get the transport matrix of a relativistic electron beam – REB. Such a matrix was obtained only in the non-relativistic case [51, 164], but standard linearization fails in the relativistic case. The transport matrix could be set up both by LEM differential recurrence formulae and by the LEM representation introduced in the second section. A programming code for the transport matrix was built up, allowing the rapid calculation of a great number of trajectories. This leads to the accurate computation of the distribution of the current density and to determine other REB physical magnitudes of interest, such as the cross-over. The obtained results were fully confirmed experimentally.

The second section concerns the time and space oscillatory chemical reaction of Belousov and Zhabotinskij [25, 165, 167], a so called chemical clock. Field and Noyes set up the nonlinear model of this reaction, about which Troy showed that it allows solitary travelling wave solutions only if the main parameter of the reaction – the stoichiometric factor – takes values contained in certain real intervals. Using asymptotic expansions in terms of LEM, it was proved that the stoichiometric factor must necessarily satisfy an algebraic equation in order to favour chemical oscillations [138, 149].

The third application regards Troesch's plasma model, describing the confinement of plasma by radiation pressure. This is a numerically unstable nonlinear two-point problem [41, 44] depending on a parameter n, physically related to the plasma density. From the physical point of view, only high values of n are of interest. The appearance of a pole of the solution that could not be spotted adds to these difficulties. The application of LEM resulted in an asymptotic approximating formula for the solution y and in a parametric formula for $y'(0)$.

It should be specified that, during the last several years, the interest in applying LEM to various mechanical and technical problems was continuously increasing. This is why the section 7.4 is devoted to the results obtained by a scientific research group of the Institute of Mechanics of Solids of the Romanian Academy (IMSAR), who applied LEM to a series of problems of great scientific interest. They extended the application of LEM, e.g., to modern high-tech modelling for shape memory alloys [16], for nonlinear mesoscopic materials [18] and to domains like damping in machine tools [17].

Let us also mention the perspectives of applying LEM to the study and determination of solitons [57] for equations like sine-Gordon, Korteweg de Vries, to models in nano-mechanics [51] or to inverse problems.

The next section concerns the LEM representation for the second Painlevé equation.

In the last section, a new connection is established between the dynamics of the nonlinear rigid pendulum and that of the motion of a solid about a fixed point, in the particular case due to Euler and Poinsot. The equations of motion are solved by LEM via the nonlinear pendulum equation. The LEM formulae for the solution are then compared numerically with the corresponding Runge-Kutta solution.

7.1. A TRANSPORT MATRIX FOR REBs BY USING LEM

The analysis of the optoelectronic systems – particularly, of the particle accelerators – needs accurate information about the particle trajectories; these are, in fact, described by means of a nonlinear initial problem [162].

The non-relativistic model is easily tackled; actually, it was possible to set up a transport matrix for the beam in this case, by using some convenient simplifications of the physical model.

But in the relativistic case, the model cannot be linearized without introducing high errors. The lack of a relativistic transport matrix, generally valid without the above mentioned simplifications, imposed the calculation of a high number of trajectories, in order to get, as accurately as possible, the aspect of the relativistic beam. This solution, however, is not acceptable, because it is not economic, from the computational point of view.

So, to find a satisfactory issue, physical models were set up in order to emphasize a small number of significant trajectories; but these simplified models are not always true.

From the mathematical point of view, the difficulty consists of the lack of a representation of the solution separating the contribution of the coefficients from that of the initial data, as in the linear case. This is the reason why any standard numerical algorithm for REB trajectories – most of them based on the Runge-Kutta method – must be entirely repeated for any set of initial data, with huge costs of time and memory. Eventually, this does not result in a qualitative analysis of the beam, as it happens in the linear case.

In this paragraph, we shall apply LEM to the nonlinear operator corresponding to the relativistic particle trajectory equations, obtaining its local inverse; this leads to the transport matrix of the REB. As it depends only on the field configuration and not on the initial data, this transport matrix represent a characteristic of the considered opto-electronic system.

The advantage of LEM in this case consists firstly of getting rapidly a great number of trajectories.

But more important than a rapid computation is the qualitative fact that the transport matrix allows the analysis of other physical quantities of interest, such as the cross-over or the REB density distribution in a plane of the accelerating system; this analysis forms the object of the last subsection.

7.1.1. THE PHYSICAL MODEL

The equations of motion of relativistic charged particles in axially symmetric fields leads to the following nonlinear ODE [51, 164]

$$\frac{\mathrm{d}^2 r}{\mathrm{d}z^2} = -\frac{e(E - e\,\Phi)}{(E - e\,\Phi)^2 - m^2 c^4}\left[1 + \left(\frac{\mathrm{d}r}{\mathrm{d}z}\right)^2\right]\left(\frac{\partial \Phi}{\partial r} - \frac{\mathrm{d}r}{\mathrm{d}z}\frac{\partial \Phi}{\partial z}\right), \tag{7.1.1}$$

that must be solved in the Cauchy conditions

$$r(z_0) = r_0; \quad \frac{\mathrm{d}r}{\mathrm{d}z}(z_0) = r_0', \qquad (r_0, r_0') \in H \tag{7.1.2}$$

The set H depends on the electron distribution at the cathode.

In the equation (7.1.1), E, e, m are known physical constants and the function $\Phi(r, z)$ represents the electric potential, which is the solution of a mixed problem – Dirichlet and Neumann – treated in numerous papers, due to its importance.

Knowing the values of Φ in a network (r, z), it can be approximated along the symmetry axis of the system by a set of polynomials of the form

$$\Phi(r,z) = \sum_{j=0}^{p} \sum_{i=0}^{2p+1} \varphi_{ij} z^i r^{2j} = \sum_{j=0}^{p} \varphi_j(z) r^{2j}, \tag{7.1.3}$$

where z will be firstly replaced with $(z - z_0)$, and then by some $(z - z_s)$, according to the position along the symmetry axis of the accelerating system.

Let us note that the expression

$$c_\Phi = -\frac{e(E - e\Phi)}{(E - e\Phi)^2 - m^2 c^4} \tag{7.1.4}$$

does not change on intervals of length up to the chosen step. Hence c_Φ may be replaced on each step by its mean value $\tilde{c}_\Phi$. The equation (7.1.1) may be thus conveniently approximated by

$$\frac{d^2 r}{dz^2} = -\tilde{c}_\Phi \left[1 + \left(\frac{dr}{dz}\right)^2\right]\left[\frac{dr}{dz}\sum_{j=0}^{p} \frac{d\varphi_j(z)}{dz} r^{2j} - \sum_{j=0}^{p-1} (j+1)\varphi_{j+1}(z) r^{2j+1}\right] \tag{7.1.5}$$

which is a polynomial second order ODE, that might be written in the form of a first order polynomial ODS

$$\mathcal{P}\begin{pmatrix} r \\ u \end{pmatrix} \equiv \begin{pmatrix} \dfrac{dr}{dz} - u \\ \dfrac{du}{dz} - P(z,r,u) \end{pmatrix} = \mathbf{0} \tag{7.1.6}$$

In (7.1.6), P is the $(2p+3)$-th degree polynomial in r and u

$$P(z,r,u) \equiv -\tilde{c}_\Phi \left(1 + u^2\right)\left[u \sum_{j=0}^{p} \frac{d\varphi_j(z)}{dz} r^{2j} - \sum_{j=0}^{p-1} (j+1)\varphi_{j+1}(z) r^{2j+1}\right] \tag{7.1.7}$$

We see that the ODS (7.1.6) is defined by $\mathcal{P}$ – a $(2p+3)$-degree differential polynomial operator.

The initial conditions (7.1.2) become

$$r(z_0)=r_0; \quad u(z_0)=r_0', \quad (r_0,r_0')\in H. \tag{7.1.8}$$

To conclude, the model of the trajectories in a REB is formed by the polynomial ODS (7.1.6) and the initial conditions (7.1.8).

7.1.2. AN APPROACH BY USING THE LEM INVERSE MATRIX

We shall apply LEM to the mathematical model (7.1.6), (7.1.8) for the trajectories of a REB.

This is precisely a system of the same type as the general system (1.1.5), treated in the first chapter. To apply LEM, we consider the exponential mapping depending on two parameters, σ and ξ

$$v(z,\sigma,\xi)=\mathrm{e}^{\sigma r+\xi u}. \tag{7.1.9}$$

The first LEM equivalent is the PDE

$$\mathfrak{L}\begin{pmatrix} r \\ u \end{pmatrix} \equiv \frac{\partial v}{\partial z}-P\left(z,\frac{\partial}{\partial \sigma},\frac{\partial}{\partial \xi}\right)v=0 \tag{7.1.10}$$

that must be solved, provided

$$v(z_0,\sigma,\xi)=\mathrm{e}^{\sigma r_0+\xi r_0'}. \tag{7.1.11}$$

Considering for v a development of the form

$$v(z,\sigma,\xi)=1+\sum_{j+k=1}^{\infty} v_{jk}(z)\frac{\sigma^j}{j!}\frac{\xi^k}{k!}, \tag{7.1.12}$$

we obtain the second LEM equivalent

$$\begin{aligned}\frac{\mathrm{d}v_{jk}}{\mathrm{d}z} &= jv_{j-1,k+1}+k\sum_{s=0}^{p}\frac{\mathrm{d}\varphi_s}{\mathrm{d}z}\left(v_{j+2s,k}+v_{j+2s,k+2}\right)- \\ &-k\sum_{s=0}^{p-1}\varphi_{s+1}(s+1)\left(v_{j+2s+1,k-1}+v_{j+2s+1,k+1}\right), \quad (j+k)\in \mathfrak{N}^*,\end{aligned} \tag{7.1.13}$$

that may be also written in matrix form

$$\mathfrak{S}\mathbf{V}\equiv\frac{\mathrm{d}\mathbf{V}}{\mathrm{d}z}-\mathbf{A}(z)\mathbf{V}=\mathbf{0}, \quad \mathbf{V}(z)=\left(\mathbf{V}_{2j-1}(z)\right)_{j\in\mathfrak{N}}. \tag{7.1.14}$$

The initial conditions (7.1.8) read for **V**

$$\mathbf{V}(z_0)=\left(\mathbf{V}_{2j-1}(z_0)\right)_{j\in\mathfrak{N}}, \quad \mathbf{V}_{2j-1}(z_0)=\left(r_0^{2j-k-1}r_0'^{k}\right)_{k=\overline{0,2j-1}}. \qquad (7.1.15)$$

The polynomial operator is odd and, in this case, the cells $\mathbf{A}_{2k,2k+2s}(z)$ are all of them null. So, we can "contract" **A**, taking into account only the nonzero cells; this also explains the form of **V** in (7.1.15). The contraction does not affect the calculation by blocks. The polynomial ODS (7.1.6) is homogeneous and, obviously, the associated LEM system (7.1.14) turned out homogeneous. Finally, let us note that its coefficients are polynomials in z.

Due to the above properties, we can apply either theorem 1.6, or theorem 3.4, to get the inverse **Π** of the polynomial operator $\mathcal{P}$.

a) The formula of **Π,** according to theorem 1.6 is

$$\mathbf{\Pi}(z-z_0)\equiv\sum_{k\geq 0}\mathbf{A}^{(k)}(z_0)\frac{(z-z_0)^k}{k!}. \qquad (7.1.16)$$

The matrices $\mathbf{A}^{(k)}$ are determined by the recurrence

$$\mathbf{A}^{(k)}(z)=\frac{\mathrm{d}\mathbf{A}^{(k-1)}}{\mathrm{d}z}(z)+\mathbf{A}^{(k-1)}(z)\mathbf{A}(z), \quad \mathbf{A}^{(0)}(z)=\mathbf{E}, \qquad (7.1.17)$$

E being the infinite unit matrix.

The first two rows of **Π**, which give the solution r of the polynomial initial problem and its derivative r' accordingly, are consistent on the interval

$$\mathrm{I}_0=\left\{z\in\mathrm{I};|z-z_0|<\frac{1}{2CQ(2p+4)}\right\}; \qquad (7.1.18)$$

in this case,

$$Q=\left(1+|r_0'|^2\right)\sum_{j=1}^{p}\left[|r_0|+(j+1)|r_0'|\right]\left|r_0^{2j}\right|. \qquad (7.1.19)$$

b) The same inverse **Π** can be obtained by applying theorem 3.4 [148,153,153,162].

As $\frac{\mathrm{d}\varphi_s}{\mathrm{d}z}, \varphi_s$ are all of them polynomials of degree p at most, we can write the LEM matrix $\mathbf{A}(z)$ in the form

$$\mathbf{A}(z)=\mathbf{A}_0+(z-z_0)\mathbf{A}_1+(z-z_0)^2\mathbf{A}_2+ \\ +\ldots+(z-z_0)^{2p+1}\mathbf{A}_{2p+1}\,. \tag{7.1.20}$$

By theorem 3.4, we find that r coincides with the first and $\mathrm{d}r/\mathrm{d}z$ coincides with the second component of the vector

$$\mathbf{V}(z)=\left[\mathbf{E}+\sum_{l=1}^{\infty}\mathbf{A}^{(l)}(z-z_0)\right]\mathbf{V}(z_0)\,, \tag{7.1.21}$$

where the infinite matrices $\mathbf{A}^{(l)}$ are computed by formulae (3.5.10), (3.5.11).

The convergence holds true at least on the corresponding to (7.1.18) interval.

The first two rows of the infinite matrix

$$\mathbf{\Pi}(z-z_0)\equiv\mathbf{E}+\sum_{l=1}^{\infty}\mathbf{A}^{(l)}(z-z_0) \tag{7.1.22}$$

represent the inverse of the polynomial operator $\mathcal{P}$. By virtue of the series convergence, we can write

$$\mathbf{\Pi}(z-z_0)=\lim_{m\to\infty}\mathbf{\Pi}_m(z-z_0), \\ \mathbf{\Pi}_m(z-z_0)\equiv\mathbf{E}+\sum_{l=1}^{m}\mathbf{A}^{(l)}(z-z_0)\,. \tag{7.1.23}$$

There are some improvements and advantages of the computation of the inverse matrix $\mathbf{\Pi}$ following b), among which we mention the following

- all the involved matrices in the matrix products of (7.1.21) are constant;
- to go as far as $(z-z_0)^{11}$ by means of theorem 1.6, one must perform 11 steps, while by theorem 3.4 the same number of iterations, easier performed, leads to terms containing $(z-z_0)^{11(2p+1)}$;
- many subroutines/procedures of the programming code from [162], theoretically founded on theorem 1.6, e.g., matrix multiplication by block-partitioning, the generation of the LEM matrices, may be effectively used in a program set up by theorem 3.4.

7.1.3. THE TRANSPORT MATRIX OF A REB

The transport matrix of the REB is precisely the LEM transport matrix of the solution of the polynomial system (7.1.6). We shall set it up following the pattern exposed in subsection 2.2.2.

Let us denote the first two rows of $\mathbf{\Pi}_m(z-z_0)$ by $\mathbf{\Pi}_m^1(z-z_0)$, $\mathbf{\Pi}_m^2(z-z_0)$.

If we split the distance I between the emission plane and the other end of the accelerating system in equal subintervals $\mathrm{I}_j = \lfloor z_j, z_{j+1})$, $j=\overline{0,N}$, such that the inequality (7.1.18) be valid on each of them, then we obtain for each $j=\overline{0,N}$ two rows defined by

$$\mathbf{T}(2j-1)=\mathbf{\Pi}_m^1(z_j), \qquad \mathbf{T}(2j)=\mathbf{\Pi}_m^2(z_j), \qquad j=\overline{1,N}. \tag{7.1.24}$$

The transport matrix will be composed of the row-vectors $\mathbf{T}(2j-1)$, $\mathbf{T}(2j)$, written one after another for $j=\overline{1,N}$; $\mathbf{T}(2j-1)$ serves to set up approximately $r(z)$ and $\mathbf{T}(2j)$ leads to $r'(z)$ on I_j. In order to get a trajectory, we see that for every interval $\mathrm{I}_j, j=\overline{0,N}$, the initial vector will be $\mathbf{V}(z_j)$, i.e.

$$\begin{aligned} &\mathbf{V}(z_j)=\left(\mathbf{V}_{2k-1}(z_j)\right)_{k\in\mathfrak{N}}, \\ &\mathbf{V}_{2k-1}(z_j)=\left(r_j^{2k-s-1} r_j'^{\,s}\right)_{s=\overline{0,2j-1}}, \end{aligned} \tag{7.1.25}$$

where

$$r_j=r(z_j), \quad r_j'=r'(z_j). \tag{7.1.26}$$

Yet, we do not need all its components. From the form of the LEM matrices, it follows that it is enough to truncate up to level $(2p+3)$ in order to emphasize the contribution of all the coefficients of the considered polynomial operator. Hence, the number of the involved components of **V** is

$$\widetilde{p}=2+4+\ldots+(2p+4)=(p+2)(p+3). \tag{7.1.27}$$

Denote by

$$\mathbf{V}_{\widetilde{p}}(z_j)=\left(\mathbf{V}_{2k-1}(z_j)\right)_{k=\overline{1,p+1}}, \tag{7.1.28}$$

and by $\mathfrak{F}_j : \mathfrak{R}^2 \rightarrow \mathfrak{R}^2$

$$\mathfrak{F}_j\left(r_j, r_j'\right) = \left(T(2j-1)\mathbf{V}_{\tilde{p}}\left(z_j\right), T(2j)\mathbf{V}_{\tilde{p}}\left(z_j\right)\right). \tag{7.1.29}$$

All the above matrix products are finite. We note that if we wish to get the values of r, r' at an arbitrary position z_l along the acceleration system, we must compose the operators $\mathfrak{F}_j$, i.e.

$$\left(r_l, r_l'\right) = \mathop{\text{o}}_{j=0}^{l\text{-}1} \mathfrak{F}_j\left(r_j, r_j'\right) \equiv \tau_l\left(r_0, r_0'\right). \tag{7.1.30}$$

In the programming code set up in [162] in the case of a given field configuration, by using theorem 1.6, we went as far as $m = 11$. A comparison was made for various developments of Φ, namely, for $p = 1, 2, 3, 4$.

The study of the trajectories starting with the same initial conditions pointed out several important aspects. For instance, we found out that the precision increases considerably with *p*. The case $p = 1$ corresponds to the so called *paraxial hypothesis*, that forms the foundation of some relativistic models. This study strongly suggests a careful examination of the results obtained as a consequence of this simplifying hypothesis.

We also observed that the precision does not increase linearly with *p*. Indeed, the trajectories obtained for $p = 1$ and $p = 2$ almost coincide; the precision increases at $p = 3$ and it is the same for $p = 4$.

7.1.4. THE CURRENT DENSITY DISTRIBUTION

Once set up, the transport matrix of an electronic system becomes one of its most important characteristics. It offers the possibility to study a motion of a set of charged particles through this system, if its input mechanical properties are known. The short time necessary to built up a trajectory allows the calculation of a great number of trajectories, representing a statistic set rich enough to give quantitative and qualitative information about the density distribution of the beam in real time.

As mentioned in the introduction, the lack of a relativistic transport matrix led to the idea of considering a small number of preferred trajectories; following this, the beam distribution in the phase space is described by the so called ellipse of emittance.

LEM allows a different treatment of this problem.

Consider a beam emitted by a plane cathode, situated at $z = z_0$ and injected in the first section of an accelerator. The probability that a particle emitted from a point (x_0, y_0) of the cathode $z = z_0$ had the components of its velocity situated within the intervals $(\dot{x}_0, \dot{x}_0 + \mathrm{d}\dot{x}_0)$, $(\dot{x}_0, \dot{x}_0 + \mathrm{d}\dot{x}_0)$, $(\dot{z}_0, \dot{z}_0 + \mathrm{d}\dot{z}_0)$ is

$$P(\dot{x}_0, \dot{y}_0, \dot{z}_0)\mathrm{d}\dot{x}_0 \,\mathrm{d}\dot{y}_0 \,\mathrm{d}\dot{z}_0 \ , \tag{7.1.31}$$

where

$$P(\dot{x}_0, \dot{y}_0, \dot{z}_0) = 2\lambda^3 \mathrm{e}^{-\lambda^2 \left(\dot{x}_0^2 + \dot{y}_0^2 + \dot{z}_0^2\right)}, \tag{7.1.32}$$

and

$$\lambda = \left(\frac{m}{2KT}\right)^{\frac{1}{2}}, \tag{7.1.33}$$

with m, K, T physical constants.

Considering the axial symmetry of the beam, the total number of the particles emitted by the surface element $\mathrm{d}x_0 \mathrm{d}y_0$ is $N_T(r_0)\mathrm{d}x_0 \,\mathrm{d}y_0$. Hence, the distribution at $z = z_0$ in the space of the variables $(x_0, y_0, \dot{x}_0, \dot{y}_0, \dot{z}_0)$ is

$$N(x_0, y_0, \dot{x}_0, \dot{y}_0, \dot{z}_0) = 2\lambda^3 \frac{\pi}{2} N_T \,\mathrm{e}^{-\lambda^2 \left(\dot{x}_0^2 + \dot{y}_0^2 + \dot{z}_0^2\right)}. \tag{7.1.34}$$

The relation between N and the current density at the cathode reads

$$j_c(r_0) = e \int_0^{+\infty} \dot{z}_0 \,\mathrm{d}\dot{z}_0 \int_{-\infty}^{+\infty} \int_{-\infty}^{+\infty} N(x_0, y_0, \dot{x}_0, \dot{y}_0, \dot{z}_0) \,\mathrm{d}\dot{x}_0 \,\mathrm{d}\dot{y}_0 \,; \tag{7.1.35}$$

by integration, it is found

$$j_c(r_0) = \frac{e}{\lambda\sqrt{\pi}} N_T(r_0), \tag{7.1.36}$$

and, eventually,

$$N(x_0, y_0, \dot{x}_0, \dot{y}_0, \dot{z}_0) = \frac{2\lambda^4}{\pi e} j_c(r_0) \,\mathrm{e}^{-\lambda^2 \left(\dot{x}_0^2 + \dot{y}_0^2 + \dot{z}_0^2\right)}. \tag{7.1.37}$$

Passing to cylindrical coordinates, the distribution function is NJ, where J is the Jacobian of the transform, therefore

$$N(r_0, \theta_0, \dot{r}_0, \dot{\theta}_0, \dot{z}_0) = \frac{2\lambda^4}{\pi e} j_c(r_0) \,\mathrm{e}^{-\lambda^2 \left(\dot{r}_0^2 + r_0^2 \dot{\theta}_0^2 + \dot{z}_0^2\right)}. \tag{7.1.38}$$

Taking into account the axial symmetry and the energy conservation principle, and mainly due to the REB transport matrix, previously set up, we can establish the relations between the initial variables $(r_0, \theta_0, \dot{r}_0, \dot{\theta}_0, \dot{z}_0)$ and the current ones, $(r, \theta, \dot{r}, \dot{\theta}, \dot{z})$

$$\begin{aligned} & r = \tau_l^1(r_0, r_0'), \quad r_0' = \frac{\dot{r}_0}{\dot{z}_0}, \\ & \dot{r} = \dot{z}r' = \dot{z}\tau_l^2(r_0, r_0'), \\ & \dot{z} = \left(1 + r'^2\right)^{-\frac{1}{2}} \left[c^2 \left(1 - \frac{m^2 c^4}{\psi^2} \right) - r^2 \dot{\theta}^2 \right]^{\frac{1}{2}}, \end{aligned} \tag{7.1.39}$$

where

$$\Psi = mc^2 \left(1 - \frac{v_0^2}{c^2} \right)^{-\frac{1}{2}} + e(\Phi_0 - \Phi). \tag{7.1.40}$$

Let us set up a network in the domain of variation of $(r_0, \theta_0, \dot{r}_0, \dot{\theta}_0, \dot{z}_0)$, namely $(r_{0i}, \theta_0, \dot{r}_{0j}, \dot{\theta}_{0k}, \dot{z}_{0l})$ for $i = \overline{1, m_1}, j = \overline{1, m_2}$., $k = \overline{1, m_3}, l = \overline{1, m_4}$. We thus obtain a set (i, j) of spatial trajectories, distinct for each pair (i, j), the indices k and l corresponding to the velocities. This choice involves the partition of the cathode in annuli. The relations (7.1.39) then read

$$\begin{aligned} & r_{ijk} = \tau_l^1\left(r_{0i}, r'_{0jk}\right) \equiv \tau_l^1\left(r_{0i}, \frac{\dot{r}_{0j}}{\dot{z}_{0k}}\right), \\ & \dot{r}_{ijkl} = \dot{z}_{ijkl} r'_{ijk} = \dot{z}_{ijkl}\tau_l^2\left(r_{0i}, r'_{0jk}\right), \\ & \dot{z}_{ijkl} = \left(1 + r'^2_{ijk}\right)^{-\frac{1}{2}} \left[c^2 \left(1 - \frac{m^2 c^4}{\psi^2_{jkl}} \right) - r^2_{ijk}\dot{\theta}^2_{0l} \right]^{\frac{1}{2}}, \end{aligned} \tag{7.1.41}$$

with

$$\Psi_{jkl} = mc^2 \left(1 - \frac{v_{jkl}^2}{c^2} \right)^{-\frac{1}{2}} + e(\Phi_0 - \Phi). \tag{7.1.42}$$

For every set of indices (i, j, k, l), we can define an average weight given by

$$r_{ijkl} = \int_{r_{0i}}^{r_{0,i+1}} \mathrm{d}r_0 \int_0^{2\pi} \mathrm{d}\theta_0 \int_{\dot{r}_{0j}}^{\dot{r}_{0,j+1}} \mathrm{d}\dot{r}_0 \int_{\dot{\theta}_{0l}}^{\dot{\theta}_{0,l+1}} \mathrm{d}\dot{\theta}_0 \int_{\dot{z}_{0k}}^{\dot{z}_{0,k+1}} \mathrm{d}\dot{z}_0 \,. \tag{7.1.43}$$

We can now obtain numerically the velocity distribution at a point $\widetilde{r}_\alpha \in (r_{\alpha-1}, r_\alpha)$ from the plane of interest z, perpendicular on the symmetry axis Oz, where we denoted by α index of partitioning following r

$$n(\widetilde{r}_\alpha, z, \dot{r}_{ijkl}, \dot{\theta}, \dot{z}_{ijkl}) = \sum_{(i,j)\in\mathcal{G}_\alpha} n_{ijkl} \,, \tag{7.1.44}$$

$\mathcal{G}_\alpha$ being the set of indices

$$\mathcal{G}_\alpha = \{(i, j) \mid r_{\alpha-1} \le r_{ij} \le r_\alpha \}. \tag{7.1.45}$$

The current density in the plane z will be

$$j_c(\widetilde{r}_\alpha, z) = \sum_{(i,j)\in\mathcal{G}_\alpha} \dot{z}_{ijkl}\, n_{ijkl} \,. \tag{7.1.46}$$

It should be mentioned that formula (7.1.46) gives the radial distribution of the current density in an arbitrary plane z. Let us also note that, due to the particular mode in which the initial velocity shows up in the constant $\widetilde{c}_\Phi$, one can get distinct trajectories only for distinct pairs (r_{0i}, r'_{0j}); this is why the sum in (7.1.44) and (7.1.46) is taken only with respect to i and j.

The method can be obviously adapted for other optoelectronic systems; one also can easily study the slit effect on the distribution of the current density.

The transport matrix was obtained by reducing the nonlinear ODE (7.1.1) to the polynomial ODS (7.1.6), such that the considered physical model be not affected. But one can easily apply LEM to more complex models, e.g., it may be extended to analytic Φ and to models taking into account the space charge.

Due to the transport matrix, we could obtain numerically the velocity distribution and, as a consequence, the radial density distribution in an arbitrary plane $z = \text{const}$ along the symmetry axis of the accelerating system. For the field configuration we dealt with, a hollow beam of several millimeters was obtained numerically; this theoretical conclusion was confirmed by the experiment. We also had good information about the cross-over. Indeed, numerous trajectories crossed the symmetry axis at apparently the same spot.

Yet, by using the equation (7.1.1), one cannot obtain trajectories spiraling along the symmetry axis; to get them, one must consider a more general mathematical model.

7.2. THE OSCILLATORY BELOUSOV-ZHABOTINSKIJ CHEMICAL REACTION

The study of the long-term behavior of the solutions of dynamical systems, as well as their dependence on parameters with physical significance, are some of the main problems of mechanics, since this was firstly noticed by Poincaré. Linear ODS are easier studied, but the nonlinear case requires specific methods, not always generally applicable.

LEM makes this study possible for nonlinear ODS too, by using the current techniques of the linear.

In this section, we shall treat by LEM the nonlinear model of a well-known chemical oscillator, the Belousov-Zhabotinskij reaction, which sometimes exhibits solitary travelling wave solutions, this phenomenon depending essentially on the values of the stoichiometric factor, the basic parameter of the reaction. We consider asymptotic expansions in vector form for the second LEM equivalent, eventually pointing out those values of the stoichiometric factor that make possible the appearance of a chemical oscillation.

In a primary form, the Belousov-Zhabotinskij reaction consists in the metal ion catalyzed oxidation by Bromate ion (BrO_3) of easily brominated organic materials. It was observed that this chemical reaction shows both temporal and spatial structure oscillations, by the appearance of the solitary travelling waves; this is the first known chemical reaction with these properties.

7.2.1. THE MATHEMATICAL MODEL

Field and Noyes [29] introduced a simple model of the reaction

$$
\begin{aligned}
A + Y &\to X, \\
X + Y &\to P, \\
B + X &\to 2X + Z, \\
2X &\to Q, \\
Z &\to fY,
\end{aligned}
\tag{7.2.1}
$$

where X, Y, Z represent the concentrations of the bromus acid ($HBrO_2$), bromide ion (Br^-) and Ce(IV) respectively. A and B denote the concentrations of the reactant (BrO_3^-), P and Q are products and f is the stoichiometric factor. Applying the law of the mass action to these reactions, the kinetic behaviour of these reactions in a continuously stirred solution could be described by a first order ODS, of polynomial type, brought to the final form [165]

$$
\begin{aligned}
\frac{\mathrm{d}x}{\mathrm{d}\tau} &= s\left(y + x - xy - qx^2\right) \equiv sF(x, y), \\
\frac{\mathrm{d}y}{\mathrm{d}\tau} &= \frac{1}{s}(-y + fz - xy) \equiv \frac{1}{s}G(x, y, z), \\
\frac{\mathrm{d}z}{\mathrm{d}\tau} &= w(x - z),
\end{aligned}
\tag{7.2.2}
$$

where the following notations were used

$$
\begin{aligned}
&x = X\frac{k_2}{k_1 A}; \qquad y = Y\frac{k_2}{k_3 B}; \quad z = Z\frac{k_2 k_5}{k_1 k_3}; \\
&\tau = \frac{\zeta}{k_1 k_2 AB} \cong 0.161\zeta; \qquad s = \sqrt{\frac{k_3 B}{k_1 A}} \cong 77.27; \\
&w = \sqrt{\frac{k_5}{k_1 k_3 AB}} \cong 0.161 k_5; \qquad q = \frac{2k_1 k_4 A}{k_2 k_3 B} \cong 8.375 \times 10^{-6}.
\end{aligned}
\tag{7.2.3}
$$

Here, ζ denotes the time, $k_1 - k_5$ are the corresponding reaction rates for the reaction (7.2.1) and the numerical values were computed under Field and Noyes assumption that $A = B = 0.06M$. Troy and Field [165] established analytically that for $f < 0.5$ or $f > 1 + \sqrt{2}$, the steady state, given by

$$x_0(f) = \frac{1}{2q}\left(1 - f - q + \left[(1 - f - q)^2 + 4q(1 + f)\right]^{\frac{1}{2}}\right),$$

$$y_0(f) = \frac{fx_0(f)}{1 + x_0(f)}, \tag{7.2.4}$$

$$z_0(f) = x_0(f),$$

$$\pi_0(f) \equiv (x_0(f), y_0(f), z_0(f)),$$

is globally stable to any initial perturbations. This means that for $f \in (-\infty, 0.5) \cup \left(1 + \sqrt{2}, +\infty\right)$ there can be no periodic solutions for the system (7.2.2), so that the travelling waves do not appear and the reagent remains red [25, 167].

Let us also introduce the notation

$$\pi_0(f) \equiv (x_0(f), y_0(f), z_0(f)). \tag{7.2.5}$$

The case in which the unstirred agent is spread in a thin layer over a flat surface was modelled by Troy [165], in the form of a system of nonlinear PDE

$$\frac{\partial x}{\partial \tau} = sF(x, y) + D_x \frac{\partial^2 x}{\partial u^2},$$

$$\frac{\partial y}{\partial \tau} = \frac{1}{s} G(x, y, z), \tag{7.2.6}$$

$$\frac{dz}{d\tau} = w(x - z).$$

A travelling wave of (7.2.6) is a nonconstant solution of the form

$$\pi(u/\alpha + \tau) = (x(u/\alpha + \tau),\ y(u/\alpha + \tau),\ z(u/\alpha + \tau)), \tag{7.2.7}$$

which also satisfies

$$\lim_{t \to \pm\infty} \pi(t) = \pi_0. \tag{7.2.8}$$

So, if we perform the change of variable $t = u/\alpha + \tau$, suggested by (7.2.7), we get

$$\frac{dx}{dt} = sF(x, y) + \frac{1}{\theta}\frac{d^2x}{dt^2},$$
$$\frac{dy}{dt} = \frac{1}{s}G(x, y, z), \tag{7.2.9}$$
$$\frac{dz}{dt} = w(x - z),$$

where $\theta = \alpha^2 / D_x$. This is a second order ODS, of polynomial type.

In addition, one must require that $x > 0$, $y > 0$, $z > 0$ for $t > 0$, as chemical concentrations cannot be negative.

Troy [165] has proved that there exists an open interval I, included in $\left(1 + \sqrt{2}, +\infty\right)$, such that for every $f \in \mathrm{I}$ and w between $\left(0, w_f\right)$, the system (7.2.9) does not allow solitary travelling wave solutions.

It is seen that, in this case, the asymptotic behavior plays an important part in getting the solutions. Our aim is to give f as accurately as possible, such that a solitary travelling wave be possible.

We firstly put the system (7.2.9) in the form of a first order ODS

$$\frac{dx}{dt} = r,$$
$$\frac{dr}{dt} = \theta r - \theta s\left(y + x - xy - qx^2\right)$$
$$\frac{dy}{dt} = \frac{1}{s}(-y + fz - xy), \tag{7.2.10}$$
$$\frac{dz}{dt} = w(x - z).$$

Clearly, (7.2.10) is a first order polynomial ODS of degree 2.

7.2.2. CRITICITY CONDITIONS BY LEM

The LEM exponential transform v depends here on 4 parameters, $\sigma, \xi, \eta, \lambda$ and satisfies the linear PDE

$$\frac{\partial v}{\partial t} = \sigma\frac{\partial v}{\partial \xi} + \xi\theta\left(\frac{\partial v}{\partial \xi} - s\left(\frac{\partial v}{\partial \sigma} + \frac{\partial v}{\partial \eta} - q\frac{\partial^2 v}{\partial \sigma^2} - \frac{\partial^2 v}{\partial \sigma \partial \eta}\right)\right) +$$
$$+ \eta \cdot \frac{1}{s}\left(-\frac{\partial v}{\partial \eta} + f\frac{\partial v}{\partial \lambda} - \frac{\partial^2 v}{\partial \sigma \partial \eta}\right) + \lambda w\left(\frac{\partial v}{\partial \sigma} - \frac{\partial v}{\partial \lambda}\right). \tag{7.2.11}$$

This is the first LEM equivalent.

With these preparations, we can prove

Theorem 7.1 [131, 133, 149] *A necessary condition for the appearance of a solitary travelling wave solution for* (7.2.9) *is that the stoichiometric factor f be one of the real roots of the algebraic equation*

$$\begin{aligned} & f^4+(4q-7)f^3+\left(5q^2-19q+17\right)f^2+ \\ & +\left(2q^3-8q^2+30q-17\right)f+q^2+2q+6=0, \end{aligned} \tag{7.2.12}$$

where q is given by

$$q=\frac{2k_1k_4A}{k_2k_3B}\cong 8.375\times 10^{-6}\,\frac{A}{B}. \tag{7.2.13}$$

Proof. As we must investigate the asymptotic behavior of the solution, we shall consider for v a series expansion of the form

$$v(t,\sigma,\xi,\eta,\lambda)=\sum_{i,j+k+l+m=0}^{\infty} v_{jklm}^{i}\,\frac{1}{t^i}\,\sigma^j\,\xi^k\,\eta^l\,\lambda^m. \tag{7.2.14}$$

Let us introduce the infinite vector

$$\mathbf{V}^i=\left(v_{jklm}^{i}\right)_{(j+k+l+m)\in\mathfrak{N}}. \tag{7.2.15}$$

Introducing the expansion (7.2.14) in the linear PDE (7.2.11), identifying the coefficients of the same negative powers of t and using the notation (7.2.15), we see that the second LEM equivalent may be written in the matrix form

$$-i\,\mathbf{V}^i=\mathbf{A}\mathbf{V}^{i+1},\quad i\in\mathfrak{N}. \tag{7.2.16}$$

The infinite LEM matrix **A** has the form

$$\mathbf{A}=\begin{pmatrix} \mathbf{A}_1 & \mathbf{B}_1 & \mathbf{0} & \mathbf{0} & \dots \\ \mathbf{0} & \mathbf{A}_2 & \mathbf{B}_2 & \mathbf{0} & \dots \\ \dots & \dots & \dots & \dots & \dots \end{pmatrix}. \tag{7.2.17}$$

The diagonal cells $\mathbf{A}_j$ are finite square matrices and they depend only on the coefficients of the linear part of (7.2.10); the matrix $\mathbf{A}_1$ is in fact the Jacobian matrix of the system

$$\mathbf{A}_1 = \begin{pmatrix} 0 & 1 & 0 & 0 \\ -\theta s & \theta & -\theta s & 0 \\ 0 & 0 & -1/s & f/s \\ w & 0 & 0 & -w \end{pmatrix}. \tag{7.2.18}$$

The rectangular matrices $\mathbf{B}_j$ depend only on the coefficients of second degree monomials in the right side of (7.2.10), i.e., on $\theta qs, \theta s, -1/s$. To be more specific, the matrix $\mathbf{B}_1 \equiv (b_{ij}) \in \mathfrak{M}(4\times 10)$ has only three nonzero elements, namely $b_{21} = \theta qs, b_{24} = \theta s, b_{34} = -1/s$.

From (7.2.16) one gets $\mathbf{AV}^1 = \mathbf{0}$, which yields the following algebraic system

$$\begin{aligned} &\beta = 0, \\ &-\theta s \alpha - \theta s \gamma + 2q\theta s x_0 \alpha + \theta s (y_0 \alpha + x_0 \gamma) + \theta \beta = 0, \\ &-\alpha + f\mu - (y_0 \alpha + x_0 \gamma) = 0, \\ &\alpha - \mu = 0 . \end{aligned} \tag{7.2.19}$$

In the above system,

$$\alpha = v^1_{1000}, \qquad \beta = v^1_{0100}, \qquad \gamma = v^1_{0010}, \qquad \mu = v^1_{0001}, \tag{7.2.20}$$

and

$$\begin{aligned} &v^1_{2000} = v^0_{1000} \cdot v^1_{1000} = \alpha x_0(f), \\ &v^1_{1010} = v^0_{0010} \cdot v^1_{1000} + v^0_{1000} \cdot v^1_{0010} = \alpha y_0(f) + \gamma x_0(f). \end{aligned} \tag{7.2.21}$$

The last formulae were obtained from the asymptotic condition, with $x_0(f), y_0(f), z_0(f)$ given by (7.2.4). The homogeneous algebraic system (7.2.19) allows nonzero solution if

$$x_0(2qx_0 + y_0 - 1) - (x_0 - 1)(f - 1 - y_0) = 0. \tag{7.2.22}$$

Taking now (7.2.4) into account, we get for f the fourth degree algebraic equation (7.2.12). The other conditions resulting from (7.2.16) for $i = 0$ are satisfied if f satisfies (7.2.12), as a result of the special form of the LEM matrix **A** and also because of (7.2.20).□

Remark. The equation (7.2.12) may be discussed with respect to q as a parameter, that is, immaterial Field and Noyes's assumption $A = B = 0.06M$. However, if this assumption is valid, then f satisfies

$$f^4 - 7f^3 + 17f^2 - 17f + 6 = 0, \qquad (7.2.23)$$

with an approximation of order 10^{-5}. This equation allows $f_1 = 1$ as a double root. Other roots are $f_3 = 2, f_4 = 3$.Only f_1, f_3 should be considered, as f_4 is within the interval $\left(1+\sqrt{2}, +\infty\right)$.

We see that studying by LEM the Belousov-Zhabotinskij reaction, we obtained information on the qualitative behaviour of the solution of Troy's model, carried out by equations, not by inequalities. Indeed, equation (7.2.23) provides exact values for the stoichiometric factor f ensuring a certain type of behaviour of the solution, unlike other methods, by which one obtains only intervals. By using the normal LEM representation for the polynomial ODS with constant coefficients (7.2.10), one can also study the time evolution of the reactants.

7.3. TROESCH'S PLASMA PROBLEM

Troesch's plasma problem concerns the confinement of plasma by radiation pressure; it is an unstable two-point problem. To solve numerically this problem, a combination of multipoint, shooting and perturbation, as well as parameter mapping techniques were applied [41, 44, 74].

The equation – or, equivalently, the two-point conditions – depend on a parameter n. Other elements further add to the difficulty of solving a nonlinear b.v.p.: the appearance of a pole within the domain of integration for some reasonable choice of $y'(0)$, which eventually turns out to be wrong, and the occurrence in the same domain of divergent integrations, that must be rejected.

The problem is the more so difficult, as only high values of n are of interest, from the physical point of view.

From the above considerations, it follows that parametric representations for both the solution $y(x)$ and $y'(0)$ should be useful. This is the chief goal of the section and, by using LEM techniques, we will achieve it.

7.3.1. THE MATHEMATICAL MODEL

Troesch's equation

$$\frac{d^2 w}{dt^2} = n \sinh (nw) \tag{7.3.1}$$

must be solved under the boundary conditions

$$w(0) = 0, \quad w(1) = 1. \tag{7.3.2}$$

Using the changes of function and variable

$$x = nt, \quad y(x) = nw, \tag{7.3.3}$$

the equation (7.3.1) is mapped into

$$\frac{d^2 y}{dx^2} = \sinh (y), \tag{7.3.4}$$

and the conditions (7.3.2) become

$$y(0) = 0, \quad y(n) = n. \tag{7.3.5}$$

The new problem (7.3.4), (7.3.5) is also a nonlinear two-point b.v.p. It is not polynomial, but we can still use LEM extension, as shown in chapter 3.

We consider a Cauchy problem for Troesch's equation by adding to the first condition (7.3.5) a Cauchy condition for the derivative

$$y(0) = 0, \quad y'(0) = \beta, \tag{7.3.6}$$

where we introduced the parameter β.

7.3.2. THE LEM SOLUTION

The LEM exponential mapping depends here on two parameters

$$v(x, \sigma, \xi) = e^{\sigma y + \xi z}. \tag{7.3.7}$$

The first LEM equivalent of (7.3.4) is

$$\frac{\partial v}{\partial x} = \sigma \frac{\partial v}{\partial \xi} + \xi \sinh (D_\sigma) v, \tag{7.3.8}$$

where the formal operator

$$\sinh (D_\sigma) \equiv \sum_{k=1}^{\infty} \frac{1}{(2k+1)!} D_\sigma^k v, \qquad D_\sigma^k \equiv \frac{\partial^k v}{\partial \sigma^k}, \tag{7.3.9}$$

has sense on $\mathcal{C}_2^0([0, n])$ and, so much the more, on $\mathcal{C}_2^1([0, n])$.

The initial conditions (7.3.6) become

$$v(0,\sigma,\xi)=\mathrm{e}^{\xi\beta}. \tag{7.3.10}$$

We consider now for v an expansion of the form

$$v(x,\sigma,\xi)=1+\sum_{j+k=1}^{\infty} v_{jk}(x)\frac{\sigma^j}{j!}\frac{\xi^k}{k!}, \tag{7.3.11}$$

that, introduced in (7.3.8), leads to the second LEM equivalent

$$v'_{jk}=jv_{j-1,k}+k\sum_{m=1}^{\infty}\frac{1}{(2m-1)!}v_{2m-1-k,k}. \tag{7.3.12}$$

This linear infinite system can be written in matrix form

$$\frac{\mathrm{d}\mathbf{V}}{\mathrm{d}x}=\mathbf{A}\mathbf{V},\quad \mathbf{V}=(\mathbf{V}_{2m-1})_{m\in\mathfrak{N}},\quad \mathbf{V}_{2m-1}=(v_{jk}(x))_{j+k=2m-1}, \tag{7.3.13}$$

and it is clearly a first order ODS with constant coefficients. Its associated LEM matrix has the form

$$\mathbf{A}=\begin{bmatrix}\mathbf{A}_{11} & \mathbf{A}_{13} & \mathbf{A}_{15} & \ldots\\ \mathbf{0} & \mathbf{A}_{33} & \mathbf{A}_{35} & \ldots\\ \mathbf{0} & \mathbf{0} & \mathbf{A}_{55} & \ldots\\ \ldots & \ldots & \ldots & \ldots\end{bmatrix}. \tag{7.3.14}$$

The LEM matrix is only column finite in this case. Yet, all the involved computation is finite, as we shall see. In (7.3.14), the diagonal cells $\mathbf{A}_{2m-1,2m-1}\in\mathfrak{M}(2m\times 2m)$ are square and depend only on the linear part of the operator. In fact, $\mathbf{A}_{11}$ is the Jacobian matrix

$$\mathbf{A}_{11}=\begin{bmatrix}0 & 1\\ 1 & 0\end{bmatrix}. \tag{7.3.15}$$

Its eigenvalues are $\lambda_1=1, \lambda_2=-1$. As it was shown in theorem 1.8, the eigenvalues of every diagonal cell $\mathbf{A}_{2m-1,2m-1}$ are

$$(2m-1-k)\cdot(+1)+k\cdot(-1),\quad k=\overline{0,2m-1}. \tag{7.3.16}$$

Each of the super-diagonal matrices $\mathbf{A}_{2m-1,2(m+l)-1}$ with $2m$ rows and $2(m+l)$ columns is set up by using only the entry $1/(2l+1)!$.

Each of them has only $(2m-1)$ nonzero entries. More precisely,

$$A_{2m-1,2(m+l)-1} = \frac{1}{(2l+1)!}\begin{bmatrix} 0 & 0 & \dots & 0 & 0 & 0 & \dots & 0 \\ 1 & 0 & \dots & 0 & 0 & 0 & \dots & 0 \\ 0 & 2 & \dots & 0 & 0 & 0 & \dots & 0 \\ 0 & 0 & \dots & 2m-2 & 0 & 0 & \dots & 0 \\ 0 & 0 & \dots & 0 & 2m-1 & 0 & \dots & 0 \end{bmatrix}. \qquad (7.3.17)$$

Note that (7.3.14) has only finite cells with odd indices. This is not casual; it comes from the oddness of the ODS (7.3.4). Indeed, the development of $\sinh(y)$ contains only odd powers of y and this leads to null cells with even indices, which can be suppressed. Finally, **A** will contain only cells with odd indices, and obviously, so will **V**.

The initial conditions (7.3.6) become for **V**

$$\mathbf{V}(0,\beta) = \left(\mathbf{V}_{2j-1}(0,\beta)\right)_{j\in\mathfrak{N}}, \quad \mathbf{V}_{2j-1}(0,\beta) = \left(\beta^{2j-1}\delta_q^{2j-1}\right)_{q=\overline{0,2j-1}}. \qquad (7.3.18)$$

The auxiliary Cauchy problem (7.3.13), (7.3.18) will serve to turn back to the nonlinear Troesch's two-point problem. The most efficient way to solve the Cauchy problem to this aim seems to be considering truncations of the system (7.3.13), by introducing the projectors P_m on $\mathcal{B}_2^1([0,n])$

$$P_m\mathbf{V} = \mathbf{V}^{(m)}, \quad \mathbf{V}^{(m)} = \left(\mathbf{V}_{2j-1}\right)_{j=\overline{1,m}} . \qquad (7.3.19)$$

Therefore, the finite truncated systems may be written in the form

$$\frac{\mathrm{d}\mathbf{V}^{(m)}}{\mathrm{d}x} = \mathbf{A}^{(m)}\mathbf{V}^{(m)}, \qquad (7.3.20)$$

$\mathbf{A}^{(m)}$ being truncated matrices, up to order m inclusive

$$\mathbf{A}^{(m)} = \begin{bmatrix} \mathbf{A}_{11} & \mathbf{A}_{13} & \cdots & \mathbf{A}_{1,2m-1} \\ \mathbf{0} & \mathbf{A}_{33} & \cdots & \mathbf{A}_{3,2m-1} \\ \cdots & \cdots & \cdots & \cdots \\ \mathbf{0} & \mathbf{0} & \cdots & \mathbf{A}_{2m-1,2m-1} \end{bmatrix}. \qquad (7.3.21)$$

Moreover, $\mathbf{V}^{(m)}$ must satisfy

$$\mathbf{V}^{(m)}(0,\beta)=\left(\mathbf{V}_{2j-1}(0,\beta)\right)_{j=\overline{1,m}},$$
$$\mathbf{V}_{2j-1}(0,\beta)=\left(\beta^{2j-1}\delta_q^{2j-1}\right)_{q=\overline{0,2j-1}}. \tag{7.3.22}$$

We can write the solution of the linear finite initial problem (7.3.20), (7.3.22) in the well known classic form

$$\mathbf{V}^{(m)}(x,\beta)\cdots=\mathrm{e}^{\mathbf{A}^{(m)}x}\,\mathbf{V}^{(m)}(0,\beta)\,. \tag{7.3.23}$$

With these preparations, by using the techniques established in chapter 2, one can prove

Theorem 7.2 [134,138] *The sequence* $\left\{{}_{10}^{(m)}(x,\beta)\right\}_{m\in\mathfrak{N}}$ *converges with m to the solution y of Troesch's equation* (7.3.4), *which satisfies the initial conditions* (7.3.6).

Proof. As $\sinh(y)\in\mathcal{A}_1$, we can consider the polynomial ODS defined by the partial sums of the McLaurin's expansion of $\sinh(y)$. Theorem 2.3 ensures the convergence of the associated linear equivalents to (7.3.8). By virtue of theorems 1.6 and 2.5, the first components of the solutions of the approximating Cauchy problems locally tend to the LEM solution of the nonlinear Cauchy problem (7.3.4),(7.3.6). ◘

Corollary 7.1 [134,138] *Let* $\beta^{(m)}$ *satisfying*

$$v_{10}^{(m)}\left(n,\beta^{(m)}\right)=n\,. \tag{7.3.24}$$

Then, if the sequence $\left\{v_{10}^{(m)}\left(x,\beta^{(m)}\right)\right\}_{m\in\mathfrak{N}}$ *is convergent, its limit is the solution of Troesch's problem* (7.3.4), (7.3.5).

These results will be used to compute effectively the solution of Troesch's problem. We must solve the finite linear ODS (7.3.20) in the initial conditions (7.3.22) . This can be easily done by block partitioning, as we know the eigenvalues of the diagonal cells. Clearly, the solution will be expressed in terms of hyperbolic functions and polynomials. For $n>10$, we retain only terms containing positive powers of exponentials, the other terms being negligible; we do so because only high values of *n* are of interest. Thus, we obtained the following asymptotic expression

$$v_{10}^{(m)}\left(n,\beta^{(m)}\right)\cong u+\sum_{k=0}^{m-1}\frac{u^{2k+1}}{(2k+1)2^{5k+1}}\,,\qquad u=\beta\mathrm{e}^{n}\,. \tag{7.3.25}$$

Formula (7.3.25) holds, as for $n > 10$ one may take $\sinh n \cong e^n / 2$. Obviously, in computing the coefficients of $v_{10}^{(m)}\left(n, \beta^{(m)}\right)$, only the term in $\sinh(2k+1)n$ is significant; this leads to the introduction of u.

These preparations allow the proof of the following constructive existence

Theorem 7.3 [134,138] *For $n > 10$, the smooth solution of Troesch's plasma problem allows the (approximating) representation formula in closed form*

$$y(x) \cong y'(0)e^{\frac{x}{4}} + \sqrt{2}\ln\frac{1 + y'(0)e^x / 2^{\frac{5}{2}}}{1 - y'(0)e^x / 2^{\frac{5}{2}}}, \tag{7.3.26}$$

where $u = y'(0)e^n$ satisfies

$$\frac{u}{2} + 2\sqrt{2}\ln\frac{1 + u / 2^{\frac{5}{2}}}{1 - u / 2^{\frac{5}{2}}} = 2n. \tag{7.3.27}$$

Proof. By theorem 7.2, the sequence $\left\{v_{10}^{(m)}(x, \beta)\right\}_{m \in \mathfrak{N}}$ locally approximates the solution of the nonlinear initial problem (7.3.4), (7.3.6). This approximation may be extended to the whole $[0, n]$ for a bounded $y(x)$, in which case the series

$$\sum_{k=0}^{\infty} \frac{u^{2k+1}}{(2k+1)2^{5k+1}} \tag{7.3.28}$$

must be convergent. This occurs for $u^2 < 2^5$; computing the series sum and imposing the condition $y(n) = n$, we get (7.3.27). The representation (7.3.26) is the corresponding sum of the similar series expansion obtained in the case of $v_{10}^{(m)}$.□

7.3.3. THE PHYSICAL BEHAVIOUR OF THE SOLUTION WITH RESPECT TO THE PARAMETER

Formulae (7.3.26) and (7.3.27) emphasize the existence of an asymptote for $y(x)$, and, accordingly, of a singular point for $y'(0)$. This fully explains the "strange" behavior of $y(x)$, namely, the occurrence of divergent integrations for particular choices of $y'(0)$, as mentioned above.

The vertical asymptote of $y(x,n)$ is given by

$$x = \ln \frac{\sqrt{32}}{y'(0)} \tag{7.3.29}$$

and it is placed, as it is seen, outside the interval $[0, n]$, but, however, closer to n as n increases; this is why no numerical treatment could overcome this difficulty.

The condition (7.3.27) may be transformed, in order to obtain an approximating formula for $y'(0)$

$$y'(0) \cong 4\sqrt{2}\,\mathrm{e}^{-n}\,\frac{\mathrm{e}^{\frac{n}{\sqrt{2}}-1} - 1}{\mathrm{e}^{\frac{n}{\sqrt{2}}-1} + 1}. \tag{7.3.30}$$

Table 7.1 contains a comparison between the values of $y'(0)$ obtained by using formula (7.3.30), denoted by LEM, and those obtained by using parameter mapping techniques from [41, 44], denoted by PMT.

Table 7.1

The values of $y'(0)$ by (a) LEM and (b) PMT

n	(a)	(b)
15.84	$0.746995 \cdot 10^{-6}$	10^{-6}
18.20	$0.705357 \cdot 10^{-7}$	10^{-7}
20.50	$0.707193 \cdot 10^{-8}$	10^{-8}
22.80	$0.709023 \cdot 10^{-9}$	10^{-9}
25.10	$0.710859 \cdot 10^{-10}$	10^{-10}
27.41	$0.705607 \cdot 10^{-11}$	10^{-11}
29.71	$0.707434 \cdot 10^{-12}$	10^{-12}

Note that the values in the column (b) emphasize n as a function of $y'(0)$, while the column (a), i.e., formula (7.3.30) established by LEM, gives the values of $y'(0)$ as a function of n.

In [44], an approximating formula for $y'(0)$ as a function of n was deduced empirically, by direct observation of their pointwise correspondence, obtained numerically

$$y'(0) \cong 8\mathrm{e}^{-n}. \tag{7.3.31}$$

We can simplify formula (7.3.30) for values of n greater that 30, obtaining a more accurate than (7.3.31) approximating formula

$$y'(0) \cong 4\sqrt{2}\mathrm{e}^{-n}\left(1+\mathrm{e}^{1-\frac{n}{\sqrt{2}}}\right)^{-1}. \tag{7.3.32}$$

From the above considerations, it is seen that, unlike pure numerical treatments, LEM provided both numerical and qualitative information about the solution of Troesch's plasma problem, and also asymptotic approximations in closed form.

We shall reconsider this problem in chapter 8, emphasizing an unexpected aspect of Troesch's plasma equation.

7.4. OTHER APPLICATIONS OF LEM IN MECHANICS AND ENGINEERING

In this section, we present several applications of LEM realized by other authors, who set up and solved models in mechanics of solids; these models, together with their LEM solutions, could be successfully applied in engineering problems.

7.4.1. A SHAPE RECOGNITION MODEL BY USING SENSOR DATA FROM TOUCH

In [15], Veturia Chiroiu, Ligia Munteanu and Cornel Nicolescu considered the problem of the shape recognition of a three dimensional object from tactile sensing by a dexterous hand. Multiple fingers slide along the object surface, providing a number of 3D contact points, belonging to object's surface. The unknown surface Γ is then determined by using Bonnet's n-ellipsoid model [13]; the parameters defining Γ are determines such that the n-ellipsoid be as close as possible to the data points.

An n-ellipsoid is defined by 10 shape parameters $d_i, i=\overline{1,10}$: arbitrary centre co-ordinates (x_G, y_G, z_G), principal axes a, b, c, the principal directions given by Euler's angles ξ, ψ, ζ and the exponent n. The advantage of the model consists of the small number of parameters required in order to efficiently represent a shape. The surface Γ is considered as a deformation of the n-unit sphere S, of equation

$$x^n + y^n + z^n = 1, \tag{7.4.1}$$

by an affine transformation; for $n = 2$, S is the unit sphere from $\Re^3$, and for $n = \infty$, the unit cube of vertices $(\pm 1, \pm 1, \pm 1)$. By the affine transformations, the unit sphere and the unit cube are mapped into ellipsoids and boxes of arbitrary centres, orientations and sizes. The tridimensional data representation is then refined by using a global deformation volumetric technique, called FFD. The FFD principle is the following: the surface Γ is embedded in a parallelipipedic box within which a three-dimensional network is considered, linking the box and the object by the trivariate polynomial defining the deformation function; this connection may be described in matrix form. The box is then deformed with respect to a parameter γ by displacing its lattice and the position of a point of the real object is computed by using an iterative procedure

$$\begin{aligned} &\frac{\mathrm{d}x_k}{\mathrm{d}\gamma} = p_{k+1}, \\ &\frac{\mathrm{d}p_k}{\mathrm{d}\gamma} = -\lambda^2 \sin x_{k+1}, \qquad \lambda^2 = a^2 + b^2 + c^2. \end{aligned} \tag{7.4.2}$$

The equations (7.4.2) must be solved in various distinct Cauchy conditions.

The geometry of the fingertip tactile sensors, as well as the configuration parameters of the manipulator, are supposedly known.

The algorithm used by the authors for the surface recognition was realized in four steps:

1. The surface Γ is tracked by three fingers of the dexterous robotic hand and the information necessary to determine the contact point is recorded; M contact points are probed with each finger, therefore there are obtained $3M$ data.
2. These $3M$ data are transformed in points of co-ordinates $(x_i, y_i, z_i), i = \overline{1,3M}$.
3. The 10 shape parameters defining the surface Γ are computed such that the n-ellipsoid and the data be best possibly fitted, by using FFD and a genetic algorithm

4. The best approximating shape of the unknown object is determined.

LEM was applied at the third step, in the frame of FFD. Indeed, the iterative equations (7.4.2), which are essential for the process of shape recognition, are similar to nonlinear pendulum's equation and could be treated by LEM.

The method proposed by the authors for shape recognition was simulated by means of a virtual manipulator equipped with a force/moment sensor to detect the contact position; it shows a satisfactory accuracy. Indeed, even with perturbed data (simulating coarse sensor resolution) there obtained satisfactory results in shape recognitions of complex objects.

The method can be successfully applied to other domains, such as non-destructive testing, identification of cavities and obstacles of unknown shape and location in acoustic media, transient elasto-dynamics, etc.

7.4.2. THE CANTILEVER BAR WITH AN EMBEDDED RIBBON OF SHAPE MEMORY ALLOY

The bending of a cantilever bar with a ribbon of shape memory alloy was studied by Veturia Chiroiu, Ligia Munteanu and Călin Chiroiu in [16]. They analysed the thermo-dynamical coupling between a cantilever bar and an embedded ribbon, made of a Ni-Ti alloy with shape memory (SMA). The ribbon is heated at a temperature higher than the transition temperature, thus yielding the rod bending. The authors prove that the rod thus treated presents the essential functions of an active system, i.e., it returns to the initial configuration.

The thermodynamic model set up by the authors is complex; the thermic process is described by the heat equation and the displacements – by Lamé-Navier type equations, written for both the rod and the SMA ribbon. To these equations, one associates corresponding initial and boundary value conditions; among them, there are put into evidence the conditions at the interface rod-SMA ribbon.

To solve this mathematical model, the authors use the Bäcklund transformation; the corresponding Monge-Ampère equations are solved by using LEM and tested with FEMLAB; the results are concordant.

By using these results, there are emphasized hysteretic curves and other mechanical characteristics, showing how the structure returns to its original shape.

Thus, the SMA NiTi alloy acts as an actuator transforming the electrical energy in mechanical energy, annihilating the deformation of the rod.

The conclusion of this study is that that shape memory alloys show properties which are not present in traditionally used in engineering materials; therefore, they constitute a basis for future high-tech innovations.

7.4.3. THE DYNAMICS OF A SYSTEM OF RIGID BODIES TREATED BY LEM

The motion of mechanical systems of rigid bodies with dry friction is thoroughly studied, due to its importance both in theory and applications. Thus, Jean and Pratt [40], as well as Moreau [54, 55], established the general equations of motion of a system of rigid bodies subjected to standard constraints and forces, and to punctual contacts with dry friction between some bodies of the system.

The authors of [58] solved by LEM the problem of Jean and Pratt; in this section, we shall present their original results, using the notations and explanations from [58].

7.4.3.1. The equations of motion

Consider a finite system of rigid bodies, of generalized coordinates $q \in \Re^n$, for which there are known the initial conditions

$$q(t_0) = q_0,\ \dot{q}(t_0) = v_0,\ \ q_0, v_0 \in \Re^n,\ t_0 \in \Re; \tag{7.4.3}$$

here, $\| q - q_0 \| \leq \varepsilon$, $\| v - v_0 \| \leq \varepsilon$, for $\varepsilon > 0$ sufficiently small. Jean and Pratt consider the motion defined on an interval $t \in [t_0, t_0 + T]$, where $0 < T \leq T_0$, $t_0, T_0 \in \Re$; T is one of the unknowns.

Definition. We say that (7.4.3) is a *noncritical initial conditions* if there exist strictly positive real numbers r_1, r_2, $0 < r_1 \leq r_2$, such that for any $t, q, \dot{q}$ and every $i = \overline{1, m}$ the inequality

$$r_1 \leq M_i(t, q, \dot{q}) \leq r_2, \tag{7.4.4}$$

is fulfilled, where $M_i(t, q, \dot{q})$ represent the normal components of the reaction in the case of a perfect sleep.

The unknowns of the problem are:

- T,

- the mapping $t \to q(t)$, describing the system motion,
- the mapping $t \to f(q(t)) \in S$, $S = S_1 \times ... \times S_m$, $S_j \in \Re^2$, which describes the pair $f_j(q(t))$ of the components of the tangential reaction force at a contact point P_j, characterized by dry friction, and
- the mapping $t \to N(q(t)) \in \Re^n$, which describes the components $N_j(q(t))$ of the normal reaction force.

All the above mappings are considered defined on $[t_0, t_0 + T]$.

If (7.4.3) is a non-critical initial condition, the usual frictionless holonomic, bilateral constraints is satisfied if there exists $0 < k < 1$ such that

$$-k/(1-k)r_1 + r_2 > 0\,,\ \bar{L}\bar{d}R \le k/m\,, \tag{7.4.5}$$

where $\bar{d}R$ is an upper bound of the friction coefficients. The kinetic and potential energies of the system are given by

$$T = 1/2\dot{q}(t)A(t,q(t))\dot{q}(t)\,,\ V = 1/2q(t)B(t,q(t))q(t)\,, \tag{7.4.6}$$

the mappings $(t,q) \to A(t,q)$, $(t,q) \to B(t,q)$ representing the kinetic, respectively, potential energy matrix, both with range in the Hilbert space of measurable, square integrable on $[t_0, t_0 + T]$ functions. The Lagrange-ian is expressed as $L = T - V$, so the general Lagrange equation of the system subjected to holonomic frictionless constraints, time-dependent or not, reads

$$A(t,q(t))\ddot{q}(t) = R(t,q(t))f(q(t)) + F(t,q(t))\,, \tag{7.4.7}$$

where $R(t,q(t))f(q(t))$ are the generalized reaction forces.

To this equation, one associates friction Coulomb's law, in generalized formulation.

7.4.3.2. The model proposed by Jean and Pratt

Consider now as an example the problem proposed by Jean and Pratt [40]. The mass m moves with respect to the mass M along the rightline C of slope $\tan\alpha$, with frictionless bilateral constraints. The mass m is also sliding along a horizontal line B of a fixed body. The friction at the contact point P obeys Coulomb's law. The mass m is subjected to a horizontal constant force of magnitude $-Mg\tan\alpha$. The mass M moves vertically along the vertical line A with a frictionless bilateral constraint (figure 7.1). The generalized coordinate $q(t)$ is the abscissa of P. We denote by $f(q(t))$ the horizontal component of the reaction force. The motion equation (7.4.7) becomes

$$A\ddot{q}(t) = f(q(t)), \qquad A = m + M\tan^2\alpha . \tag{7.4.8}$$

Applying Coulomb's law, without insisting on modelling details, the equation of (7.4.8) reads

$$\ddot{q}(t) = B\tanh(Bq(t)), \tag{7.4.9}$$

where

$$B = RN_0 / A, \quad N_0 = \frac{Ag(m+M)}{A \mp RM\tan\alpha}; \tag{7.4.10}$$

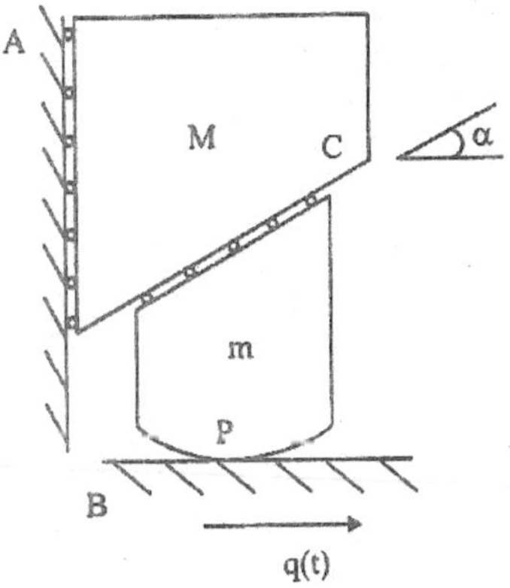

Figure 7.1. The system proposed by Jean and Pratt

This equation must be solved in the initial conditions

$$q(t_0) = 0, \; \dot{q}(t_0) = v_0 . \tag{7.4.11}$$

7.4.3.3. The LEM solution of Jean and Pratt problem

The nonlinear initial problem (7.4.8), (7.4.11) was solved in [58] by using LEM.

The authors first introduced the change of function and variable

$$x = Bt, \; y(x) = Bq , \tag{7.4.12}$$

which, applied to (7.4.8), yields

$$\frac{\mathrm{d}^2 y}{\mathrm{d}x^2} - \tanh y = 0 ; \tag{7.4.13}$$

The above equation was written in the form of a nonlinear first order ODS

$$\begin{aligned} y' &= z, \\ z' &= \tanh y; \end{aligned} \tag{7.4.14}$$

the initial conditions corresponding to (7.4.11) being

$$y(0) = 0, \quad z(0) = v_0 . \tag{7.4.15}$$

The LEM mapping depends on two parameters

$$v(x, \sigma, \xi) = \mathrm{e}^{\sigma y + \xi z} ; \tag{7.4.16}$$

the first LEM equivalent reads

$$\frac{\partial v}{\partial x} = \sigma \frac{\partial v}{\partial \xi} + \xi \tanh(\mathrm{D}_\sigma) v , \tag{7.4.17}$$

where

$$\tanh(\mathrm{D}_\sigma) \equiv \sum_{k=1}^{\infty} \frac{2^{2k}\left(2^{2k} - 1\right) B_{2k}}{(2k)!} \mathrm{D}_\sigma^{2k-1}, \quad \mathrm{D}_\sigma^k \equiv \frac{\partial^k}{\partial \sigma^k}, \tag{7.4.18}$$

B_{2k} being the Bernoulli numbers [3].

Taking for v the expansion

$$v(x, \sigma, \xi) = 1 + \sum_{j+k=1}^{\infty} v_{jk}(x) \frac{\sigma^j}{j!} \frac{\xi^k}{k!} , \tag{7.4.19}$$

the equation (7.4.17) leads to the second LEM equivalent

$$v'_{jk} = j v_{j-1,k} + k \sum_{m=1}^{\infty} \frac{2^m (2^{2m-2} - 1) B_{2m-2}}{(m-1)!} v_{2m-1-k,k} , \tag{7.4.20}$$

that can be written in matrix form

$$\frac{\mathrm{d}\mathbf{V}}{\mathrm{d}x} = \mathbf{A}\mathbf{V}, \quad \mathbf{V} = (\mathbf{V}_{2m-1})_{m \in \mathfrak{N}}, \quad \mathbf{V}_{2m-1} = \left[v_{jk}(x)\right]_{j+k=2m-1} . \tag{7.4.21}$$

As the ODS (7.4.14) is odd, we can drop the cells with even indices of the LEM matrix **A**. Hence the “contracted” **A** reads

$$\mathbf{A} = \begin{bmatrix} \mathbf{A}_{11} & \mathbf{A}_{13} & \mathbf{A}_{15} & \dots \\ \mathbf{0} & \mathbf{A}_{33} & \mathbf{A}_{35} & \dots \\ \mathbf{0} & \mathbf{0} & \mathbf{A}_{55} & \dots \\ \dots & \dots & \dots & \dots \end{bmatrix}, \tag{7.4.22}$$

where the diagonal cells are

$$\mathbf{A}_{11} = \begin{bmatrix} 0 & 1 \\ 1 & 0 \end{bmatrix}, \quad \mathbf{A}_{2j-1,2j-1} = \begin{bmatrix} 0 & 2j-1 & \dots & 0 & 0 \\ 1 & 0 & \dots & 0 & 0 \\ \dots & \dots & \dots & \dots & \dots \\ 0 & 0 & \dots & 0 & 1 \\ 0 & 0 & \dots & 2j-1 & 0 \end{bmatrix}. \tag{7.4.23}$$

The initial conditions (7.4.15) become for $\mathbf{V}$

$$\mathbf{V}(0) = \left(\mathbf{V}_{2j-1}\right)_{j \in \mathfrak{N}}, \quad \mathbf{V}_{2j-1} = z_0^{2j-1}\, \mathbf{e}_{2j-1}. \tag{7.4.24}$$

The initial LEM problem is solved by truncating the system up to an order m. After some elementary computation, the authors obtain for Jean and Pratt's problem the following analytical approximating solution

$$y(x) \cong z_0 x - \frac{1}{2}x^2 + \frac{4}{5z_0}x^3 - \frac{3}{5z_0^2}x^4 + \frac{32}{75z_0^3}x^5. \tag{7.4.25}$$

7.4.3.4. Conclusions

For m, R and α fixed up, the authors of [58] determined a LEM approximation of the (unique) solution of the problem (7.4.9), (7.4.11) of the form

$$q(t) \cong v_0 t - \frac{B}{2}t^2 + \frac{4}{5}\frac{B^2}{v_0}t^3 - \frac{3}{5}\frac{B^3}{v_0^2}t^4 + \frac{32}{75}\frac{B^4}{v_0^3}t^5, \tag{7.4.26}$$

where

$$B = \frac{RN_0}{m + M\tan^2\alpha}, \tag{7.4.27}$$

for $t \in [0, \tau_0]$, and

$$\tau_0 = \frac{v_0}{B} = \frac{v_0\left(m + M\tan^2\alpha\right)}{RN_0}. \tag{7.4.28}$$

In the above formula,

$$N_0 = \frac{g(m+M)\left(m + M\tan^2\alpha\right)}{m + M\tan^2\alpha - RM\tan\alpha}. \tag{7.4.29}$$

The LEM solution can be qualitatively studied. In [58], there are put into evidence several interesting aspects of the behaviour of the solution.

Thus, for $t = \tau_0$ it results that $q = 0.42\frac{v_0^2}{B}$. If $t > \tau_0$, then $v_0 = 0$, $q(t) = 0, f(t) = 0, N(t) = 0$.

For $R > \tan\alpha$, $\frac{RM\tan\alpha}{m + M\tan^2\alpha} \to 1$, $N_0 \to \infty$ and $\tau_0 \to 0$. This may result in shocks in the limit process, as suggested in [40]. The solution (7.4.26) exists only if the condition (7.4.5) is satisfied. Indeed, for $v_0 > 0$ this condition is satisfied because f is negative, N is positive, and $f = -RN$, which implies $\frac{RM\tan\alpha}{m + M\tan^2\alpha} < 1$. If $v_0 < 0$, we have

$$N_0 = \frac{g(m+M)\left(m + M\tan^2\alpha\right)}{m + M\tan^2\alpha + RM\tan\alpha}, \tag{7.4.30}$$

and (7.4.5) may be satisfied or not. The unique solution of the problem is in this case

$$q(t) \cong v_0 t + \frac{B}{2}t^2 + \frac{4}{5}\frac{B^2}{v_0}t^3 - \frac{3}{5}\frac{B^3}{v_0^2}t^4 + \frac{32}{75}\frac{B^4}{v_0^3}t^5, \tag{7.4.31}$$

for $t \in [0, \tau_0]$. For $t > \tau_0$, one has $q(t) = 0, f(t) = 0$ and $N(t) = mg + Mg$.

We conclude this section by mentioning also other applications of LEM, realized by the research group from IMSAR, led by Veturia Chiroiu: active control damping of machine tools [17], study of quasi-periodic oscillations of a weakly nonlinear coupled oscillator [22], damping effects on the stability of dynamical systems, etc.

7.5. LEM SOLUTIONS FOR THE SECOND PAINLEVÉ EQUATION

Painlevé [61] was the first to notice the existence of ODEs whose solutions allow possible singularities not depending on the Cauchy data. The classification according to this criterion pointed out six such equations, later known as Painlevé I-VI.

While obtained in a pure mathematical frame, these equations turned out to be of the greatest importance in applications in various domains. Among the most pertinent applications, let us quote general relativity, quantum field theory, nonlinear waves, nonlinear optics and fiber optics, quantum gravity, plasma physics, polyelectrolytes, superconductivity, etc.[1,2]. The interest in studying the Painlevé equations is recently growing, as they were also obtained as reductions of soliton equations solvable by inverse scattering.. Thus, the second Painlevé equation was obtained in the study of solitons for PDEs as Korteweg-de Vries, Schrödinger, sine-Gordon, Boussinesq, Davey-Stewartson. Clarkson [2] mentioned that some exact solutions of the Painlevé equations may expressed in terms of special functions, such as Airy or Bessel functions.

In this subsection, we shall use the above LEM representations to get locally the solutions of the Cauchy problem with arbitrary data in the particular case of the second Painlevé equation (PII)

$$\ddot{y} + ty = 2y^3 . \tag{7.5.1}$$

Let us associate to PII some arbitrary Cauchy conditions

$$y(0) = \alpha, \quad \dot{y}(0) = \beta . \tag{7.5.2}$$

The nonlinear polynomial ODE (7.5.1) may be also written in the form of the polynomial ODS

$$\mathcal{P}y \equiv \begin{bmatrix} \dot{y} - z \\ \dot{z} - ty + 2y^3 \end{bmatrix} = \mathbf{0}, \tag{7.5.3}$$

and the initial conditions (7.5.2) are

$$y(0) = \alpha, \quad z(0) = \beta . \tag{7.5.4}$$

Let us apply LEM to the above problem. Considering the LEM mapping

$$v(t, \sigma, \xi) = e^{\sigma y + \xi z}, \tag{7.5.5}$$

we get the first LEM equivalent

$$\mathfrak{L}v \equiv \frac{\partial v}{\partial t} - \sigma\frac{\partial v}{\partial \xi} - \xi\left(-t\frac{\partial v}{\partial \sigma} - 2\frac{\partial^3 v}{\partial \sigma^3}\right). \qquad (7.5.6)$$

Proceeding as specified before, we use in this PDE an expansion of v in the form

$$v(t,\sigma,\xi) = 1 + \sum_{i+j=1}^{\infty} v_{ij}(t)\frac{\sigma^i}{i!}\frac{\xi^j}{j!} \qquad (7.5.7)$$

to get the second LEM equivalent of the system (7.5.3)

$$\mathfrak{S}\mathbf{V} \equiv \frac{\mathrm{d}\mathbf{V}}{\mathrm{d}t} - \mathbf{A}(t)\mathbf{V} = \mathbf{0},$$
$$\mathbf{V}(t) = (\mathbf{V}_{2k-1})_{k\in\mathfrak{N}}, \quad \mathbf{V}_{2k-1}(t) = \lfloor v_{2k-1-j,j} \rfloor_{j=\overline{0,2k-1}}, \qquad (7.5.8)$$

whose associated LEM matrix is

$$\mathbf{A}(t) = \begin{bmatrix} \mathbf{A}_{11}(t) & \mathbf{A}_{13} & \mathbf{0} & \mathbf{0} & \cdots \\ \mathbf{0} & \mathbf{A}_{33}(t) & \mathbf{A}_{35} & \mathbf{0} & \cdots \\ \mathbf{0} & \mathbf{0} & \mathbf{A}_{55}(t) & \mathbf{A}_{57} & \cdots \\ \cdots & \cdots & \cdots & \cdots & \cdots \end{bmatrix}. \qquad (7.5.9)$$

This matrix is compressed because of the oddness of the polynomial operator.

The initial conditions are firstly mapped to

$$v(0,\sigma,\xi) = \mathrm{e}^{\sigma\alpha+\xi\beta}, \quad \sigma,\xi \in \mathfrak{R}, \qquad (7.5.10)$$

being associated to the PDE (7.5.6) and then, by (7.5.7), they become

$$\mathbf{V}(0) = (\mathbf{V}_{2k-1}(0))_{k\in\mathfrak{N}}, \quad \mathbf{V}_{2k-1}(0) = \left\lfloor \alpha^{2k-1-j}\beta^j \right\rfloor_{j=\overline{0,2k-1}}, \qquad (7.5.11)$$

indicating an initial condition for the system (7.5.8).

The diagonal matrices are

$$\mathbf{A}_{2k-1,2k-1}(t)=\begin{bmatrix} 0 & 2k-1 & 0 & \dots & 0 & 0 \\ -t & 0 & 2k-2 & \dots & 0 & 0 \\ 0 & -2t & 0 & \dots & 0 & 0 \\ \dots & \dots & \dots & \dots & 0 & 0 \\ 0 & 0 & 0 & \dots & 0 & 1 \\ 0 & 0 & 0 & \dots & -(2k-1)t & 0 \end{bmatrix} \tag{7.5.12}$$

and the rectangular ones read

$$\mathbf{A}_{2k-1,2k+1}=2\begin{bmatrix} 0 & 0 & \dots & 0 & 0 & 0 & 0 \\ 1 & 0 & \dots & 0 & 0 & 0 & 0 \\ 0 & 2 & \dots & 0 & 0 & 0 & 0 \\ \dots & \dots & \dots & \dots & \dots & 0 & 0 \\ 0 & 0 & \dots & 0 & 2k-1 & 0 & 0 \end{bmatrix}. \tag{7.5.13}$$

Let us note that the diagonal cells depend on t, while the rectangular ones are all of them constant.

The linear part of the polynomial ODS (7.5.3) is

$$\dot{y}=z, \quad \dot{z}=-ty, \tag{7.5.14}$$

or else

$$\ddot{y}+ty=0. \tag{7.5.15}$$

A fundamental system of solutions for the linear part is therefore

$$\sqrt{t}J_{\frac{1}{3}}\left(\frac{2}{3}t^{\frac{3}{2}}\right); \quad \sqrt{t}J_{-\frac{1}{3}}\left(\frac{2}{3}t^{\frac{3}{2}}\right), \tag{7.5.16}$$

where J represents the Bessel functions of first order; we thus see that the LEM representations will use as a basis the Bessel functions.

Taking into account the expansion of Bessel functions, one can consider for (7.5.14) the following fundamental system

$$\begin{aligned} p(t)&=t\sum_{k=0}^{\infty}\frac{(-1)^k}{k!(3k+1)(3k-2)..4\cdot 1}\left(\frac{t^3}{3}\right)^k, \\ q(t)&=1+\sum_{k=1}^{\infty}\frac{(-1)^k}{k!(3k-1)(3k-4)..5\cdot 2}\left(\frac{t^3}{3}\right)^k. \end{aligned} \tag{7.5.17}$$

They even form a normal system, as

$$q(0)=1, \quad \dot{q}(0)=0, \quad p(0)=0, \quad \dot{p}(0)=1. \tag{7.5.18}$$

The solution of the Cauchy problem (7.5.14), (7.5.4) is

$$y(t)=\alpha q(t)+\beta p(t), \quad z(t)=\alpha \dot{q}(t)+\beta \dot{p}(t). \tag{7.5.19}$$

Denote by

$$f(t)=\alpha q(t)+\beta p(t). \tag{7.5.20}$$

and let us truncate the LEM system up to the fifth order. The associated truncated system, giving the approximate LEM solution, is

$$\begin{aligned} \frac{\mathrm{d}\mathbf{V}_1^{(5)}}{\mathrm{d}t} &= \mathbf{A}_{11}(t)\mathbf{V}_1^{(5)} + \mathbf{A}_{13}\mathbf{V}_3^{(5)}, \\ \frac{\mathrm{d}\mathbf{V}_3^{(5)}}{\mathrm{d}t} &= \mathbf{A}_{33}(t)\mathbf{V}_3^{(5)} + \mathbf{A}_{35}\mathbf{V}_5^{(5)}, \\ \frac{\mathrm{d}\mathbf{V}_5^{(5)}}{\mathrm{d}t} &= \mathbf{A}_{55}(t)\mathbf{V}_5^{(5)}. \end{aligned} \tag{7.5.21}$$

The initial conditions associated to this system are, obviously,

$$\mathbf{V}_{2k-1}^{(5)}(0)=\left[\alpha^{2k-1-j}\beta^j\right]_{j=\overline{0,2k-1}}, \quad k=\overline{1,3}. \tag{7.5.22}$$

We can solve it by block partitioning, starting with the last block.

Let us notice that the fundamental matrices associated to the linear systems

$$\frac{\mathrm{d}\mathbf{V}_{2k-1}}{\mathrm{d}t} = \mathbf{A}_{2k-1,2k-1}(t)\mathbf{V}_{2k-1}, \quad k=\overline{1,2}, \tag{7.5.23}$$

are, accordingly,

$$\begin{bmatrix} q & p \\ \dot{q} & \dot{p} \end{bmatrix}, \quad \begin{bmatrix} q^3 & 3pq^2 & 3p^2q & p^3 \\ q^2\dot{q} & q(2p\dot{q}+\dot{p}q) & p(2p\dot{q}+\dot{p}q) & p^2\dot{p} \\ q\dot{q}^2 & \dot{q}(2\dot{p}q+p\dot{q}) & \dot{p}(2\dot{p}q+\dot{p}q) & p\dot{p}^2 \\ \dot{q}^3 & 3\dot{p}\dot{q}^2 & 3\dot{p}^2\dot{q} & \dot{p}^3 \end{bmatrix}; \tag{7.5.24}$$

they present a certain symmetry and this leads to the conclusion that one can obtain them for any k. The same remark holds for their inverses that are needed in order to find the solution of the involved non-homogeneous systems.

Indeed, for the matrices (7.5.24) these inverses are, accordingly,

$$\begin{bmatrix} \dot{p} & -p \\ -\dot{q} & q \end{bmatrix}, \quad \begin{bmatrix} \dot{p}^3 & -3p\dot{p}^2 & 3p^2\dot{p} & -p^3 \\ -\dot{p}^2\dot{q} & \dot{p}(2p\dot{q}+\dot{p}q) & -p(2\dot{p}q+p\dot{q}) & p^2\dot{p} \\ \dot{p}\dot{q}^2 & -\dot{q}(2\dot{p}q+p\dot{q}) & q(2p\dot{q}+\dot{p}q) & p\dot{p}^2 \\ -\dot{q}^3 & 3q\dot{q}^2 & -3q^2\dot{q} & q^3 \end{bmatrix}. \tag{7.5.25}$$

The solution of the linear and homogeneous Cauchy problem

$$\begin{aligned} &\frac{\mathrm{d}\mathbf{V}_5^{(5)}}{\mathrm{d}t} = \mathbf{A}_{55}(t)\mathbf{V}_5^{(5)}, \\ &\mathbf{V}_5^{(5)}(0) = \left[\alpha^{5-j}\,\beta^j\right]_{j=\overline{0,5}}, \end{aligned} \tag{7.5.26}$$

is, by theorem 2.2,

$$\mathbf{V}_5^{(5)}(t) = \left[f^5, f^4\dot{f}, f^3\dot{f}^2, f^2\dot{f}^3, f\dot{f}^4, \dot{f}^5\right]^{\mathrm{T}}. \tag{7.5.27}$$

The next block is non-homogeneous. The solution of the associated homogeneous system, also satisfying the initial conditions (7.5.22) for $k=2$ is, again by theorem 2.2,

$$\mathbf{V}_3^{hom}(t) = \left[f^3, f^2\dot{f}, f\dot{f}^2, \dot{f}^3\right]^{\mathrm{T}}. \tag{7.5.28}$$

To find a particular solution of the second block of (7.5.21), we shall use the inverse (7.5.25) of the associated fundamental system. After getting $\mathbf{V}_3^{(5)}(t)$, we solve the first block of (7.5.21). We get the final form of the fifth approximation in vector form

$$\begin{bmatrix} y(t) \\ z(t) \end{bmatrix} \cong \begin{bmatrix} q(t) & p(t) \\ \dot{q}(t) & \dot{p}(t) \end{bmatrix} \begin{bmatrix} \alpha \\ \beta \end{bmatrix} + 2\begin{bmatrix} pq_1 - qp_1 \\ \dot{p}q_1 - \dot{q}p_1 \end{bmatrix} + 12\begin{bmatrix} pq_2 - qp_2 \\ \dot{p}q_2 - \dot{q}p_2 \end{bmatrix}, \tag{7.5.29}$$

where we used the notations

$$\begin{aligned} &p_1(t) = \int_0^t p(\tau) f^3(\tau)\,\mathrm{d}\tau, \quad q_1(t) = \int_0^t q(\tau) f^3(\tau)\,\mathrm{d}\tau, \\ &p_2(t) = \int_0^t p(\tau) f^2(\tau)\left[p(\tau)q_1(\tau) - q(\tau)p_1(\tau)\right]\mathrm{d}\tau, \\ &q_2(t) = \int_0^t q(\tau) f^2(\tau)\left[p(\tau)q_1(\tau) - q(\tau)p_1(\tau)\right]\mathrm{d}\tau. \end{aligned} \tag{7.5.30}$$

Obviously, all the involved particular solutions satisfy null initial conditions.

A similar to (7.5.29) representation may be immediately obtained for the following variant of (7.5.1)

$$\ddot{y} - ty = 2y^3. \tag{7.5.31}$$

Indeed, a fundamental system of solutions of the linear part of (7.5.31) is [3, 42]

$$\sqrt{t}J_{\frac{1}{3}}\left(\frac{2}{3}\mathrm{i}t^{\frac{3}{2}}\right);\quad \sqrt{t}J_{-\frac{1}{3}}\left(\frac{2}{3}\mathrm{i}t^{\frac{3}{2}}\right). \tag{7.5.32}$$

As previously, we shall use instead of this the normal system

$$\begin{aligned} P(t) &= t\sum_{k=0}^{\infty}\frac{1}{k!\,(3k+1)(3k-2)..4\cdot 1}\left(\frac{t^3}{3}\right)^k, \\ Q(t) &= 1+\sum_{k=1}^{\infty}\frac{1}{k!\,(3k-1)(3k-4)..5\cdot 2}\left(\frac{t^3}{3}\right)^k. \end{aligned} \tag{7.5.33}$$

Thus, the LEM fifth approximate solution of (7.5.31) is precisely (7.5.29), in which p and q are replaced by P and Q accordingly.

One can estimate the above approximating formulae (7.5.29) to see how close they are to the LEM series that give the exact solution of the nonlinear ODS.

Suppose that $|t| < M$. Then, from (7.5.29) it follows that the inequalities

$$|p(t)| \le P(|t|) \le P(M),\quad |q(t)| \le Q(|t|) \le Q(M) \tag{7.5.34}$$

hold true for both equations (7.5.1) and (7.5.31).

Starting from (7.5.34), we can also use the rough approximations

$$\begin{aligned} &|p(t)| \le M\mathrm{e}^{\frac{M^3}{3}},\quad |q(t)| \le \mathrm{e}^{\frac{M^3}{3}},\quad |f(t)| \le (|\alpha| + M|\beta|)\mathrm{e}^{\frac{M^3}{3}}, \\ &|p(t)q_1(t) - q(t)p_1(t)| \le 2M^2\mathrm{e}^{\frac{5M^3}{3}}(|\alpha| + M|\beta|)^3, \\ &|p(t)q_2(t) - q(t)p_2(t)| \le 4M^5\mathrm{e}^{\frac{7M^3}{3}}(|\alpha| + M|\beta|)^5, \end{aligned} \tag{7.5.35}$$

for both the couples p, q and P, Q.

Thus, if, e.g. $M < 0.5$, $|\alpha| + M|\beta| < 0.5$, then $|f(t)| < 0.521$ and

$$\begin{aligned} &|p(t)q_1(t) - q(t)p_1(t)| < 0.1525, \\ &|p(t)q_2(t) - q(t)p_2(t)| < 0.005234. \end{aligned} \tag{7.5.36}$$

Naturally, the rate of convergence of the LEM solutions (7.5.29) of both second Painlevé equations (7.5.1), (7.5.31) depends on that of the involved Bessel functions.

7.6. NORMAL LEM SOLUTIONS FOR EULER-POINSOT'S SOLID

In some previous papers [128,158], a new and rather surprising connection was established between the dynamics of the nonlinear rigid pendulum and that of the motion of a solid about a fixed point, in the particular case due to Euler and Poinsot. Using the LEM solution for the pendulum, it was possible to get an analytical approach of the solution of the equation of motion, that, compared numerically with the Runge-Kutta solution, lead to small relative errors on large intervals.

The motion of Euler-Poinsot's solid was thoroughly studied and its solution is well known (see, e.g., [89]). I tackled this problem despite this, because the classical solution is based on first integrals, implicit with respect to the unknown functions; there is a good chance that analytical approximations of the explicit solution give more insight into the problem.

7.6.1. EULER-POINSOT'S SOLID

The frictionless motion of a rigid solid about one of its points, fixed up with respect to an inertial reference system, represents one of the basic problems of mechanics, of both theoretical and practical importance. This problem was firstly considered by d'Alembert in 1749, but the equations of motion were established in their present form by Euler (1758). Later on, the problem was tackled by other scientists, like J.L.Lagrange, L. Poinsot, S.D. Poinsot, C.G.C. Jacobi, Ch. Hermite and Sonia Kovalevskaia; it is still attracting numerous searchers.

Euler-Poinsot's is one of the cases of integrability. It is the case in which the moment of the external forces given with respect to a fixed pole is null; e.g., the resultant of the external forces is null or passes always through the fixed pole. The particular case considered here is that in which the solid is fixed up right in its center of mass, being acted upon by its own weight. The equations of motion then have the form

$$\begin{aligned} I_1\dot{\omega}_1 &= (I_2 - I_3)\omega_2\omega_3, \\ I_2\dot{\omega}_2 &= (I_3 - I_1)\omega_3\omega_1, \\ I_3\dot{\omega}_3 &= (I_1 - I_2)\omega_1\omega_2, \end{aligned} \tag{7.6.1}$$

where $\omega_1, \omega_2, \omega_3$ are the angular velocities and I_1, I_2, I_3 are the moments of inertia with respect to the fixed pole, ordered as $I_1 > I_2 > I_3$, without lost of generality. These inequalities justify the following notations

$$a^2 = \frac{I_2 - I_3}{I_1}, \quad b^2 = \frac{I_1 - I_3}{I_2}, \quad c^2 = \frac{I_1 - I_2}{I_3}, \tag{7.6.2}$$

due to which the system (7.6.1) becomes

$$\begin{aligned} \dot{\omega}_1 &= a^2\omega_2\omega_3, \\ \dot{\omega}_2 &= -b^2\omega_3\omega_1, \\ \dot{\omega}_3 &= c^2\omega_1\omega_2. \end{aligned} \tag{7.6.3}$$

7.6.2. A LEM APPROACH

Let us associate to this system the arbitrary initial conditions

$$\omega_j(0) = \omega_{j0}, \quad j = \overline{1,3}. \tag{7.6.4}$$

I. From the first two equations, by using a well-known integral combination, it straighforwardly follows that one can search for the unknown functions ω_1, ω_2 in the form

$$\omega_1 = aC\cos\theta, \quad \omega_2 = bC\sin\theta, \tag{7.6.5}$$

Each of the first two equations of (7.6.3) give

$$\omega_3 = -\frac{1}{ab}\dot{\theta}. \tag{7.6.6}$$

From the third, it results that $x = 2\theta$ satisfies the equation

$$\ddot{x} + a^2b^2c^2C^2\sin x = 0, \quad C^2 = \frac{\omega_{10}^2}{a^2} + \frac{\omega_{20}^2}{b^2}, \tag{7.6.7}$$

in which we recognize the equation of the ***nonlinear rigid pendulum***.

II. Using the same pattern, we deduce that in the last two equations of (7.6.3) one can search for ω_2, ω_3 in the form

$$\omega_2 = bA\sin\varphi, \quad \omega_3 = cA\cos\varphi; \tag{7.6.8}$$

then ω_1 is

$$\omega_1 = -\frac{1}{bc}\dot{\varphi}. \tag{7.6.9}$$

From the first equation of (7.6.3) it results that $y = 2\varphi$ also satisfies the ***pendululum equation***

$$\ddot{y} + a^2b^2c^2A^2 \sin y = 0, \quad A^2 = \frac{\omega_{20}^2}{b^2} + \frac{\omega_{30}^2}{c^2}, \tag{7.6.10}$$

III. The last variant concerns the first and the last equation of (7.6.3); here, one must take into account the sign of the expression $E \equiv \frac{\omega_{10}^2}{a^2} - \frac{\omega_{30}^2}{c^2}$.

a) If $E > 0, \quad E \equiv B^2$, then we can take

$$\omega_1 = aB\cosh\psi, \quad \omega_3 = cB\sinh\psi, \tag{7.6.11}$$

b) If $E < 0, \quad E \equiv -B^2$, then we must take

$$\omega_1 = aB\sinh\psi, \quad \omega_3 = cB\cosh\psi. \tag{7.6.12}$$

Both cases a) and b) lead to

$$\omega_2 = \frac{1}{ac}\dot{\psi}; \tag{7.6.13}$$

this time, $z = 2\psi$ satisfies an equation complementary to that of the pendulum

$$\ddot{z} + a^2b^2c^2B^2 \sinh z = 0, \quad B^2 = \frac{\omega_{10}^2}{a^2} - \frac{\omega_{30}^2}{c^2}. \tag{7.6.14}$$

So, we put into evidence three paths; by using any of them, one can get the motion of the solid about a fixed point, in the above considered case.

Applying the results of chapter 3, especially those obtained in subsection 3.4.2, one obtains the normal LEM representations of x, y, z. Let us note that we already obtained the LEM formulae for the pendulum in chapter 6. One can write all these three expressions in a common form $S(t)$, for any of the equations (7.6.7), (7.6.10), (7.6.14). Attributing to S the initial data α, β, deduced from the arbitrary Cauchy conditions (7.6.4), we obtain, going as far as thirds order effects

- for (7.6.7) and (7.6.10), the normal LEM solution for the pendulum equation

$$S(t) \cong \frac{1}{K}(\alpha K \cos Kt + \beta \sin Kt) + \\ + \frac{(-1)^j}{2^6}\left[\alpha^3 \varphi_1(Kt) + \alpha^2\beta\, \varphi_2(Kt) + \alpha\beta^2\, \varphi_3(Kt) + \beta^3\, \varphi_4(Kt)\right] \quad (7.6.15)$$

where $j = 2$, K is respectively replaced by $abcC, abcB$ and the functions φ_j, $j = \overline{1,4}$ are given by the formulae

$$\begin{aligned} K^2\varphi_1(\tau) &= \cos 3\tau - \cos\tau - 12\tau\sin\tau, \\ K^3\varphi_2(\tau) &= 3\sin 3\tau - 21\sin\tau + 12\tau\cos\tau, \\ K^4\varphi_3(\tau) &= -3\cos 3\tau + 3\cos\tau - 12\tau\sin\tau, \\ K^5\varphi_4(\tau) &= -\sin 3\tau - 9\sin\tau + 12\tau\cos\tau; \end{aligned} \quad (7.6.16)$$

- for (7.6.14), the same LEM representation (7.6.15), where $j = 1$, K is replaced by $abcA$, and φ_j, $j = \overline{1,4}$ are given by the same formulae (7.6.16).

The above normal LEM representations efficiently approximate x, y, z. These functions are differentiated and then multiplied by constants, according to (7.6.6), (7.6.9), (7.6.13), thus giving $\omega_3, \omega_1, \omega_2$. The remaining two functions are determined on each path from the formulae (7.6.5), (7.6.8), (7.6.12).

We see that the pendulum equation, as well as its complementary (7.6.14), are essential. The LEM solution of the pendulum equation was analysed in the previous sections of this chapter from the numerical point of view.

In this case, we set up a table in which the Runge-Kutta solution of (7.6.3) is compared with the solution of the same system, by using the analytic LEM formula (7.6.15), applied to equation (7.6.7). We took $a^2 = 0.4, b^2 = 1.3333, c^2 = 2$ and made the comparison on the large enough interval $\left[0, 10\pi\right]$. This comparison showed relative errors around $5 \cdot 10^{-2}$ [158].

Table 7.2

Comparison between the Runge-Kutta and LEM solution

$\omega_1(0)$	$\omega_2(0)$	$\omega_3(0)$	Relative error
1	0	0.2	0.0609
0.5	0.2	0.2	0.0703
0.5	0.1	0.1	0.0319
1.5	0.1	0.1	0.0673
1.5	–0.1	–0.1	0.1030
–1	–0.1	0.1	0.0250

Let us mention that a complete numerical study should also include the other two above specified paths, having LEM representations as a kernel. It is possible that the Euler-Poinsot solid should present a "mathematical kernel" in the form of pendulum equation in other cases too; one could thus obtain analytical representations of the solution based on LEM. Finally, it should be very interesting to identify the physical significance of the magnitudes θ, φ, ψ and, if possible, to explain the role they play in the motion of the solid around a fixed point.

Chapter 8

INTRINSIC EQUATIONS AND LEM SOLUTIONS

The scope of this chapter is to unify some of the models studied by LEM under the sign of a mathematical kernel represented by intrinsic equations. The attribute "intrinsic" refers to the independence of these equations of physical factors: they are purely mathematical.

In the first section, an ODE is put into evidence that governs from the abstract both phenomena from the mechanics of solids – the B.-E. bar deflection, the motion of the nonlinear rigid pendulum, the nonlinear two-bar frame – and the plasma confinement by radiation pressure, modelled by Troesch. The LEM solution of the intrinsic equation can be applied to all these models.

We notice that the intrinsic equation can be brought to a particular case of Düffing's equation. This equation is treated by LEM in section 2.

The next – and last – section describes a model and solution analogy, based on Voinea's analogy. The terms of this analogy are the relativistic motion of a particle acted upon by an arbitrary force and the nonlinear bending of a straight bar. To these terms, a third one is added: the deflection of a relativistic particle beam under the influence of a magnetic field. These phenomena obey the same abstract mathematical law, expressed in form of another intrinsic equation, different from those in the previous sections, whose solution is then discussed for each of the considered cases.

8.1. AN UNEXPECTED MATHEMATICAL CONNECTION BETWEEN THE SOLID AND PLASMA STATE

In this section, there are presented several mechanical and physical models governed by the same intrinsic polynomial ODE with constant coefficients, purely abstract.

The first case for which the intrinsic equation was deduced was the nonlinear pendulum [112,145]. By simple transformations, the pendulum equation was reduced to a fourth degree polynomial ODE, completely independent of the physical magnitudes involving the pendulum motion; we called this equation *intrinsic* precisely because it represents the "hard core"– the archetype – of the phenomenon. Further researches [116,117,147] showed that this polynomial ODE also governs other models, such as the B.-E. bar, the nonlinear elastic frame, belonging to the solid state, but also Troesch's plasma model, treated in section 7.

We show here that the normal LEM solution obtained for the intrinsic equation, valid for all these models, no matter if they belong to the solid or plasma state, permits a qualitative study, which was realized for each model.

8.1.1. THE MODELS

The solid state models allowing the intrinsic equation as mathematical pattern are: the nonlinear pendulum, the B.-E. bar and the nonlinear frame. To these models one adds Troesch's model, belonging to the plasma state.

Solid state:

1. The nonlinear free rigid pendulum, already presented in chapter 6. The equation of motion is (6.1.2), where we take $A = 0$. Putting θ instead of y, we obtain the nonlinear homogeneous equation

$$\ddot{\theta} + \omega^2 \sin \theta = 0, \qquad \omega^2 = \frac{g}{l}. \tag{8.1.1}$$

2. The Bernoulli-Euler (B.-E.) bar, presented in section 5.2. Using the same notations, we obtain as a model the following nonlinear ODS

$$\begin{aligned} &\frac{\mathrm{d}y}{\mathrm{d}s} = \sin \theta, \\ &\frac{\mathrm{d}\theta}{\mathrm{d}s} = \alpha^2 (f - y), \quad \alpha^2 = \frac{P}{EI}. \end{aligned} \tag{8.1.2}$$

3. The elastic frame. The buckling of an elastic frame composed of two bars, one vertical and one horizontal, a fundamental problem of the statics of constructions, was studied in section 5.4; there, we specified that we shall re-consider the method of getting the normal LEM solution

(5.4.10). With the notations of this section, the nonlinear model of the frame, established by P.P.Teodorescu, is formed by the following two nonlinear ODSs

$$x_1' = \cos\theta_1,$$
$$y_1' = \sin\theta_1, \tag{8.1.3}$$
$$\theta_1' = \widetilde{X}x_1 - \left(\alpha^2 - \omega^2\right)y_1,$$

and

$$x_{21}' = \cos\theta_2,$$
$$y_2' = \sin\theta_2, \tag{8.1.4}$$
$$\theta_2' = -\omega^2 x_2 + \widetilde{X}y_2,$$

that must be solved in the conditions

$$x_j(0) = 0, \quad y_j(0) = 0, \quad j = 1,2,$$
$$y_1(l) = f, \quad l - x_1(l) = y_2(l), \quad l - x_2(l) = f, \tag{8.1.5}$$
$$\theta_1(l) = \theta_2(l) = -\theta_0, \quad \theta_0 > 0.$$

Plasma state:

Troesch's model, describing the confinement of plasma by radiation pressure, was already treated in section 7.3. Let us recall that Troesch's problem is composed of the nonlinear EDO

$$y'' = \sinh y, \tag{8.1.6}$$

to be solved in the two-point conditions

$$y(0) = 0, \quad y(n) = n; \tag{8.1.7}$$

the parameter n has the physical signification of plasma density.

8.1.2. THE INTRINSIC EQUATION AND THE NORMAL LEM SOLUTION

The intrinsic equation is obtained by successive differentiation of the above equations and by replacing the trigonometric, accordingly hyperbolic, sine by its second derivative, developed from the corresponding equation. We obtain the polynomial ODE

$$z_j^{IV} z_j' - z_j''' z_j'' = (-1)^j z_j'' z_j'^3, \quad j = 1,2, \tag{8.1.8}$$

where $j=1$ corresponds to the solid and $j=2$ – to the plasma state.

As the intrinsic equation is representative for all the above models, we shall denote by x its independent variable, re-considering this notation and its significance when we will treat each case apart.

Consequently, the unknown functions from all the above models satisfy the ODE (8.1.8); its coefficients are purely abstract, they do not depend on physical data or magnitudes and this is why we called it the *intrinsic equation.*

Being the mathematical hard core of all the considered models, it will obviously lead to an unified solution of the four problems.

We shall establish the normal LEM solution for the unknown function z_j. Integrating once the intrinsic equation, we obtain a new ODE of polynomial type, of third degree

$$z_j''' = Kz_j' + \frac{(-1)^j}{2} z_j'^3, \tag{8.1.9}$$

where K is an integration constant. Putting $z_j' = u_j$, we deduce

$$u_j'' = Ku_j + \frac{(-1)^j}{2} u_j^3, \tag{8.1.10}$$

i.e., a polynomial ODE of second order and third degree.

As in the general case, we associate to (8.1.10) the arbitrary Cauchy conditions

$$u_j(0) = \beta, \quad u_j'(0) = \gamma. \tag{8.1.11}$$

The LEM mapping depends on two parameters, σ and ξ

$$v(x, \sigma, \xi) = e^{\sigma u + \xi u'}; \tag{8.1.12}$$

applying it to the intrinsic equation, we get its first linear equivalent. Taking then for v the expansion

$$v(x, \sigma, \xi) = 1 + \sum_{j+k=1}^{\infty} v_{jk}(x) \frac{\sigma^j}{j!} \frac{\xi^k}{k!}, \tag{8.1.13}$$

we obtain the linear equivalent ODS. Let us note that the ODE (8.1.10) is odd, therefore the associated LEM matrix can be compressed to cells with odd indices.

The LEM system associated to the intrinsic equation is

$$\frac{d\mathbf{V}}{dx} = \mathbf{AV}, \quad \mathbf{V} = [\mathbf{V}_{2m-1}]_{m\in\mathfrak{N}}, \quad \mathbf{V}_{2m-1} = [v_{jk}(x)]_{j+k=2m-1}. \tag{8.1.14}$$

The corresponding LEM matrix reads

$$\mathbf{A} = \begin{bmatrix} \mathbf{A}_{11} & \mathbf{A}_{13} & \mathbf{0} & \mathbf{0} & \dots & \dots & \dots \\ \mathbf{0} & \mathbf{A}_{33} & \mathbf{A}_{35} & \mathbf{0} & \dots & \dots & \dots \\ \mathbf{0} & \mathbf{0} & \mathbf{A}_{55} & \mathbf{A}_{57} & \dots & \dots & \dots \\ \dots & \dots & \dots & \dots & \dots & \dots & \dots \\ \mathbf{0} & \mathbf{0} & \mathbf{0} & \dots & \mathbf{A}_{2m-1,2m-1} & \mathbf{A}_{2m-1,2m+1} & \dots \\ \dots & \dots & \dots & \dots & \dots & \dots & \dots \end{bmatrix}, \tag{8.1.15}$$

where the diagonal cells are given by

$$\mathbf{A}_{2j-1,2j-1} = \begin{bmatrix} 0 & 2j-1 & 0 & \dots & 0 & 0 \\ K & 0 & 2j-2 & \dots & 0 & 0 \\ 0 & 2K & 0 & \dots & 0 & 0 \\ \dots & \dots & \dots & \dots & \dots & \dots \\ 0 & 0 & 0 & \dots & 0 & 1 \\ 0 & 0 & 0 & \dots & (2j-1)K & 0 \end{bmatrix}. \tag{8.1.16}$$

To be more explicit, we also specified the form of the rectangular cells $\mathbf{A}_{2j-1,2j+1}$, $j = 1,2$, in order to emphasize fifth order effects in the LEM solution

$$\mathbf{A}_{13} = -\frac{1}{2}\begin{bmatrix} 0 & 0 & 0 & 0 \\ 1 & 0 & 0 & 0 \end{bmatrix}, \ \mathbf{A}_{35} = -\frac{1}{2}\begin{bmatrix} 0 & 0 & 0 & 0 & 0 & 0 \\ 1 & 0 & 0 & 0 & 0 & 0 \\ 0 & 2 & 0 & 0 & 0 & 0 \\ 0 & 0 & 3 & 0 & 0 & 0 \end{bmatrix}. \tag{8.1.17}$$

Taking into account theorem 3.3, one can easily prove

Theorem 8.1 [116, 147]. *The intrinsic equation allows the following normal LEM solution*

$$u(x) = \sum_{m=1}^{\infty} \sum_{j=0}^{2m-1} u_{2m-1-j,j}(x)\beta^{2m-1-j}\gamma^{j}, \tag{8.1.18}$$

where $u_{2k-1-j,j}(x)$ *are components of the vectors*

$$\mathbf{U}_{2m-1}(x)=\left\lfloor u_{2m-1-j,j}(x)\right\rfloor_{j=\overline{0,2m-1}}, \tag{8.1.19}$$

satisfying the linear ODS with constant coefficients

$$\begin{aligned}
&\frac{\mathrm{d}\mathbf{U}_1}{\mathrm{d}x}=\mathbf{A}_{11}^{\mathrm{T}}\mathbf{U}_1,\\
&\frac{\mathrm{d}\mathbf{U}_3}{\mathrm{d}x}=\mathbf{A}_{33}^{\mathrm{T}}\mathbf{U}_3+\mathbf{A}_{13}^{\mathrm{T}}\mathbf{U}_1,\\
&\cdots\cdots\cdots\cdots\cdots\cdots\cdots\cdots\cdots\cdots\\
&\frac{\mathrm{d}\mathbf{U}_{2m-1}}{\mathrm{d}x}=\mathbf{A}_{2m-1,2m-1}^{\mathrm{T}}\mathbf{U}_{2m-1}+\mathbf{A}_{2m-1,2m+1}^{\mathrm{T}}\mathbf{U}_{2m-3}
\end{aligned} \tag{8.1.20}$$

and the Cauchy conditions

$$\mathbf{U}_1(0)=\begin{bmatrix}1 & 0\end{bmatrix}^{\mathrm{T}},\quad \mathbf{U}_{2k-1}(0)=\mathbf{0},\quad k=\overline{2,m}. \tag{8.1.21}$$

Note that, if we wish to study $(2k+1)$-effects, we must take $m=2k-1$ in the normal LEM solution (8.1.18). For fifth order effects, for instance, we must take $m=3$.

Let us also mention that, by virtue of theorem 1.8, we know the eigenvalues of the diagonal cells $\mathbf{A}_{2j-1,2j-1}$, which are given by formula (1.4.50) as linear combinations of the eigenvalues of $\mathbf{A}_{11}$

$$\mathbf{A}_{11}=\begin{bmatrix}0 & 1\\ K & 0\end{bmatrix}. \tag{8.1.22}$$

These eigenvalues depend on the sign of *K*; for $K<0$ they are purely imaginary and if $K>0$, they are real and opposite. Hence, if $K=-a^2$ the normal LEM solution contains only trigonometric functions, while for $K=a^2$, it only contains hyperbolic functions, in both cases occasionally multiplied by polynomials in *x*.

The finite LEM system (8.1.20) can be solved by blocks, starting with the first one; adding new vectors **U** does not modify the preceding computation, as mentioned on the occasion of the proof of theorem 3.3.

Applying the Laplace transformation to the linear Cauchy problem (8.1.20), (8.1.21), we get, going as far as to fifth order effects,

1. for $K = -a^2$, the formulae

$$\begin{aligned} u(x) \cong & \frac{1}{a}\left(\beta a \cos ax + \gamma \sin ax\right) + \\ & + \frac{1}{2^6}\left[\beta^3 \varphi_1(ax) + \beta^2 \gamma \varphi_2(ax) + \beta\gamma^2 \varphi_3(ax) + \gamma^3 \varphi_4(ax)\right] + \\ & + \frac{1}{2^{12}}\left[\beta^5 \Phi_1(ax) + \beta^4 \gamma\, \Phi_2(ax) + \beta^3 \gamma^2\, \Phi_3(ax) + \beta^2 \gamma^3\, \Phi_4(ax) + \right. \\ & \left. + \beta\gamma^4\, \Phi_5(ax) + \gamma^5\, \Phi_6(ax)\right], \end{aligned} \tag{8.1.23}$$

where

$$\begin{aligned} a^2 \varphi_1(\tau) &= \cos 3\tau - \cos\tau - 12\tau \sin\tau, \\ a^3 \varphi_2(\tau) &= 3 \sin 3\tau - 21 \sin\tau + 12\tau \cos\tau, \\ a^4 \varphi_3(\tau) &= -3 \cos 3\tau + 3 \cos\tau - 12\tau \sin\tau, \\ a^5 \varphi_4(\tau) &= -\sin 3\tau - 9 \sin\tau + 12\tau \cos\tau, \end{aligned} \tag{8.1.24}$$

$$\begin{aligned} a^4 \Phi_1(\tau) &= \cos 5\tau - 24 \cos 3\tau + 23 \cos\tau - 36\tau \sin 3\tau \\ & \quad + 96\tau \sin\tau - 72\tau^2 \cos\tau, \\ a^5 \Phi_2(\tau) &= 5 \sin 5\tau - 132 \sin 3\tau + 599 \sin\tau + 108\tau \cos 3\tau \\ & \quad - 336\tau \cos\tau - 72\tau^2 \sin\tau, \\ a^6 \Phi_3(\tau) &= -10 \cos 5\tau + 180 \cos 3\tau - 170 \cos\tau + 72\tau \sin 3\tau \\ & \quad + 528\tau \sin\tau - 144\tau^2 \cos\tau, \end{aligned} \tag{8.1.25}$$

$$\begin{aligned} a^7 \Phi_4(\tau) &= -10 \sin 5\tau + 12 \sin 3\tau + 854 \sin\tau + 72\tau \cos 3\tau \\ & \quad + 912\tau \cos\tau - 144\tau^2 \sin\tau, \\ a^8 \Phi_5(\tau) &= 5 \cos 5\tau + 108 \cos 3\tau - 113 \cos\tau + 108\tau \sin 3\tau \\ & \quad + 240\tau \sin\tau - 72\tau^2 \cos\tau, \\ a^9 \Phi_6(\tau) &= \sin 5\tau + 48 \sin 3\tau + 271 \sin\tau - 36\tau \cos 3\tau \\ & \quad + 384\tau \cos\tau - 72\tau^2 \sin\tau; \end{aligned} \tag{8.1.26}$$

2. for $K = a^2$, the formulae

$$
\begin{aligned}
u(x) \cong & \frac{1}{a}\left(\beta a \cosh ax + \gamma \sinh ax\right) + \\
& + \frac{1}{2^6}\left[\beta^3 \psi_1(ax) + \beta^2 \gamma\, \psi_2(ax) + \beta \gamma^2 \psi_3(ax) + \gamma^3 \psi_4(ax)\right] + \\
& + \frac{1}{2^{12}}\left[\beta^5\, \Psi_1(ax) + \beta^4 \gamma\, \Psi_2(ax) + \beta^3 \gamma^2\, \Psi_3(ax) + \beta^2 \gamma^3\, \Psi_4(ax) + \right. \\
& \quad \left. + \beta \gamma^4\, \Psi_5(ax) + \gamma^5 \Psi_6(ax)\right],
\end{aligned}
\tag{8.1.27}
$$

where

$$
\begin{aligned}
a^2 \psi_1(\tau) &= \cosh 3\tau - \cosh \tau - 12\tau \sinh \tau, \\
a^3 \psi_2(\tau) &= -3 \sinh 3\tau + 21 \sinh \tau - 12\tau \cosh \tau, \\
a^4 \psi_3(\tau) &= 3 \cosh 3\tau - 3 \cosh \tau + 12\tau \sinh \tau, \\
a^5 \psi_4(\tau) &= - \sinh 3\tau - 9 \sinh \tau + 12\tau \cosh \tau,
\end{aligned}
\tag{8.1.28}
$$

$$
\begin{aligned}
a^4 \Psi_1(\tau) &= \cosh 5\tau - 24 \cosh 3\tau + 23 \cosh \tau + 36\tau \sinh 3\tau - \\
& \quad - 96\tau \sinh \tau + 72\tau^2 \cosh \tau, \\
a^5 \Psi_2(\tau) &= 5 \sinh 5\tau - 132 \sinh 3\tau + 599 \sinh \tau + 108\tau \cosh 3\tau - \\
& \quad - 336\tau \cosh \tau - +72\tau^2 \sinh \tau, \\
a^6 \Psi_3(\tau) &= 10 \cosh 5\tau - 180 \cosh 3\tau + 170 \cosh \tau + 72\tau \sinh 3\tau + \\
& \quad + 528\tau \sinh \tau - 144\tau^2 \cosh \tau, \\
a^7 \Psi_4(\tau) &= 10 \sinh 5\tau - 12 \sinh 3\tau - 854 \sinh \tau - 72\tau \cosh 3\tau + \\
& \quad + 912\tau \cosh \tau - 144\tau^2 \sinh \tau, \\
a^8 \Psi_5(\tau) &= 5 \cosh 5\tau + 108 \cosh 3\tau - 113 \cosh \tau - 108\tau \sinh 3\tau - \\
& \quad - 240\tau \sinh \tau + 72\tau^2 \cosh \tau, \\
a^9 \Psi_6(\tau) &= \sinh 5\tau + 48 \sinh 3\tau + 271 \sinh \tau - 36\tau \cosh 3\tau - \\
& \quad - 384\tau \cosh \tau + 72\tau^2 \sinh \tau .
\end{aligned}
\tag{8.1.29}
$$

Integrating directly the above normal LEM representations, we obtain efficient approximations for z_j in the case of the solid state.

Similar formulae can be obtained for the plasma state ($j = 2$), changing the signs of the rectangular matrices (8.1.17).

8.1.3. THE ANALYSIS BY LEM OF THE MODELS

We shall apply the normal LEM solution to each of the models governed by the intrinsic equation, also specifying the particularities of each model.

Solid state

1. In the case of the *rigid pendulum*, $z_1 = \theta, K = -\omega^2$ and the independent variable is the time t. We thus apply formula (8.1.23) for $a = \omega$; the obtained results do not differ from those obtained in section 6.2.

2. For a critical and postcritical study of the *cantilever bar*, we observe that the rotation θ of the bar cross section satisfies the pendulum equation

$$\frac{\mathrm{d}^2\theta}{\mathrm{d}s^2} + \alpha^2\theta = 0\,, \tag{8.1.30}$$

as a consequence of Kirchhoff's analogy; we still apply formula (8.1.23), for $z_1 = \theta$, $K = -\alpha^2$ and $\gamma = 0$, the independent variable being the arc s. in this case too, the obtained solution is analogous to the LEM representations deduced in chapter 5, starting from other equivalents of the B.-E equation; therefore it leads to the postcritical formulae from section 5.2.

3. For the elastic frame, we note that both θ_j, $j = 1{,}2$ satisfy the intrinsic equation. But for the horizontal bar we have

$$K = \omega^2 \sin\theta_2(0) + \tilde{X}\cos\theta_2(0)\,, \tag{8.1.31}$$

which is positive, while for the vertical bar we get

$$K = -\tilde{X}\sin\theta_1(0) - \left(\alpha^2 - \omega^2\right)\cos\theta_1(0)\,, \tag{8.1.32}$$

which is negative. The signs of K in the two cases are established according to physical hypothesis and to the initial conditions

It results that for the horizontal bar the LEM solution (8.1.27) holds true, while for the vertical one we must use the LEM formula (8.1.23). In both cases $\beta = 0$.

We therefore conclude that, no matter how complicated the bar system, in the physical hypotheses in which the nonlinear model for the two-bar frame was established, only two LEM representations are possible: one, expressed by trigonometric functions (for vertical bars) and one – by hyperbolic functions (for horizontal bars). These representations result in an important fact: the most responsible for the loss of the structure stability is the vertical bar.

Plasma

In this case, the intrinsic equation is valid for $j = 2$. The normal LEM representation also permits to deduce asymptotic formulae; which are similar to (7.3.29) and (7.3.30), found in section 7.3, starting directly from Troesch's equation.

The analysis of the above models shows that the intrinsic equation plays a fundamental part in the behaviour of the solutions, therefore it is important for all the phenomena it is governing; with good reason, the intrinsic equation can be considered the mathematical hard core of all these mechanical and physical phenomena, completely distinct.

The idea of finding an intrinsic equation valid for several models belonging to completely distinct physical phenomena proved workable in [98,99], where an analogy between the bending of a cantilever bar and the deflection of an electron beam under the influence of a magnetic field was emphasized. This forms the subject of section 8.3.

8.2. NORMAL LEM SOLUTIONS FOR DÜFFING'S OSCILLATOR

Düffing's oscillator is mathematically modelled as [26]

$$\ddot{x} + \delta\dot{x} + \beta x + \alpha x^3 = A\cos\omega t\,, \qquad (8.2.1)$$

with a positive damping constant δ. For positive values of β, this can be physically interpreted as a forced oscillator with a spring of nonlinear restoring force; for positive α, one has a hardening, while for negative α – a softening spring. For $\beta < 0$, it can be regarded as describing the dynamics of a point mass in a double well potential [129,88,175].

Let us also note that Düffing's model is an algebraically simple equation involving time-dependent acceleration (jerks) that have chaotic solutions, as previously shown by Ueda [166].

In what follows, we take $\alpha > 0,\ \beta \ge 0,\ \delta \ge 0$.

Gottlieb pointed out [34] that the simplest ODE in a single variable exhibiting chaos is third order, following Poincaré-Bendixson theorem; Düffing's equation with $\alpha = \omega = 1$ may be written as a fourth order homogeneous polynomial equation depending on an unique parameter, δ [81]. The fourth derivative is, in fact, the time derivative of the jerk; it is also called spasm, jounce or sprite, because of its behaviour. J. Sprott deduced numerically polynomial jerks allowing chaotic solutions [81] and simple first order polynomial ODS with three equations and three unknown functions allowing solutions with chaotic behaviour; among them, one can recognize Lorenz's system and Rössler type ODSs, both with the corresponding chaotic attractors [82].

In this section, we present the results obtained by using the normal LEM representations. There are established the parametric LEM solutions emphasizing third order effects; a numerical comparison with the Runge-Kutta method is then provided. For wide ranges of the involved parameters, it is shown that the LEM formulae can be applied on large time intervals, thus putting into evidence qualitatively the long-term behaviour of the solution.

8.2.1. THE NORMAL LEM SOLUTIONS

We set up here the normal LEM solutions for Düffing's oscillator in two different cases: a) free undamped and b) forced damped.

a) ***Free undamped oscillator***

In this case, Düffing's equation becomes

$$\ddot{x} + \beta x + \alpha x^3 = 0, \qquad (8.2.2)$$

and coincides with the *intrinsic equation*, found as a mathematical hard core of several physical and mechanical phenomena, completely distinct, both mathematically and physically; all this forms the subject of the previous two sections. After a convenient change of variable, to the purpose of bringing equation (8.2.1) to the form (8.1.10), also taking into account that β is positive, we claim that the LEM solution up to fifth order effects is given by formulae (8.1.23)–(8.1.26).

b) ***Forced damped oscillator***

Let us take $\beta = 0$ with Ueda. Introducing three auxiliary functions $y = \dot{x}, u = \cos \omega t, v = \sin \omega t$, Düffing's equation may be written in the form of a homogeneous polynomial first order ODS

$$\begin{aligned} \dot{x} &= y\,, \\ \dot{y} &= -\,\delta y - \alpha x^3 + Au\,, \\ \dot{u} &= -\,\omega v\,, \\ \dot{v} &= \omega u\,. \end{aligned} \tag{8.2.3}$$

The transposed of the associated LEM matrix is then

$$\mathbf{A} = \begin{bmatrix} \mathbf{A}_{11}^{\mathrm{T}} & \mathbf{0} & \mathbf{0} & \mathbf{0} & \ldots \\ \mathbf{A}_{13}^{\mathrm{T}} & \mathbf{A}_{33}^{\mathrm{T}} & \mathbf{0} & \mathbf{0} & \ldots \\ \mathbf{0} & \mathbf{A}_{35}^{\mathrm{T}} & \mathbf{A}_{55}^{\mathrm{T}} & \mathbf{0} & \ldots \\ \ldots & \ldots & \ldots & \ldots & \ldots \end{bmatrix} \tag{8.2.4}$$

where

$$\mathbf{A}_{11}^{\mathrm{T}} = \begin{bmatrix} 0 & 0 & 0 & 0 \\ 1 & -\delta & 0 & 0 \\ 0 & A & 0 & \omega \\ 0 & 0 & -\omega & 0 \end{bmatrix}, \tag{8.2.5}$$

$$\mathbf{A}_{13}^{\mathrm{T}} = \left[a_{jk}\right]_{\substack{j=\overline{1,20},\\ k=1,4,}} \qquad a_{jk} = -\,\alpha\delta_1^2\,.$$

If we stick to third order effects, then we truncate the LEM matrix to

$$\mathbf{A}_3 = \begin{bmatrix} \mathbf{A}_{11}^{\mathrm{T}} & \mathbf{0} \\ \mathbf{A}_{13}^{\mathrm{T}} & \mathbf{A}_{33}^{\mathrm{T}} \end{bmatrix}. \tag{8.2.6}$$

The eigenvalues of $\mathbf{A}_{11}^{\mathrm{T}}$ are

$$0,\ \delta,\ \pm\,\mathrm{i}\omega\,; \tag{8.2.7}$$

according to theorem 1.8, the eigenvalues of $\mathbf{A}_{33}^{\mathrm{T}}$ are

$$\begin{aligned} &0,\ 0, -\delta,\ -\delta,\ -2\delta,\ -3\delta,\ \pm\,\mathrm{i}\omega,\ \pm\,\mathrm{i}\omega,\ \pm\,2\,\mathrm{i}\omega, \\ &\pm\,3\mathrm{i}\omega,\ \ \delta \pm \mathrm{i}\omega,\ \ \delta \pm 2\mathrm{i}\omega,\ \ 2\delta \pm \mathrm{i}\omega\,. \end{aligned} \tag{8.2.8}$$

The associated LEM ODS for up to third order effects has two blocks; solving it by using the Laplace transformation, we find the normal LEM solution for null Cauchy data in the form

$$\begin{aligned} x(t) \cong {} & \frac{A}{\delta^2+\omega^2}\left(\mathrm{e}^{-\delta t}-\cos\omega t+\frac{\delta}{\omega}\sin\omega t\right)-6\alpha A^3\left[c_0+c_1\mathrm{e}^{-\delta t}+\right. \\ & +c_2\, t\,\mathrm{e}^{-\delta t}+c_3\,\mathrm{e}^{-3\delta t}+c_4\cos\omega t+c_5\sin\omega t+c_6\cos 3\omega t \\ & +c_7\sin 3\omega t+\mathrm{e}^{-\delta t}\left(c_8\cos 2\omega t+c_9\sin 2\omega t\right)+ \\ & \left.+\mathrm{e}^{-2\delta t}\left(c_{10}\cos\omega t+c_{11}\sin\omega t\right)\right], \end{aligned} \tag{8.2.9}$$

where, with the notation $\sigma^2=\delta^2+\omega^2$, the coefficients $c_j, j=\overline{1,10}$ have the following expressions

$$c_0=\frac{4\delta^2+11\omega^2}{9\delta^2\omega^4\left(\delta^2+4\omega^2\right)\left(4\delta^2+\omega^2\right)},$$

$$\begin{aligned} & c_1=-\frac{3}{\delta^2\sigma^4\left(\delta^2+9\omega^2\right)},\quad c_2=-\frac{1}{4\delta\omega^2\sigma^4},\quad c_3=\frac{1}{36\delta^2\sigma^6}, \\ & c_4=-\frac{\delta^2-\omega^2}{8\omega^4\sigma^6},\quad c_5=-\frac{2\omega\delta}{8\omega^4\sigma^6}, \\ & c_6=\frac{\delta^4-12\delta^2\omega^2+3\omega^4}{72\omega^4\sigma^6\left(\delta^2+9\omega^2\right)},\quad c_7=\frac{2\delta\omega\left(3\delta^2-5\omega^2\right)}{72\omega^4\sigma^6\left(\delta^2+9\omega^2\right)}, \end{aligned} \tag{8.2.10}$$

$$\begin{aligned} & c_8=\frac{-1}{4\sigma^6\left(\delta^2+4\omega^2\right)},\qquad c_9=\frac{\delta\left(\delta^2+3\omega^2\right)}{8\omega^3\sigma^6\left(\delta^2+4\omega^2\right)}, \\ & c_{10}=\frac{1}{2\sigma^6\left(4\delta^2+\omega^2\right)},\qquad c_{11}=\frac{\delta}{\omega\sigma^6\left(4\delta^2+\omega^2\right)}. \end{aligned} \tag{8.2.11}$$

8.2.2. NUMERICAL COMPARISON

We compared the values given by the two above LEM formulae with the corresponding numerical solutions obtained by using the Runge-Kutta method, for various ranges of the involved parameters, also establishing the intervals of concordance (denoted by I) of both solutions. Let us note that, immaterial the previously established intervals of convergence of the LEM representations, such a comparison is more realistic, as it can show significant enlargements of the domain of validity of the LEM solutions.

Tables 1 and 2 show this comparison for the free undamped Düffing's oscillator (formula (8.1.23)) and for the damped forced oscillator (formula (8.2.9)) accordingly.

Table 8.1

Comparison between the LEM formula (8.1.23) and the numerical solution

β	x_0	$\dot{x}_0$	relative error/step	Interval of concordance (I)
1	0.1	0.1	0.0354	[0,100]
5	0.1	0.1	0.0871	[0, 500]
10	0.1	0.1	0.0867	[0,700]
100	0.1	0.1	0.0710	[0,2300]
1	0.01	0.01	0.0108	[0, 4000]
5	0.01	0.01	0.0584	[0,10000]
10	0.01	0.01	0.0240	[0, 30000]
100	0.01	0.01	0.0543	[0, 1000]

In table 8.1, there are also considered the initial values $x_0, \dot{x}_0$; they are taken around the equilibrium point (0,0) in the phase space. We observe that, the greater the value of β, the larger the interval of concordance; yet for large β we note that I is smaller, because the influence of secular terms in the LEM formula is significantly larger in this case. Hence, one cannot yet speak of long term concordance.

Table 8.2 contains on its first row the Ueda values of the parameters [166]; we see that I is small, even if it contains around 1500 Runge-Kutta steps. For large ω, I is large enough to yield long term concordance; note that, unlike (8.1.23), formula (8.2.9) contains only terms bounded at infinity; the arrows on the right mean that I can still be larger.

Larger damping coefficients seem to have less effect on I. The relative error per step was taken to be no greater that 9% , while in many of the cases it does not exceed 5%.

Table 8.2

Comparison between the LEM formula (8.2.9) and the numerical solution

δ	ω	*A*	α	relative error/step	Interval of concordance (I)
0.05	1	7.5	1	0.0883	[0, 1.25]
0.05	5	7.5	0.05	0.0833	[0, 10]
0.05	5	7.5	0.5	0.0727	[0, 3]
0.05	10	7.5	0.5	0.0855	[0, 15]
0.05	10	7.5	1	0.0401	[0,7]
0.05	50	7.5	1	0.0546	[0, 6000]
0.05	100	7.5	1	0.0274	[0, 9000]
0.05	1	5	1	0.0183	[0, 1.3]
0.05	1	1	1	0.0468	[0, 2.5]
0.05	10	1	1	0.08869	[0, 150]
0.05	100	1	1	0.0284	[0, 10000] →
0.5	5	7.5	1	0.0686	[0, 1.3]
0.5	10	7.5	1	0.0725	[0, 40]
0.5	100	1	1	0.0436	[0, 100000] →
10	1	7.5	1	0.0051	[0, 3]

We can conclude that the comparison of LEM representations with the numerical solutions obtained by using the Runge-Kutta method showed that the two solutions are concordant on large time intervals for certain ranges of the parameters. This emphasizes the normal LEM solution for Düffing's oscillator as a qualitative tool for studying the long term behaviour of the phenomenon.

8.3. A MODEL AND SOLUTION ANALOGY

In [11], R. Voinea emphasized an analogy between two completely different physical phenomena: the rectilinear displacement in the relativistic frame under a constant force and the large deformations of a straight bar for a constant bending moment and constant rigidity. He showed that the corresponding governing equations differ by a sign and both the solutions for null Cauchy data may be put under a common form of a conic, depending on a parameter a. The case $a < 0$ yields a hyperbola and represents the implicit solution of the relativistic Cauchy problem, while $a > 0$ corresponds to an ellipse (or circle) and gives the implicit solution of the standard cantilever bar problem.

In what follows, we consider the relativistic model for time-dependent forces on the one hand, and the Bernoulli-Euler bar acted upon by variable bending moments and rigidities on the other hand. We firstly try to reduce each model class to an intrinsic equation, which does not depend on the physical data, and then find the corresponding solutions for associated Cauchy problems with arbitrary data. It should be mentioned that this problem, solved by LEM, served as a common frame for several typical bar problems: cantilever, simply supported and hyperstatic [6], [9].

In the last section, a third term of comparison is emphasized: the deflection of a relativistic electron beam – REB – under a magnetic field, previously associated to the Bernoulli-Euler bar deflection [7], [8].

8.3.1. THE MODELS

The general relativistic movement of a material particle $x = x(t)$ of rest mass m_0 in the case of an arbitrary force $F = F(t)$ is given by the second order ODE [1], [11]

$$\frac{\mathrm{d}^2 x}{\mathrm{d}t^2} = \frac{F(t)}{m_0}\left[1 - \frac{1}{c^2}\left(\frac{\mathrm{d}x}{\mathrm{d}t}\right)^2\right]^{\frac{3}{2}}, \tag{8.3.1}$$

where c is the speed of light in vacuum.

The deformation $y = y(x)$ of a Bernoulli-Euler bar [3], [4] is given by

$$\frac{d^2 y}{dx^2} = \frac{M(x)}{E(x)I(x)}\left[1 + \left(\frac{dy}{dx}\right)^2\right]^{\frac{3}{2}}, \tag{8.3.2}$$

where $M = M(x)$ is the bending moment and $E(x)I(x)$ is the rigidity (product between the modulus of longitudinal elasticity and the moment of inertia of the cross section with respect to the neutral axis), both considered as functions of x, varying on an interval $[0, l]$ taken along the rest position of the bar, l being the bar length.

8.3.2. THE INTRINSIC EQUATIONS

We get now the corresponding intrinsic equations for the above two models.

Let us first take equation (8.3.1). Performing the change of variable

$$\tau = ct, \tag{8.3.3}$$

we get the equation

$$\frac{d^2 x}{d\tau^2} = f(\tau)\left[1 - \left(\frac{dx}{d\tau}\right)^2\right]^{\frac{3}{2}}, \tag{8.3.4}$$

where

$$f(\tau) = \frac{F\left(\frac{\tau}{c}\right)}{m_0 c^2}. \tag{8.3.5}$$

Now, if we denote by

$$z = \frac{dx}{d\tau}, \tag{8.3.6}$$

the equation (8.3.1) becomes

$$\frac{dz}{d\tau} = f(\tau)\left(1 - z^2\right)^{\frac{3}{2}}. \tag{8.3.7}$$

Introducing in (8.3.7) the new variable

$$h(\tau)=\int_{\tau_0}^{\tau} f(\theta)\mathrm{d}\theta\,, \tag{8.3.8}$$

we get a new equation

$$\frac{\mathrm{d}z}{\mathrm{d}h}=\left(1-z^2\right)^{\frac{3}{2}}. \tag{8.3.9}$$

We may call this an *intrinsic equation*, as it does not formally depend on any physical data.

Let us consider now the Bernoulli-Euler bar equation (8.3.2). Here also we can denote by

$$z=\frac{\mathrm{d}y}{\mathrm{d}x}; \tag{8.3.10}$$

thus (8.3.2) becomes

$$\frac{\mathrm{d}z}{\mathrm{d}x}=f(x)\left(1+z^2\right)^{\frac{3}{2}}, \tag{8.3.11}$$

where

$$f(x)=\frac{M(x)}{E(x)I(x)}. \tag{8.3.12}$$

Introducing in (8.3.11) the new variable

$$h(x)=\int_{x_0}^{x} f(x')\mathrm{d}x'\,, \tag{8.3.13}$$

we get in this case another equation

$$\frac{\mathrm{d}z}{\mathrm{d}h}=\left(1+z^2\right)^{\frac{3}{2}}, \tag{8.3.14}$$

for which one has any reason to call it *intrinsic*, as it does not depend on the physical data.

So, we see that both physical phenomena, different as they are, have a similar mathematical core, differing by a sign only.

8.3.3. THE SOLUTIONS

To get the solutions of the above models, we add some arbitrary Cauchy conditions

$$x(t_0) = \alpha, \quad \frac{dx}{dt}(t_0) = c\beta \tag{8.3.15}$$

to equation (8.3.1) and the arbitrary Cauchy conditions

$$y(x_0) = \alpha, \quad \frac{dy}{dx}(x_0) = \beta \tag{8.3.16}$$

to equation (8.3.2)

With these specifications, we try to solve the above models, starting from the corresponding intrinsic equations. We begin with the Cauchy problem (8.3.1), (8.3.15), thus starting from (8.3.9), in which we perform the change of function

$$z = \sin u, \tag{8.3.17}$$

which leads to the ODE

$$\frac{du}{dh} = \cos^2 u, \tag{8.3.18}$$

allowing the general solution

$$\tan u = h + k, \tag{8.3.19}$$

or, in terms of z,

$$z = \frac{h + k}{\sqrt{1 + (h + k)^2}}. \tag{8.3.20}$$

From (8.3.15) and (8.3.20), we immediately get

$$k = \frac{\beta}{\sqrt{1 - \beta^2}}. \tag{8.3.21}$$

So, the general solution of the Cauchy problem (8.3.1), (8.3.15) is expressed as

$$x(t) = \alpha + \int_{ct_0}^{ct} \frac{h(\tau') + k}{\sqrt{1 + [h(\tau') + k]^2}} d\tau'. \tag{8.3.22}$$

In [5] we deduced by similar techniques the solution of the second Cauchy problem (8.3.2), (8.3.16), also starting from the corresponding intrinsic equation (8.3.14). Yet, in this case, we used no more the trigonometric change of function (8.3.17), but the change

$$z = \sinh u\,, \tag{8.3.23}$$

leading to the ODE

$$\frac{\mathrm{d}u}{\mathrm{d}h} = \cosh^2 u\,, \tag{8.3.24}$$

whose general solution is

$$\tanh u = h + k\,, \tag{8.3.25}$$

and from (8.3.14) we find

$$z = \frac{h+k}{\sqrt{1-(h+k)^2}}\,. \tag{8.3.26}$$

The constant k results from (8.3.16) and (8.3.22)

$$k = \frac{\beta}{\sqrt{1+\beta^2}}\,. \tag{8.3.27}$$

Consequently, the general solution of the Cauchy problem (8.3.2), (8.3.16) is expressed as

$$y(x) = \alpha + \int_{x_0}^{x} \frac{h(x') + k}{\sqrt{1-\left[h(x') + k\right]^2}}\,\mathrm{d}x'\,, \tag{8.3.28}$$

with k defined in (8.3.27) and α given by (8.3.16). We see that the sign difference of the two considered models is also reflected in their solutions and, even more, in the associated constants. To get more insight into this similarity, we shall write the intrinsic equations and their corresponding solutions under common formulae, by introducing the parameter sign

$$\text{sign} = \begin{cases} -1 & \text{in the relativistic case} \\ 1 & \text{in the case of the bar} \\ 0 & \text{in the limit case.} \end{cases} \tag{8.3.29}$$

Thus, the intrinsic equations (8.3.9), (8.3.14) may be written under the common form

$$\frac{\mathrm{d}z}{\mathrm{d}h} = \left(1 + \operatorname{sign} z^2\right)^{\frac{3}{2}}. \tag{8.3.30}$$

Now, if we define the function

$$k(\operatorname{sign}) = \frac{\beta}{\sqrt{1 + \operatorname{sign} \beta^2}}, \tag{8.3.31}$$

we see that both formulae (8.3.20) and (8.3.26) may be also written in a common frame

$$z = \frac{h + k(\operatorname{sign})}{\sqrt{1 - \operatorname{sign} \left[h + k(\operatorname{sign})\right]^2}}. \tag{8.3.32}$$

The limit case – sign =0 – fits in for both models.

In case of the Cauchy problem (8.3.1), (8.3.15), we admit that $f(\tau) \neq 0$, i.e. $F(t) \neq 0$ for $t \geq t_0$. Analogously, in case of the B.-E. bar we admit that $f(x) \neq 0$, hence $M(x) \neq 0$, thus excluding a point of inflexion of the deformed bar axis, that would have required a piecewise calculation. We observe that both functions $h(\tau)$ and $h(x)$ are dimensionless; this follows from the geometric and mechanic significations of the functions $f(\tau)$, $f(x)$ accordingly, as well as from those of their primitives (8.3.8) and (8.3.13). As a consequence, the function z is also dimensionless, and this is an outstanding property that should also be expected from the intrinsic character of the equations (8.3.9) and (8.3.14).

Introducing the velocity $v = \dfrac{\mathrm{d}x}{\mathrm{d}t}$, we can also write

$$v = c \tanh \theta, \tag{8.3.33}$$

where $\theta = \theta(t)$ is a dimensionless function; the second condition (8.3.15) leads to

$$v_0 = v(t_0) = c\beta = c \tanh \theta_0, \quad \theta_0 = \theta(t_0). \tag{8.3.34}$$

We can "transfer" the Cauchy conditions (8.3.15) for the intrinsic equation (8.3.9), putting $\tau_0 = ct_0$. We have

$$z\big|_{h=0} = z\big|_{\tau=\tau_0} = \left.\frac{dx}{d\tau}\right|_{\tau=\tau_0} = \tanh\theta_0,$$
$$\left.\frac{dz}{dh}\right|_{h=0} = \left.\left(1-z^2\right)^{\frac{3}{2}}\right|_{h=0} = \operatorname{sech}^3\theta_0. \tag{8.3.35}$$

From (8.3.20) we also have

$$h+k = \frac{z}{\sqrt{1-z^2}}, \tag{8.3.36}$$

hence the integration constant is

$$k = \sinh\theta_0. \tag{8.3.37}$$

Relations (8.3.20) and (8.3.36) can be written in the form

$$\frac{1}{z^2} - \frac{1}{(h+k)^2} = 1, \tag{8.3.38}$$

representing a rectangular hyperbola with respect to the variables $\frac{1}{z}$ and $\frac{1}{h+k}$. We can write

$$z = \tanh\theta, \qquad h+k = \sinh\theta, \tag{8.3.39}$$

which should be expected. We also notice that

$$-1 < z = \frac{v}{c} = \tanh\theta < 1,$$
$$-\infty < h+k = \sinh\theta < \infty, \tag{8.3.40}$$

so that from the hyperbola only two half-branches correspond, except for the points $(\pm 1, 0)$ (figure 8.1, thick line).

In the case of the B.-E. bar, we can write

$$z = \frac{dy}{dx} = \tan\theta, \tag{8.3.41}$$

where the dimensionless function $\theta = \theta(x)$ represents the rotation of the cross section of the bar. The Cauchy conditions for the equation (8.3.14) are of the form

$$z\Big|_{h=0} = z\Big|_{x=x_0} = \frac{\mathrm{d}y}{\mathrm{d}x}\Big|_{x=x_0} = \tan\theta_0, \quad \theta_0 = \theta(x_0),$$
$$\frac{\mathrm{d}z}{\mathrm{d}h}\Big|_{h=0} = \left(1 + z^2\right)^{\frac{3}{2}}\Big|_{h=0} = \sec^3\theta_0. \tag{8.3.42}$$

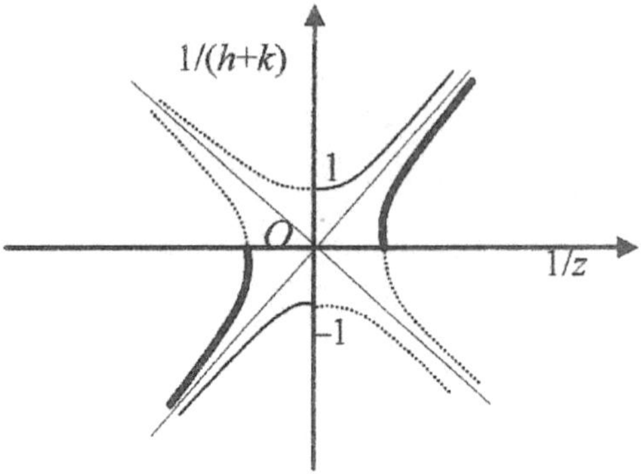

Figure 8.1. Hyperbolae's graphics

From (8.3.26) we also have

$$h + k = \frac{z}{\sqrt{1 + z^2}}, \tag{8.3.43}$$

and thus the integration constant is

$$k = \sin\theta_0. \tag{8.3.44}$$

Relations (8.3.43) lead to

$$\frac{1}{(h+k)^2} - \frac{1}{z^2} = 1, \tag{8.3.45}$$

i.e. another rectangular hyperbola, this time with respect to the variables $\frac{1}{h+k}$ and $\frac{1}{z}$, conjugate to the hyperbola (38). Observing that

$$z = \tan\theta, \quad h + k = \sin\theta, \tag{8.3.46}$$

we also notice that

$$-1 < h + k < 1, \quad -\infty < z < \infty, \tag{8.3.47}$$

which means $-\pi/2 < \theta < \pi/2$; hence

$$h < 1 - k = 1 - \sin\theta_0 . \tag{8.3.48}$$

So, in this case too, there correspond two half branches of the hyperbola, but the points $(0, \pm 1)$ (figure 8.1, thin lines).

Another outstanding property of the hyperbolae is that their graphs do not change, no matter *z*, *h* and *k*; thus, they are invariants for each corresponding problem.

8.3.4. GRAPHICAL APPROACH

The graph of the function (8.3.20) is represented in figure 8.2, where we emphasize its remarkable points and the horizontal asymptotes.

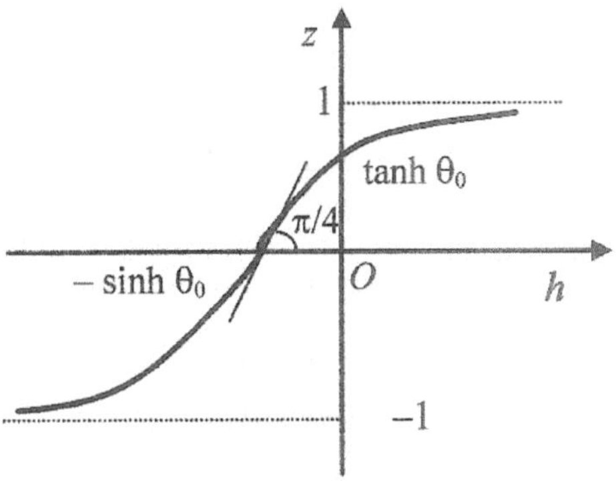

Figure 8.2. The graphic of the solution (8.3.20)

Given the initial velocity v_0, therefore $\tanh\theta_0$, and consequently θ_0, the corresponding graph allows to get the velocity *v* as a function of the new variable *h*. For homogeneous initial conditions $v_0 = 0$, we have $\theta_0 = 0$, involving $\sinh\theta_0 = 0$, $\tanh\theta_0 = 0$ and $k = 0$ and, obviously, the function (8.3.20) becomes

$$z = \frac{h}{\sqrt{1 + h^2}} . \tag{8.3.49}$$

Its graph is represented in figure 8.3.

Admitting that $F(t) > 0$, it follows $f(\tau) > 0$, whence $h(\tau) > 0$. Joining the graph of the function $h(\tau)$ with respect to an $O\tau$ – axis along the Oz – axis and of opposite sense (the function $h(\tau)$ is obtained by quadrature), we obtain a graphical approach of $\tanh\theta$ for an arbitrary $\tau = \text{const}$.

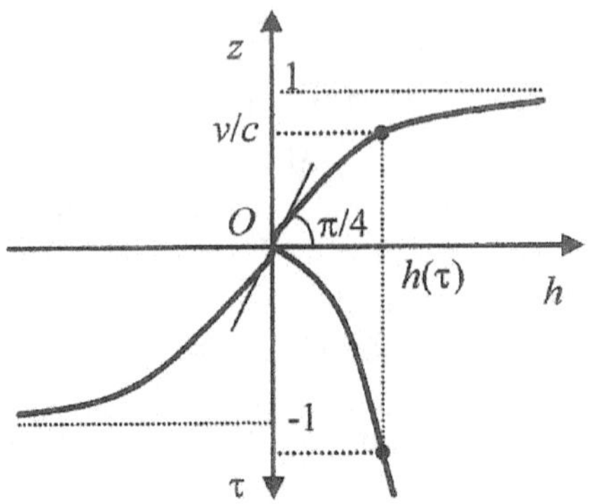

Figure 8.3. The graphic of the solution (8.3.49)

Hence we know the velocity v at any moment t. We observe that for $t > 0$ we get $h > 0$ and $z > 0$, hence $\theta > 0$ and $v > 0$. For $t \to \infty$, we get $v \to c$. It is convenient to use a dimensionless coordinate $\dfrac{\tau}{\tau_1} = \dfrac{t}{t_1}$ (t_1 being arbitrarily chosen), so that the two graphs be compatible (the coordinates be dimensionless in both cases).

Similarly, the graph of the function (8.3.26) is represented in figure 8.4, where its remarkable points are put into evidence as well as the vertical asymptotes.

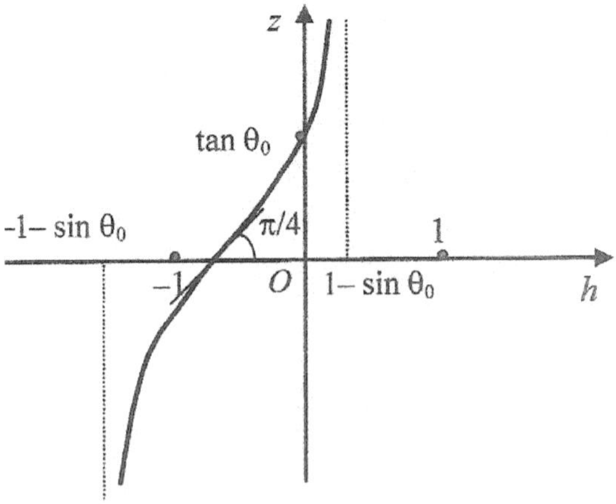

Figure 8.4. The graphic of the solution (8.3.26)

Given the rotation θ_0 at the bar left end, the corresponding graph allows us to get the rotation of the bar cross section (i.e., $\tan\theta$) as a function of the new variable h. In the case of a cantilever bar, $\theta_0 = 0$ (figure 8.5), hence $k = 0$, and thus the function (8.3.26) becomes

$$z = \frac{h}{\sqrt{1-h^2}}, \qquad h = \sin\theta < 1, \tag{8.3.50}$$

and its graph takes the form of figure 8.5. Admitting $M(x) > 0$, it follows that $f(x) > 0$, whence $h(x) > 0$. Joining the graph of the function $h(x)$ with respect to an Ox-axis along the Oz-axis and of opposite sense (the function $h(x)$ is obtained by quadrature), we obtain a graphical approach for $\tan\theta$ for an arbitrary cross-section x, hence for its rotation θ. We observe that for $x > 0$ we get $h > 0$ and $z = \tan\theta > 0$, hence $\theta > 0$. For $x = l$ (l is the bar length), we get $\theta(l) = \theta_{\max}$. In this case too, it is convenient to use a dimensionless co-ordinate x/l, so that the two graphs be compatible (the coordinates be dimensionless in both cases).

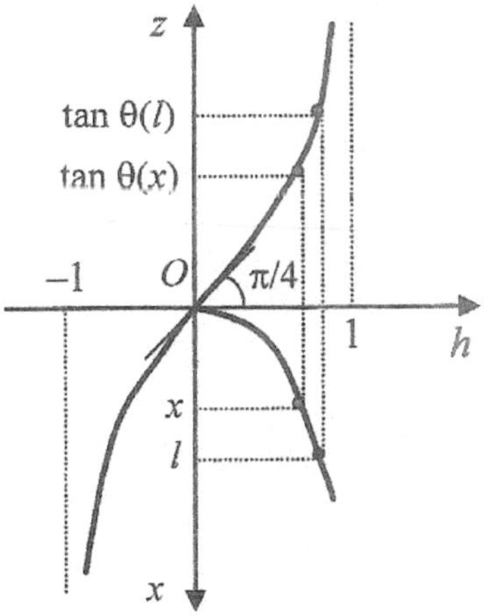

Figure 8.5. The graphic of the solution (8.3.50)

We thus emphasize a connection between the deformed neutral axis of a straight B.-E. bar, of variable rigidity, acted upon by an arbitrary bending moment, in the nonlinear case, in an Euclidean space, and the world line of a particle in relativistic rectilinear movement, acted upon by a time-dependent force, in a Minkowkij pseudo-Euclidean space.

8.3.5. THE THIRD TERM OF THE ANALOGY

In some previous papers [7,8], we put into evidence another analogy, between the Bernoulli-Euler bar deformation in the nonlinear case and the deflection of a REB under certain magnetic fields.

A REB is obtained e.g. in a linear accelerator, the magnetic field being produced by a deflecting coil. One may therefore consider the magnetic induction **B** as defined by three components with respect to a Cartesian co-ordinate system $xOyz$.

$$B_x(x,y,0)) \equiv B(x,y), \quad B_y = 0, \quad B_z = 0. \tag{8.3.51}$$

Without restrictions, we may assume that the electronic beam is contained in the plane xOy. Under these hypothesis, the parametric equations of the plane trajectories [2], together with the components v_x, v_y of the velocity, lead to the following first order ODS, where differentiation is taken with respect to time

$$\begin{aligned} \dot{x} &= v_x\,, \\ \dot{y} &= v_y\,, \\ \dot{v}_x &= -\frac{e}{m}B(x,y)v_y\,, \\ \dot{v}_y &= \frac{e}{m}B(x,y)v_x\,. \end{aligned} \tag{8.3.52}$$

In (8.3.52), e is the electron charge and m the relativistic mass.

In [8] it was proved that the system (8.3.52) may be written in the form (8.3.2). Indeed, by eliminating v_x, v_y we find

$$\begin{aligned} \ddot{x} &= -\frac{e}{m}B(x,y)\dot{y}\,, \\ \ddot{y} &= \frac{e}{m}B(x,y)\dot{x}\,. \end{aligned} \tag{8.3.53}$$

If we consider y as a function of x, then

$$\dot{y} = \dot{x}\frac{\mathrm{d}y}{\mathrm{d}x}, \quad \ddot{y} = \dot{x}^2\frac{\mathrm{d}^2y}{\mathrm{d}x^2} + \ddot{x}\frac{\mathrm{d}y}{\mathrm{d}x} \tag{8.3.54}$$

and this yields

$$\frac{e}{m}B(x,y)\dot{x} = \dot{x}^2\frac{\mathrm{d}^2y}{\mathrm{d}x^2} - \frac{e}{m}B(x,y)\dot{y}\frac{\mathrm{d}y}{\mathrm{d}x}, \tag{8.3.55}$$

or

$$\dot{x}^2 \frac{\mathrm{d}^2 y}{\mathrm{d}x^2} = \frac{e}{m} B(x, y)\left[1 + \left(\frac{\mathrm{d}y}{\mathrm{d}x}\right)^2\right]. \tag{8.3.56}$$

From (8.3.53), we get straightforwardly the classic prime integral

$$\frac{1}{2} m\left(\dot{x}^2 + \dot{y}^2\right) = K, \tag{8.3.57}$$

which, combined with (8.3.54), gives

$$\dot{x} = \sqrt{\frac{2K}{m}}\left[1 + \left(\frac{\mathrm{d}y}{\mathrm{d}x}\right)^2\right]^{-\frac{1}{2}}. \tag{8.3.58}$$

Introducing this in (8.3.56), it is immediately obtained

$$\frac{\mathrm{d}^2 y}{\mathrm{d}x^2} = \frac{e}{2K} B(x, y)\left[1 + \left(\frac{\mathrm{d}y}{\mathrm{d}x}\right)^2\right]^{\frac{3}{2}}. \tag{8.3.59}$$

If the induction B depends at most on x, then (8.3.59) is exactly of the form (8.3.2). Yet, from the physical point of view, one only has either B effectively dependent on both x and y, or B constant. A constant induction is obtained only for an infinite deflection coil; however difficult, the tendency is for such an induction in the case of an industrial accelerator.

Putting in (8.3.59) $z = \frac{\mathrm{d}y}{\mathrm{d}x}$ and

$$h(x) = \frac{e}{2K} \int_{x_0}^{x} B(x', y(x'))\mathrm{d}x', \tag{8.3.60}$$

we get again (8.3.14). But, in this case, the intrinsic equation gives no more directly the solution, as h depends on y.

The Cauchy conditions are handled somewhat more complicated, because of the constant K and the time-dependence of the unknown functions in (8.3.52). From physical reasons, the following initial conditions should be associated to the system (8.3.52)

$$x(0) = 0,\ y(0) = 0, \quad v_x(0) = v_0,\ v_y(0) = 0. \tag{8.3.61}$$

Thus, in this case, due to model restrictions, we cannot take arbitrary Cauchy data and must take $t_0 = 0$, which yields $x_0 = x(t_0) = 0$.

So far, we did not discuss the constant *K* in formula (8.3.57). From (8.3.61) we get

$$K = \frac{1}{2} m v_0^2 , \tag{8.3.62}$$

and the initial conditions (8.3.61) become, by the first formula (8.3.54), for *y* thought as a function of *x*

$$y(0) = 0, \quad \frac{dy}{dx}(0) = \frac{\dot{y}(0)}{\dot{x}(0)} = \frac{0}{v_0} = 0 . \tag{8.3.63}$$

In the ideal case of a constant *B*, the problem is completely solved, either straightforwardly, or starting from the associated intrinsic equation (8.3.30), written for $\text{sign} = 1$, whose solution (8.3.32) becomes

$$z = \frac{\frac{e}{2K} Bx}{\sqrt{1 - \left(\frac{e}{2K} Bx \right)^2}} , \tag{8.3.64}$$

as $k = 0$ and $h(x) = \frac{e}{2K} Bx$, for $x_0 = 0$. Consequently, the solution of (8.3.59) under the null Cauchy conditions (8.3.63) may be written in the implicit form of a circle $x^2 + y^2 = \frac{4K^2}{e^2}$.

REFERENCES

1. ABDULLAEV, A.S., *On the theory of Painlevé's second equation*, Dokl. Akad. Nauk SSSR, 273, 5, pp.1033-1036, 1983.

2. ABLOWITZ, M.J., CLARKSON, P.A., *Solitons, nonlinear evolution equations and inverse scattering*, Cambridge University Press, 1991.

3. ABRAMOWITZ, M., STEGUN, I. A. (eds.), *Handbook of mathematical functions*, U. S. Dept. of Commerce, 1984.

4. BAKER, G.L., GOLLUB, J.P., *Chaotic dynamics. An introduction*, Cambridge Univ. Press, 1992.

5. BARDINET, E., COHEN, L.D., AYACHE, N., *Analyzing the deformation of the left ventricle of the heart with a parametri deformale model*, Proc. Conf. on Computer Vision, Virtual Reality and Robotics in Medicine (CVRMed), Nice, France, April 1995.

6. BARTEN, H.J., *On the deflection of a cantilever beam*, Quart. Appl. Math, **3**, pp. 272, 1945.

7. BHATTACHARYYA, R., *Behavior of a rubber spring pendulum*, J. of Appl. Mech., 67. June, Transactions of ASME, 2000.

8. BEJU, I., SOÓS, E., TEODORESCU, P.P., *Euclidean tensor calculus with applications*, Abacus Press, Tunbridge Wells, Kent, England, 1983

9. BIGGS, N., *Algebraic graph theory*, Cambridge University Press, 1974.

10. BICKLEY, W.G., *The heavy elastica*, Phil. Mag. Ser. 7, 17, 603-622, 1934.

11. BODMAN, S., The industrial practice of chemical process engineering, MIT Press, Massachussetts Inst. of Technology, Cambridge, Mass., London, England, 1968.

12. BONDARENKO, N., TEODORESCU, P.P., TOMA, I., *On some generalized equations of the theory of elasticity with quasipolynomial solutions,* Rev. Roum. des Sci. Tech., série Mécanique Appliquée, **43**, *4*, pp. 439-455, 1998.

13. BONNET, M., *Shape identification problems using boundary elements and shape differentiation*, Proc. of the 2nd Conf. on Boundary and Finite Element ELFIN 2, Sibiu, pp. 35-48, 1993.

14. CHIROIU, V., TEODORESCU P.P., TOMA, I., *Cnoidal and normal LEM representations for the nonlinear pendulum,* The annual symposium of the Institute of Solid Mechanics of the Roumanian Academy of Science, SISOM 2001,Bucharest, 24-25.05, pp. 77-84 (CD-R), 2001.

15. CHIROIU, V., MUNTEANU, L., NICOLESCU, C. M., *A shape description model by using sensor data from touch*, 4h Symp. on Multibody Dynamics and Vibration at the Nineteenth Biennial Conference on Mechanical Vibration and Noise ASME

International Design Engng. Tech. Conferences, Chicago, Illinois, September 2-6, 2003, paper nr. DETC2003/VIB-48337, 2003.

16. CHIROIU, V., MUNTEANU, L., CHIROIU, C., *On the bending of a cantilever flexible beam with an embedded ribbon of shape memory alloy*, ASME 2003 International Design Eng. Tech. Conferences and the Computers and Information in Engng. Conference, Chicago, Illinois September 2-6, 2003, paper nr. DETC2003/VIB-48363, 2003.

17. CHIROIU, V., MUNTEANU, L., RUGINA, C., *On the Active Control of Damping in Machine Tools*, Proc. of MMSS'2003, 3rd International Conference on Machining and Measurement of Sculptured Surfaces, 24 - 26 September, 2003, Kraków, Poland, pp. 127, 2003.

18. CHIROIU, V., *A phenomenological multiscale acoustical approach in modelling of nonlinear mesoscopic materials*, 10th International Congress on Sound and Vibration, 7-10 July 2003, Stockholm, Sweden, pp. 1584, 2003.

19. CHIROIU, V., MUNTEANU, L., ŞTIUCĂ, P., DONESCU, Şt., *Introducere în nanomecanică* (*Introduction to nanomechanic)s,* The Publishing House of the Romanian Academy, Bucharest, 2004.

20. CHIROIU, V., MUNTEANU, L., *The wobble soliton behavior of a heavy elastica,* Rev.Roum.Sci.Techn., série Méc.Appl., **50**, 2005

21. DONESCU, Şt., *LEM representations for the coupled pendulum*, Proc. of SISOM 2000, pp. 31-36 (CD-R), 2000.

22. DONESCU, Şt., CHIROIU, V., *Studiul oscilaţiilor quasi-periodice ale unui oscilarot cuplat slab nelinear (A study of the quasi-periodic oscillations of a weakly nonlinear coupled oscillator)*, Proc. of the Sci. session of the Dept. of Math. and Computer Sci. of TUCEB, Bucharest 26.05.2001, pp. 90-92, 2001.

23. DONESCU, Şt., *Analysis of the nonlinear differential equations having algebraic nonlinearities*, Proc. of the International Conference, Greece, 2001.

24. DONESCU, Şt., *Pendulul nelinear cuplat privit în perspectiva metodelor LEM şi cnoidală* (*The nonlinear coupled pendulum in view of LEM and the cnoidal method(*, Proc. of the Sci. session of the Dept. of Math. and Computer Sci. of TUCEB , 24.05.03, Conspress, Bucharest, pp. 32-34, 2003.

25. DONESCU, St., *LEM representations of the solutions of some nonlinear equations*, Proc. of ICTCAM 2007, 20-23.06.2007.

26. G.DÜFFING, *Erzwungene Schwingungen bei verändlicher Eigenfrequenz, Vieweg*, Braunschweig, 1918.

27. EULER, L., *De curvis elasticis*, 1744.

28. FERMI, E., PASTA, J.R., ULAM, S.M., *Collected papers of Enrico Fermi,* vol.2, E. Fermi, The Univ. of Chicago Press, Chicago, 1965.

29. FIELD, R.J., NOYES, R.M., *Oscillations in chemical systems IV. Limit cycle behavior in a model of a chemical reaction,* J. Chem. Phys., **160**, pp. 1877-1884, 1975.

30. FIRICĂ, O., *Polynomial graph generated by linear equivalence*, Bull. Math. de la Soc. Sci. Math. de la Roumanie, **29(77)**, *1*, pp. 33-43, 1985.

31. FIRICĂ, O., *Operatorial graph generated by linear equivalence*, Bull. Math. de la Soc. Sci. Math. de la Roumanie, **31(79)**, *3*, pp. 210-217, 1987.

32. FIRICĂ, O., *Spectral properties of s-operatorial graphs*, Analele Univ. Bucharest, seria Matematică, XXXVII, *3*, 1988.

33. FLIESS, M., *Fonctionnelles causales non linéaires et indeterminées noncommutatives*, Bull. Soc. Math. de France, **109**, pp. 3-40, 1981.

34. GOTTLIEB, H.P.W., *Question #38. What is the simplest jerk function that gives chaos?*, Am.J. Phys., **64**, 525, 1996.

35. GRASHOF, *Theorie der Elasticität*, 1878.

36. GREENHILL, A.G., *Determination of the greatest height consistent with stability that a vertical pole or mast can be made, and of the greatest height to which a tree of given proportions can grow*, Proc. Camb. Phil. Soc., **4**, pp. 65-73, 1881.

37. HILLE, E., PHILIPS, R., *Functional analysis and semigroups*, Providence, Rhode Island, 1957.

38. ISAACSON, N., KELLER, H., *Analysis of numerical methods*, John Wiley & Sons, NY-London, 1966.

39. ISIDORI, A., *Nonlinear control systems. An introduction,* Lect. Notes in Information and Control Sci., Springer-Verlag, Berlin-Heidelberg-NY-Tokio, **72**, 1985.

40. JEAN, M., PRATT, E., *A system of rigid bodies wih dry friction*, Int. J. of Engng. Sci., **23**, *5*, pp.497-513, 1985.

41. JONES, D.J., *Solutions of Troesch's and other two-point B.V.P. by shooting techniques*, J. Comp. Phys., **12**, pp. 429-434, 1972.

42. KAMKE, E., *Differentialgleichungen Lösungsmethoden und Lösungen*, Mir, Moscova, 1951 (trad. in Russian).

43. KOITER, W.T., *Elastic stability and postbuckling behaviour*, Proc. Symp. Nonlinear Problems, Ed. By R.E. Langer, Univ. Wesconsin Press, pp. 257, 1963.

44. KUBIČEK, M., HLAVAČEK, V., *Solutions of Troesch's two-point B.V.P. by shooting techniques*, J. Comp. Phys., **17**, pp. 95-101, 1975.

45. KRON, G., *Invariant Form of the Maxwell-Lorentz Field Equations for Accelerated Systems*, J. of Applied Physics, IX, March, pp.196-208, 1938.

46. LAMBERT, J.D., *A modification of the shooting method for two-pooint boundary value problems*, Intern. Scrift. Numerische Math., Birkhäuser Verlag, Basel und Stuttgart, **19**, pp. 133-143, 1974.

47. LANDAU, L., LIFSCHITZ, E., *Théorie des champs*, Mir, Moscova, 1970.

48. LAU, J.H., *Large deflection of cantilever beam*, J. Engng. Mech. Division, Proc. ASCE, 107, pp. 259, 1981.

49. LORENZ, E. N., *Deterministic nonperiodic flow*, J. Atmos. Sci., **20,** pp. 130-141, 1963.

50. MAWHIN, J., *The forced pendulum. A paradigm for nonlinear analysis and dynamical systems*, Expositiones Mathematicae, Bibliographisches Institut & F.A. Brockhaus AG, **6**, pp. 271-286, 1988.

51. MIHĂILESCU, M., CHIROIU, V., *Advanced mechanics on shells and intelligent structures,* The Publishing House of the Romanian Academy, Bucharest, 2004.

52. MOCANU, C.I., NEMOIANU, C., TOMA, I., VASILIU, M., *Nouveau modèle des bobines de déviation pour les accélérateurs industriels,* Rev. Roum. des Sci. Techn., série . Élth. et Énerg., **29**, 4, pp.387-393, 1984.

53. MOONEY, M., *A theory of large elastic deformation*, Journal of Applied Physics, 11(9), 1940, pp. 582-592.

54. MOREAU, J.J., *On unilateral constraints, friction and plasticity*, New Variational Techniques in Mathematical Physics (eds. G. Capriz, G.Stampacchia) pp.173–322, Edizioni Cremonese, Roma, 1974.

55. MOREAU, J.J., *Application of convex analysis to some problems of dry friction*, Trends in Applications of Pure mathematics to Mechanics (ed. H. Zorski), **2**, pp.263–280, Pitman Pub. Ltd, London, 1973.

56. MUNTEANU, L., BADEA, T., CHIROIU, V., *Linear equivalence method for the analysis of the double pendulum's motion*, Complexity International Journal, **9**, pp. 26-43, 2002.

57. MUNTEANU, L., DONESCU, Ş., *Introduction to the soliton theory, Applications to mechanics*, Book Series: Fundamental Theories of Physics, Kluwer Academic Publishers, 2004.

58. MUNTEANU, L., POPA, D., SECARA, C., CHIROIU, V., *The analysis of a system of rigid bodies's dynamics by linear equivalence method,* Proc. of the Roumanian Academy, series A, **8**, *2*, pp. 145-150, 2007.

59. MUNTEANU, L., CHIROIU, V., DONESCU, Şt., *Analytical solutions for a nonlinear coupled pendulum*, WSEAS Transactions on Mathematics, **7**, *7*, pp. 503-514, 2008.

60. NEMOIANU, C., TOMA, I. , *La correction dela trajectoire des faisceaux dans les bobines de déviation des accélérateurs industriels de particules*, Rev. Roum. des Sci. Techn., série Elth. et énerg., **40**, *1*, pp. 49-53, 1995.

61. NIKOMAROV, M., *Exact bending calculus of cantilever and simply supported beams*, Azerneštr, Baku, 1965.

62. OGDEN, R., *Non- Linear Elastic Deformations*, New York, J. Wiley and Sons, 1984.

63. PAINLEVÉ, P., *Sur les équations différentielles du second ordre à point critique fixe*, C.R. Acad. Sci. (Paris), 143, pp.1111-1117, 1906.

64. PERKO, L., *Differential equations and dynamical systems*, Springer-Verlag, NY-Berlin-Heidelberg-London-Paris-Tokio-Hong Kong-Barcelona, Texts in Applied mathematics, 7, 1991.

65. PETROŞANU, D., *Sur la Déformation d'un Réservoir Sphérique,* Rev. Roum. des Sci. Techniques, Série de Mécanique Appliquée, **45**, *3*, pp. 309-320, 2000.

66. PETROŞANU, D., *Exemples de grandes déformations pour un reservoir sphérique et un tube cylindrique*, University Politehnica of Bucharest, Sci. Bull., Series A: Appl. Math. and Physics, **66**, *1*, pp. 37-46, 2004.

67. PETROŞANU, D., *Sur la déformation d'un tube cylindrique*, Rev. Roum.Sci. Techniques, Série de Mécanique Appliquée, **55**, *1*, pp. 39-50, 2010.

68. PFLÜGGER, A., *Stabilitätsprobleme des Elastostatik*, 2nd ed., Springer-Verlag, Berlin- Göttingen-Heidelberg- NY, 1964.

69. POSTON, T., STEWART, J., *Catastrophe theory and its applications*, Pitman, London-San Francisco-Melbourne, 1978.

70. RALL, L.B., *Solutions for abstract polynomial equations by iterative methods,* MRC Report, #892, August, 1968.

71. REID, W.T., *Ordinary differential equations*, John Wiley & Sons, NY-London-Sidney, 1971.

72. RIVLIN, R. S., *Large elastic deformations of isotropic materials. IV. Further developments of the general theory*, Philosophical Transactions of the Royal Society of London. Series A, Mathematical and Physical Sciences, 241(835), pp. 379-397, 1948.

73. RHODE, F.V., *Large deflection of a cantilever beam with uniformly distributed load*, Quart. Appl. Math., **11**, pp. 337, 1953.

74. ROBERTS, S.M., SHIPMAN, J.S., *Solutions of Troesch's two-point b.v.p. by a combination of techniques*, J. Comp. Phys., **10**, pp.232-241, 1972.

75. ROORDA, J., *Stability of structures with small imperfectins*, J. Engng. Mech. Div. ASCE, **91**, E1, Proc. Paper 4230, *86*, 1965.

76. SCHEIDI, R., TROGER, K., ZEMAN, K., *Coupled flutter and divergence bifurcation of a double pendulum*, Int. J. of Nonlinear Mechanics, **19**, *2*, pp. 163-176, 1983.

77. SCHNEIDER, Z., Z. Oester. Ing. Archiv Verlag, Wien, 1901.

78. SEDERBERG, T.W., PARRY, S.R., *Free-form deformationof solid geometric models*, Computer Graphics (Proc. of SIGGRAPH 86), **20**, *4*, pp.151-160, 1986.

79. SOARE, M.V., TEODORESCU, P.P., TOMA, I., *Ecuaţii diferenţiale cu aplicaţii în mecanica construcţiilor (Differential Equations with applications to the mechanics of constructions)*, Ed.Tehnică, Bucharest, 1999.

80. SOARE, M.V., TEODORESCU, P.P., TOMA, I., *Ordinary differential equations with applications to mechanics,* Springer, Dordrecht, 2007.

81. J.C. SPROTT, *Some simple chaotic jerk functions*, Am. J. Phys. **65**, 6, pp.537-543, 1997.

82. J.C.SPROTT, *Some simple chaotic flows*, Phys. Rev. E, **50**, 2, pp. R647-R650, 1994.

83. STĂNESCU, N.D., MUNTEANU, L., CHIROIU, V., PANDREA, N., *Sisteme dinamice. Teorie and aplicaţii (Dynamical systems. Theory and applications),* The Publishing House of the Romanian Academy, Bucharest, 2007.

84. STĂNESCU, NICOLAE-DORU, *Some aspects concerning the stability of motion for an automotive with neo-Hookean suspension*, 2nd Int. Conf. on Experiments/Process/System Modeling/Simulation& Optimization, Athens, 4-7.07.2007.

85. STĂNESCU, N.D.,PANDREA, M., PANDREA, N., STAN, M., *About the stability of the double hanging pendulum with one neo-Hookean rod*, Proc. of SISOM, 2005, Bucharest, 2005.

86. ŞTIUCĂ, P., CHIROIU, V., NICOLESCU, C.M., *On the Mechanical behavior of nanostructures materials,* Topics in Mechanics, eds. V. Chiroiu, T. Sireteanu, The Publishing House of the Romanian Academy, Bucharest, vol.II, pp.354-390, 2004.

87. TABOR, M., *Chaos and integrability in nonlinear dynamics*, John Wiley & Sons, NY-Chichester-Brisbane-Toronto-Singapore, 1989.

88. TAKASHI KANAMARU , *Duffing oscillator*, Scholarpedia, **3**(3):6327, 2008.

89. TEODORESCU, P.P., *Mechanical systems. Classical models*, t.I,II,III, Springer, 2006, 2009, 2013.

90. TEODORESCU, P.P., ILLE, V., *Teoria elasticitaţii şi introducere în mecanica solidelor deformabile* (*Theory of elasticity and introduction to the mechanics of deformable solids)* , **2**, Dacia Publ. House, Cluj-Napoca, 1979.

91. TEODORESCU, P.P., NICOROVICI, N.-A., *Applications of the theory of groups in mechanics and physics*, Book Series: Fundamental Physics, Kluwer Acad. Publ., Boston/Dordrecht/London, 2004.

92. TEODORESCU, P.P., STĂNESCU, N.D.,PANDREA, M., *Numerical Analysis with Applications in Mechanics and Engineering*, Wiley-IEEE Press, 2013.

93. TEODORESCU, P.P., TOMA, I., *New considerations concerning the Cauchy type problem in the nonlinear bending of a straight bar*, An. Univ. A.I.Cuza, Iaşi, suppl. **27**, seria I, pp. 247-251, 1981.

94. TEODORESCU, P.P., TOMA, I., *On the Cauchy type problem in the nonlinear bending of a straight bar*, Mech. Res. Comm., **9**, pp. 151-158, 1982.

95. TEODORESCU, P.P., TOMA, I., *Two fundamental cases in the nonlinear bending of a straight bar*, Meccanica, **19**, pp. 51-60, 1984.

96. TEODORESCU, P.P., TOMA, I., *On criticity conditions for the straight bar by the linear equivalence method*, Lucrările Conf. Naţionale de Mecanica Solidelor, 24-25.05.1985, Timişoara, pp. 89-92, 1985.

97. TEODORESCU, P.P., TOMA, I., *On the nonlinear bending of a hyperstatic bar*, Int. J. Engng. Sci., **24**, pp. 1257-1270, 1986.

98. TEODORESCU, P.P., TOMA, I., *On an analogy involving the Bernoulli-Euler equation of elastic beams*, Proc. of the IV-th Symposium of Tensometry, 24-27.09.1986, Galaţi, pp. 383-386, 1986.

99. TEODORESCU, P.P., TOMA, I., *On the analogy between the deflected elastic beams and deflected relativistic electron beams*, Mech. Res. Comm., **13**, 5, pp. 265-270, 1986.

100. TEODORESCU, P.P., TOMA, I., *The nonlinear bending of a straight bar treated by linear equivalence*, Lucrările seminarului de Mecanică, Univ. din Timişoara, fasc. **5**, 1987.

101. TEODORESCU, P.P., TOMA, I., *Bifurcation for two fundamental cases of a nonlinear straight bar*, Rev. Roum. des Sci. Tech., série Mécanique, Appl., **32**, *4*, pp. 397-493, 1987.

102. TEODORESCU, P.P., TOMA, I., *Nonlinear study of criticity for two special cases of straight bars*, Bull. Math. Soc. Sci. Math. de la Roumanie, **32(80)**, *3*, pp. 273-280, 1988.

103. TEODORESCU, P.P., TOMA, I., *On the postcritical behaviour of a cantilever bar*, Lect. Notes in Eng. Sci., A Symposium dedicated to A.C.Eringen, June 20-22 1988, Berkeley, California, Springer Verlag, **39**, pp. 233-244, 1988.

104. TEODORESCU, P.P., TOMA, I., *A postcritical study of an elastic structure*, Rev. Roum. des Sci. Tech., série Mécanique Appl., **34**, *6*, pp. 591-594, 1989.

105. TEODORESCU, P.P., TOMA, I., *An application of the linear equivalence method to the postcritical study of a Bernoulli-Euler bar*, Rev. Roum. des Sci. Tech., série Mécanique Appl., **34**, *4*, pp. 509-521, 1989.

106. TEODORESCU, P.P., TOMA, I., *Using LEM in critical and postcritical behaviour of elastic structures described by nonlinear differential systems with variable coefficients*, L Proc. of the VI-th National Symp. of Tensometry, Craiova, 24-25.09.1992, **II**, pp. 649-653, 1992.

107. TEODORESCU, P.P., TOMA, I., *Nonlinear study of stability for a Bernoulli-Euler bar by the LEM*, Proc. of the Seminary of Mechanics, Univ. of Timişoara, fasc. **34**, 1992.

108. TEODORESCU, P.P., TOMA, I., *A new formulation in the stability problem of an elastic frame*, Bul. of the XVIII-th Conf. Mech. Solids, 9-11.06.1994, Constanţa, **I**, pp. 57-64, 1994.

109. TEODORESCU, P.P., TOMA, I., *On an analogy between the nonlinear pendulum equation and the Lotka-Volterra system*, Rev. Roum. des Sci. Tech., série Mécanique Appl., **39**, *5*, pp. 443-451, 1994.

110. TEODORESCU, P.P., TOMA, I., *Asupra stabilităţii unui sistem de bare în tratare modulară (On the stability of a bar system),* Proc. of the XIX-th Conference on the Mechanics of Solids, 2-3.06.1995, Târgovişte, pp. 15-18, 1995.

111. TEODORESCU, P.P., TOMA, I., *Un studiu al pendulului nelinear dublu prin echivalenţă lineară (A study of the nonlinear double pendulum by linear equivalence)*, Proc. of the VII-th Symp. of Tensometry, 17-19.10.1996, Suceava, **III**, pp. 35-38, 1996.

112. TEODORESCU, P.P., TOMA, I., *On some elastic structures governed by the same intrinsic equation,* Bull. For Appl. & Computer Math. (BAM) 1567-1591/'98, LXXXVI-B, PC-122/'98, pp. 8-14, 1998.

113. TEODORESCU, P.P., TOMA, I., *Efecte de ordin superior în studiul pendulului nelinear amortizat (Higher order effects in the study of the nonlinear damped pendulum)*, Sci. Bull of the XXII-th Conf. Mech. Solids, 29.10.1998, Braşov, **I**, pp. 5-8, 1998.

114. TEODORESCU, P.P., TOMA, I., *Reprezentări LEM normale pentru pendulul amortizat (Normal LEM representations for the damped pendulum),* Bul. Univ. Petrol-Gaze Ploieşti, **LI**, *1*, pp. 351-356, 1999.

115. TEODORESCU, P.P., TOMA, I., *Soluţii periodice pentru pendulul nelinear supus acţiunii unei forţe perturbatoare (Periodic solutions for the nonlinear forced pendulum)*, Bul. Şt. al Univ. Tehn. a Moldovei (special issue), Chişinău, **I**, pp. 3-6, 2000.

116. TEODORESCU, P.P., TOMA, I., *A class of elastic structures with the same mathematical core,* Honorary volume dedicated to professor emeritus Ioannis D. Mittas, Aristotle Univ. of Thessaloniki, Fac. of Eng., Dept. of Math, Phys.Sci., Division of Math., pp. 499-508, 2000.

117. TEODORESCU, P.P., TOMA, I., *An unexpected mathematical connection between the solid and plasma state*, Proc. of the VIII-th International Conference of Tensometry and TEHNONAV 2000, pp. 179-184 , 2000.

118. TEODORESCU, P.P., TOMA, I., *Nonlinear damped pendulum treated by linear equivalence,* Mech. Res. Comm, **27**, *3*, pp. 373-380, 2000.

119. TEODORESCU, P.P., TOMA, I., *Formule LEM pentru perioada pendulului nenlinear (LEM formulae for the nonlinear double pendulum period)*, Bul. Univ. Petrol-Gaze Ploieşti, **LII**, seria Tehnică, *2*, pp. 11-16, 2000.

120. TEODORESCU, P.P., TOMA, I., *Normal LEM representations for the nonlinear pendulum,* An. Univ. Ovidius, Constanţa**,** Seria Matematică, **IX** (suppl), *2*, pp. 69-72, 2001.

121. TEODORESCU, P.P., TOMA, I., *Normal LEM representations for the nonlinear pendulum,* GAMM 2001, Annual Scientific Conference, 12-15.02, ETH Zürich, pp. 144 , 2001.

122. TEODORESCU, P.P., TOMA, I., *New estimations in the nonlinear bending of a straight bar,* Bul. Şt. al celei de a XXVI-a Conferinţe Naţionale de Mecanica Solidelor, 14-15.06.01, Brăila, pp. 187-190, 2002.

123. TEODORESCU, P.P., TOMA, I., *Asupra unei analogii de model şi soluţie (On a model and solution analogy)*, Bul. Şt. al Univ. din Piteşti, seria Mec. Apl, **1**(**7**), pp. 287-292, 2003.

124. TEODORESCU, P.P., TOMA, I., *On Voinea's analogy,* Proc. of the Romanian Acad., series A, **4**, *3*, pp. 157-166 , 2003.

125. TEODORESCU, P.P., TOMA, I., *New integral LEM formulae applied to the nonlinear bar*, Mech. Res. Comm., **31**, *1*, pp. 161-168, 2004.

126. TEODORESCU, P.P., TOMA, I., *Nonlinear elastic deformations treated by LEM,* in: Topics in Applied Mechanics, The Publishing House of the Romanian Academy, Bucharest, eds. V. Chiroiu, T. Sireteanu, vol.II, pp.391-442, 2004.

127. TEODORESCU, P.P., TOMA, I., *Applying LEM to Düffing's oscillator*, U.P.B. Sci. Bull. Series D, **72**, 3, 2010, pp. 3-13.

128. TEODORESCU, P.P., TOMA, I., *The pendulum - a mathematical hard core,* Acta Technica Napocensis, series: Applied Mathematics and Mechanics, 53, vol. II, pp.125-132, 2010.

129. THOMPSON, J.M.T., HUNT, G.W., *A general theory of elastic stability*, John Wiley & Sons, NY, London, Sydney, Toronto, 1973.

130. TOMA, I., *Sufficient conditions of uniqueness for the regular solutions of a special class of nonlinear boundary value problems,* Internazionale Schriftenreihe zur Numerische Mathematik, Birkhäuser Verlag, Basel und Stuttgart, **19**, pp. 203-211, 1974.

131. TOMA, I., *Necessary conditions on the values of the stoidchiometric factor allowing the appearance of the solitary travelling wave solution in a model concerning the Belousov-Zhabotinskij chemical reaction,* Proc. of the Int. Symposium of Applications of Mathematics in System Theory, 27-30.12.1979, Braşov, **I**, pp. 199-205, 1978.

132. TOMA, I., *On polynomial differential equations*, Bull. Math. Soc. Sci. Math. de la Roumanie, **24**(**72**), *4*, pp. 417-424, 1980.

133. TOMA, I., *A method for solving polynomial differential equations with applications to the mathematical model concerning the Belousov-Zhabotinskij chemical reaction,* Bull Math. de la Soc. Sci. Math. de la Roumanie, **24**(**72**), *3*, pp. 319-325, 1980.

134. TOMA, I., *Solutions of bilocal polynomial problems by linearization,* Analele Univ. Bucharest, Seria Matematică, **30**, pp. 71-80, 1981.

135. TOMA, I., *Local inversion of polynomial differential operators by linear equivalence,* Analele Univ. Bucuresti, Seria Matematică, **31**, pp. 75-80, 1982.

136. TOMA, I., *Extension of a linearization method for polynomial operators,* Rev. Roum. des Math. Pures et Appl., **31**, *6*, pp. 531-538, 1983.

137. TOMA, I., *φ-Equivalence and nonlinear operators*, Bull. Math. Soc. Sci. Math. de la Roumanie, **29(77)**, *1*, pp. 81-88, 1985.

138. TOMA, I., *On Troesch's plasma problem,* Rev. Roum. des Sci. Techn., série Mécanique Appl., **31**, *1*, pp. 13-18, 1986.

139. TOMA, I., *Techniques of computation by linear equivalence,* Bull. Math. Soc. Sci. Mat. de la Roumanie, **33**(**81**), *4*, pp. 363-373, 1989.

140. TOMA, I., *On some connections between Lie derivatives and the linear equivalence method,* Analele Univ. Bucharest,. Seria Matematică, **39**, *1-2*, pp. 72-79, 1990.

141. TOMA, I., *An Application of LEM to Fliess' series*, Proc. of the Sci. session of the Dept. of Math. and Computer Sci. of TUCEB, 11.05.1991, Bucharest, pp. 29-32, 1991.

142. TOMA, I., *Reprezentări LEM normale în cazul centrelor (Normal LEM represenetations around centers),* Proc. of the Sci. session of the Dept. of Math. and Computer Sci. of TUCEB, Bucharest, 27.05. 1995, pp. 16-17, 1995.

143. TOMA, I., *Metoda echivalenţei lineare şi aplicaţiile ei (The linear equivalence method and its applications),* Ed. Flores, Bucharest, 1995.

144. TOMA, I., *Inversul numeric matriceal al unui operator diferenţial polinomial (The numerical matrix inverse of a differential polynomial operator)*, Proc. of the

Sci. session of the Dept. of Math. and Computer Sci. of TUCEB,, 24.05.1997, pp.11-13, 1997.

145. TOMA, I., *Asupra unei ecuaţii ce guvernează o clasă de structuri elastice (On an equation governing a class of elastic structures)*, Proc. of the Sci. session of the Dept. of Math. and Computer Sci. of TUCEB, 15.05. 1999, pp. 12-14, 1999.

146. TOMA, I., *Soluţii LEM pentru pendulul nelinear liber (LEM solutions for the nonlinear free pendulum)*, Proc. of the Sci. session of the Dept. of Math. and Computer Sci. of TUCEB, 26.05. 2001, pp. 11-13, 2001.

147. TOMA, I., *Modele intrinseci tratate prin LEM (Intrinsic models treated by LEM)*, Proc. of the Sci. session of the Dept. of Math. and Computer Sci. of TUCEB, 24.05.03. 2003, pp. 102-104, 2003.

148. TOMA, I., *A transport matrix for REBs by using LEM integral formulae,* Proc. of MENP, Bucharest, 2004, pp. 144-148 , 2006.

149. TOMA, I., *Specific LEM techniques for some polynomial dynamical systems,* in: Topics in Applied Mechanics, The Publishing House of the Romanian Academy, Bucharest, eds. V. Chiroiu, T. Sireteanu, t.III, pp.427-459, 2006.

150. TOMA, I., *Normal LEM representations for the nonlinear forced pendulum,* II NNMAE, (Nonsmooth/Nonconvex Mechanics with Applications in Engineering, Proc. of the International Conference in Memoriam of Professor P.D. Panagiotopoulos, ed. C.C. Baniotopoulos, Thessaloniki, Greece, 7-8.07.2006, pp.329-332, 2006.

151. TOMA, I., *New periodic LEM solutions for the nonlinear pendulums,* SISOM 2007 and the Homagial Session of Acoustics, D28 (CD-R), 2007

152. TOMA, I., *The nonlinear pendulum from a LEM perspective,* in: Research Trends in Mechanics, The Publishing House of the Romanian Academy, Bucharest, eds. D.Popa, V. Chiroiu, I.Toma, t.I, pp. 395-422, 2007.

153. TOMA, I., *LEM solutions in mechanics and engineering,* Proc. of ICTCAM, 20-23.06.2007, pp. 123-128, 2007.

154. TOMA, I., *Extensions of LEM to non-autonomous systems,* in: Research Trends in Mechanics, The Publishing House of the Romanian Academy, Bucharest, eds. D.Popa, V. Chiroiu, I.Toma, t.II, pp. 361-378, 2008.

155. TOMA, I., *Metoda echivalenţei lineare şi aplicaţiile ei în mecanică (The linear equivalence method and its applications to mechanics)*, Ed. Tehnică, Bucharest, 2008.

156. TOMA, I., *LEM solutions for two neo-Hookean models*, in: Research Trends in Mechanics, The Publishing House of the Romanian Academy, eds. D.Popa, V. Chiroiu, L. Munteanu, t..III, pp.444-468, 2009.

157. TOMA, I., *Normal LEM solutions for Düffing's equation*, 10th workshop of the Dept. of Mathematics and Computer Science, TUCEB, Bucharest, Romania, 23 May 2009, pp.140-145, 2009.

158. TOMA, I., *An abstract pattern for some dynamical models*, Revue Roumaine des Sciences Techniques, série de Mécanique Appliquée, **55** (3), pp.267-278, 2010.

159. TOMA, I., *An Ogden type model treated by LEM*, Proc. of the Annual Symp.of the Inst. of the Mechanics if Solids of the Roumanian Academy and session of the Commission of Acoustics, SISOM 2011, 25-26.05.2011, pp. 111-116, 2011.

160. TOMA, I., *LEM solutions for a class of hyperelastic problems*, in: Inverse problems and computational mechanics, The Publishing House of the Romanian Academy, eds.L. Marin, L. Munteanu, V. Chiroiu, t.I, pp. 339-368, 2011.

161. TOMA, I., *The linear equivalence method and its applications to nonlinear dynamics*, Bul. Şt. Univ. Piteşti, seria Mec. Apl., 2(16), pp. 230-272, 2008.

162. TOMA, I., POPA, L., *Matrice de transport pentru fascicule de particule încărcate relativistei (A transport matrix for REBs)*, St. Cerc. Mat., **36**, *2*, pp. 160-168, 1984.

163. TRÈVES, J.F., *Lectures on linear PDEs with constant coefficients,* fasc. publ. Inst.de Matematica Pura e Aplicada do Conselho Nacional de Pesquisas, Rio de Janeiro, 1961.

164. TOZONI, O.V., *Mathematical models for the evaluation of electric and magnetic fields,* ILIFE, 1968.

165. TROY, W.C., *The disappearance of solitary travelling wave solutions of a model concerning Belousov-Zhabotinskij chemical reaction*, Rocky Mountains J. of Math., 7, 3, 1977.

166. UEDA, Y. , *The road to chaos*, Aerial Press, 1992.

167. VOINEA, R., *An analogy*, Proc. Ro. Acad., Series A, **1**, *1*, pp. 51-53, 2000.

168. VOINEA, R.P., STROE, I.V., *Introducere în teoria sistemelor dinamice (Introduction to the theoty of dynamical systems),* The Publishing House of the Romanian Academy, 2000.

169. VOLTERRA, V., *Variations and fluctuations of the number of individuals in animal species living together*, Animal Ecology, McGraw-Hill, 1931 (transl. by R. N. Chapman from the 1928 edition)

170. WANG, CHANG-YI, *Large deformations of a heavy cantilever*, Quarterly of Appl. Math., XXXIX, 2, pp. 261-273, 1981.

171. WASOW, W.R., *Asymptotic expansions for ordinary differential equations*, Interscience, NY, 1965.

172. WIGGINS, St., *Global bifurcations and chaos. Analytical methods*, Springer-Verlag, Berlin-Heidelberg-London-Paris-Tokio, 1988.

173. YANO, Kentaro, *The theory of Lie derivatives and its applications*, North Holland Publ. Co. Amsterdam P, Noordhaff Ltd., Groeningen, 1965.

174. ZAIKIN, A.N., ZHABOTINSKIJ, A., *Concentration wave propagation in a two-dimensional liquid phase self-oscillating system*, Nature, **225**, pp. 135-137, 1970.

175. ZEEMAN, Ch., *Düffing's equation: catastrophic jumps of amplitude and phase*, conference, Univ. of Texas at San Antonio, 31.03.2000.

ABOUT THE AUTHOR

Ileana Toma graduated the faculty of Mathematics-Mechanics of the University of Bucharest. After graduation, she became a scientific searcher at the Institute of Mathematics of the Romanian Academy (IMAR). She got a PhD degree in mathematics in 1982. Following the dissolution of IMAR, she was transferred to ICPE - the Research Institute for Electrotechnics from Bucharest. In 1991 she became a full professor at the Technical University of Civil Engineering of Bucharest. She wrote - alone or in collaboration - four scientific books, about 20 textbooks on Analysis, Advanced Calculus and Numerical Analysis and over 160 scientific papers. While having results in various domains of pure and applied mathematics, her main scientific contribution concerns LEM - the linear equivalence method, an original method that she created to the purpose of determining, both qualitatively and numerically, the solutions of nonlinear ordinary differential equations and systems in a classical linear frame. She wrote two books on LEM in Romanian and over 120 papers.